Sustainable Processes for the Removing of Heavy Metals from Aqueous Solutions

Sustainable Processes for the Removing of Heavy Metals from Aqueous Solutions

Editors

Cristina Palet
Julio Bastos-Arrieta

MDPI • Basel • Beijing • Wuhan • Barcelona • Belgrade • Manchester • Tokyo • Cluj • Tianjin

Editors

Cristina Palet
Department of Analytical Chemistry
The Autonomous University of Barcelona
Barcelona
Spain

Julio Bastos-Arrieta
Chemical Engineering and Analytical Chemistry
University of Barcelona
Barcelona
Spain

Editorial Office
MDPI
St. Alban-Anlage 66
4052 Basel, Switzerland

This is a reprint of articles from the Special Issue published online in the open access journal *Water* (ISSN 2073-4441) (available at: www.mdpi.com/journal/water/special_issues/processes_removing_heavy_metals_aqueous_solution).

For citation purposes, cite each article independently as indicated on the article page online and as indicated below:

LastName, A.A.; LastName, B.B.; LastName, C.C. Article Title. *Journal Name* **Year**, *Volume Number*, Page Range.

ISBN 978-3-0365-6975-8 (Hbk)
ISBN 978-3-0365-6974-1 (PDF)

Contents

About the Editors

Cristina Palet

Dr. Cristina Palet has been linked to the Universitat Autònoma de Barcelona since 1989, obtaining her Chemistry degree, and in 1994, achieved her doctor's degree with Cum Laude with the Prof's tutorship Manuel Valiente and Prof. Maria Muñoz. Since March 2000, she has been a Professor of Titular (Analytical Chemistry) at the Chemistry Department of Universitat Autònoma de Barcelona. Her research work focuses on:

In Basic research, Separation Systems have been developed, primarily by using membrane setups in flat and hollow fiber configurations (liquid, polymeric and composite membranes) and chromatographic materials (resins).

Furthermore, she studies the characterization and use of bio-waste materials for separation purposes, so different biomasses have been also studied, mainly for metals (human hair, dog hair, degreased wool and chicken feathers as keratin biomaterials from animal origin, and cork and algae as vegetal origin have been used). Nanoparticles (NPs) are prepared and characterized, mainly for their immobilization in different supports to improve the efficiency and/or the selectivity of heavy metal adsorption. One application of NPs (SPION) for the adsorption of PAHs has been also checked.

In a more applied field, the extraction, identification and quantification of PAHs by Gas Chromatography with a Mass Detector have been developed. In the last few years, the development of new sensor materials is one research line of her interest, mainly applied to environmental and biomedical applications. To follow many of such separation and detection systems, different Instrumental Techniques have been used and some analytical procedures have been developed Some Design of Experiments Systems have been applied to optimize different designed separation methodologies. Beamline time obtained this in 2018, in ALBA (Catalunya, Spain) and Diamond (England) Synchrotrons to determine the speciation of chromium onto different biomass systems, for the first time.

Julio Bastos-Arrieta

Dr. Julio Bastos Arrieta o got his Master's degree (September 2010) and Ph.D. in Chemistry with "Cum Laude"/International Doctor Quality honors (October 2014). In June 2015, he was hired as Postdoc Researcher for the Separation Techniques and Waste Management Group (SETRI), currently known as Resource Recovery and Environmental Management (R2EM) at Universitat Politècnica de Catalunya (UPC) under the supervision of Drs. Joan de Pablo and Ignasi Casas. There he co-directed the Master's theses of students and took part in different research projects with public and private entities, for example: Empresa Nacional de Residuos Radiactivos (ENRESA, Spain), Institute for Transuranium Elements (ITU; Germany), Universitat de Girona (UdG, Spain) and Universitat de Barcelona (UB, Spain).

In September 2017, he moved to Dresden (Germany) to enroll as a Postdoc in the Group of Prof. Eychmüller, at Technische Universität Dresden (TUD), to work on the preparation and analysis of nanoactive colloids. He came back to Spain in 2020 and carried out a research stay at UAB (June-September 2020) based on the impedimetric analysis of nanomaterials. In September 2020, he joined the Molecular and Industrial Biotechnology group (GBMI) at the UPC as Senior Postdoc, under the supervision of Dr. Tzanko Tzanov. From February to July 2021, he taught a class in UAB as a visiting Professor for the Analytical Chemistry Department.

Since November 2021 he enrolled in the Universitat de Barcelona as a Tenure Lecturer Professor and researcher of the Electroanalysis Group, of the Analytical Chemistry Department. He collaborates actively with his former colleagues in TUD, UPC and UAB. His main research interests include the preparation and characterization of nanomaterials (e.g., nanoparticles, active colloids) and their application towards the electrocatalytical enhancement of sensing systems.

Preface to "Sustainable Processes for the Removing of Heavy Metals from Aqueous Solutions"

This article collection from the Special Issue "Sustainable Processes for the Removing of Heavy Metals from Aqueous Solutions" attempts to summarize the state-of-the-art of current macro-, micro- and nanotechnologies for water purification, discussing their field of application, especially for heavy-metal ion removal. For instance, it presents the recently available information on utilizing different biomass materials for heavy metals removal, highlighting the increasing use of these materials due to their low cost, regeneration ability, high adsorption efficiency, and small chemical or biological sludge with a possibility of metal recovery.

Cristina Palet and Julio Bastos-Arrieta
Editors

water

MDPI

Editorial

Sustainable Processes for the Removal of Heavy Metals from Aquatic Systems

Julio Bastos-Arrieta [1,2,*] and Cristina Palet [3]

1 Departament d'Enginyeria Química i Química Analítica, Universitat de Barcelona (UB), Martí i Franquès 1-11, 08028 Barcelona, Spain
2 Institut de Recerca de l'Aigua (IdRA), Universitat de Barcelona (UB), 08028 Barcelona, Spain
3 Grup de Tècniques de Separació en Química, Unitat Química Analítica, Departament de Química, Facultat de Ciències, Universitat Autònoma de Barcelona, 08193 Bellaterra, Spain
* Correspondence: julio.bastos@ub.edu

Citation: Bastos-Arrieta, J.; Palet, C. Sustainable Processes for the Removal of Heavy Metals from Aquatic Systems. *Water* **2023**, *15*, 761. https://doi.org/10.3390/w15040761

Received: 29 January 2023
Revised: 2 February 2023
Accepted: 7 February 2023
Published: 15 February 2023

Water pollution is a global problem threatening the entire biosphere and affecting the life of many millions of people. It is not only one of the foremost global risk factors for illness, diseases and death, but also contributes to the continuous reduction of the available drinkable water sources worldwide. Delivering valuable solutions, which are easy to implement and affordable, often remains a challenge.

Heavy metal ions are some of the most harmful and widespread contaminants, with adverse effects to the environment. These ionic species, mainly cations, are one of many deadly contaminants in ground water across the globe. The physical-chemical properties and composition of the affected waters and sediments present strong spatial variations depending, mainly, on the proximity to the discharge point, and strong seasonal variations are detected depending on the rainfall and temperature regime.

Therefore, the use of environmentally friendly technologies and the reuse or revalorization of all waste generated is a strategy that must be widely assumed. For example, biosorption of heavy metals by metabolically inactive non-living biomass of microbial or plant origin is an innovative and alternative technology for the removal of these pollutants from aquatic systems.

This Special Issue compiled 13 different research works [1–13], including 1 review paper [13]. This publication collection covers a wide range of topics related to the enhanced removal of heavy metals during primary treatments in wastewater management including speciation [13], sorption technologies [1–7,9,11], complexation [8] and coagulation [12]. The main heavy metals addressed are: Cr(VI) [3,5,12], Cr(III) [2,6,9], Co(II) [11], iron [8], Cu(II) [2,4,7], Ni(II) [1,3], Cd(II) [1,2,6], Pb(II) [2,6] and Zn(II) [1], among others, such as alkaline earth elements [8].

Different materials such as Ni-Al alloy [12], inorganic adsorbents such as silica SBA-15 and titanosilicate ETS-10 [4], biomass [1–3,5,11], membranes (i.e., chitosan) [9], nanocellulose [8], ZnO nanoparticles [7], biosynthesized adsorbents (from oyster shells) such as hydroxyapatite [6] and nano modified coffee husk and coffee lignin [2] are presented in this volume. The strategy for their removal varies between direct (bio)adsorption, optimizing parameters such as pH, contact time, adsorption temperature and surface charge density [10]. In two cases, a complementary antibacterial effect is included during the separation process [2,6], which is related to corresponding biomass development. In one case, a strategic safe disposal of the spent adsorbents is presented [1], which is an interesting and novel approach for the sustainability of adsorption processes.

Finally, thanks to all the contributions, this Special Issue demonstrates the feasibility of sustainable materials and processes mainly for heavy metal removal from aqueous wastes.

Institutional Review Board Statement: Not applicable.

Informed Consent Statement: Not applicable.

Data Availability Statement: Not applicable.

Conflicts of Interest: Authors declare no conflict of interest.

References

1. Simón, D.; Palet, C.; Costas, A.; Cristóbal, A. Agro-Industrial Waste as Potential Heavy Metal Adsorbents and Subsequent Safe Disposal of Spent Adsorbents. *Water* **2022**, *14*, 3298. [CrossRef]
2. Guevara-Bernal, D.F.; Ortíz, M.Y.C.; Cifuentes, J.A.G.; Bastos-Arrieta, J.; Palet, C.; Candela, A.M. Coffee Husk and Lignin Revalorization: Modification with Ag Nanoparticles for Heavy Metals Removal and Antifungal Assays. *Water* **2022**, *14*, 1796. [CrossRef]
3. Villabona-Ortíz, A.; Tejada-Tovar, C.; González-Delgado, Á.D. Elimination of Chromium (VI) and Nickel (II) Ions in a Packed Column Using Oil Palm Bagasse and Yam Peels. *Water* **2022**, *14*, 1240. [CrossRef]
4. Humelnicu, D.; Zinicovscaia, I.; Humelnicu, I.; Ignat, M.; Yushin, N.; Grozdov, D. Study on the SBA-15 Silica and ETS-10 Titanosilicate as Efficient Adsorbents for Cu(II) Removal from Aqueous Solution. *Water* **2022**, *14*, 857. [CrossRef]
5. Villabona-Ortíz, A.; González-Delgado, Á.; Tejada-Tovar, C. Equilibrium, Kinetics and Thermodynamics of Chromium (VI) Adsorption on Inert Biomasses of Dioscorea rotundata and Elaeis guineensis. *Water* **2022**, *14*, 844. [CrossRef]
6. Jang, S.; Park, K.; Song, S.; Lee, H.; Park, S.; Youn, B.; Park, K. Removal of Various Hazardous Materials Using a Multifunctional Biomass-Derived Hydroxyapatite (HAP) Catalyst and Its Antibacterial Effects. *Water* **2021**, *13*, 3302. [CrossRef]
7. Leiva, E.; Tapia, C.; Rodríguez, C. Highly Efficient Removal of Cu(II) Ions from Acidic Aqueous Solution Using ZnO Nanoparticles as Nano-Adsorbents. *Water* **2021**, *13*, 2960. [CrossRef]
8. de Jesus Carvalho de Souza, V.; Caraschi, J.C.; Botero, W.G.; de Oliveira, L.C.; Goveia, D. Development of Cotton Linter Nanocellulose for Complexation of Ca, Fe, Mg and Mn in Effluent Organic Matter. *Water* **2021**, *13*, 2765. [CrossRef]
9. Zakmout, A.; Sadi, F.; Velizarov, S.; Crespo, J.G.; Portugal, C.A.M. Recovery of Cr(III) from Tannery Effluents by Diafiltration Using Chitosan Modified Membranes. *Water* **2021**, *13*, 2598. [CrossRef]
10. Bartzis, V.; Sarris, I.E. Time Evolution Study of the Electric Field Distribution and Charge Density Due to Ion Movement in Salty Water. *Water* **2021**, *13*, 2185. [CrossRef]
11. Acosta-Rodríguez, I.; Rodríguez-Pérez, A.; Pacheco-Castillo, N.C.; Enríquez-Domínguez, E.; Cárdenas-González, J.F.; Martínez-Juárez, V.-M. Removal of Cobalt (II) from Waters Contaminated by the Biomass of Eichhornia crassipes. *Water* **2021**, *13*, 1725. [CrossRef]
12. Ogata, F.; Nagai, N.; Tabuchi, A.; Toda, M.; Otani, M.; Saenjum, C.; Nakamura, T.; Kawasaki, N. Evaluation of Adsorption Mechanism of Chromium(VI) Ion Using Ni-Al Type and Ni-Al-Zr Type Hydroxides. *Water* **2021**, *13*, 551. [CrossRef]
13. Sylwan, I.; Thorin, E. Removal of Heavy Metals during Primary Treatment of Municipal Wastewater and Possibilities of Enhanced Removal: A Review. *Water* **2021**, *13*, 1121. [CrossRef]

water

MDPI

Review

Removal of Heavy Metals during Primary Treatment of Municipal Wastewater and Possibilities of Enhanced Removal: A Review

Ida Sylwan * and **Eva Thorin** *

Future Energy Center, School of Business, Society and Engineering, Mälardalen University, SE-721 23 Västerås, Sweden
* Correspondence: ida.sylwan@mdh.se (I.S.); eva.thorin@mdh.se (E.T.); Tel.: +46-(0)-21-151-767 (I.S.); +46-(0)-21-101-564 (E.T.)

Abstract: Resource reuse has become an important aspect of wastewater management. At present, use of sludge in agriculture is one of the major reuse routes. Conventional municipal wastewater treatment does not involve any designated process for removal of heavy metals, and these distribute mainly between effluent and sludge. Enhanced removal of heavy metals during primary treatment may decrease the heavy metal concentrations in both effluent and sludge from secondary treatment and promote long-term reuse of secondary sludge. This review considers heavy metal occurrence and removal during primary settling, together with possible treatment technologies for heavy metal removal in primary settlers and their theoretical performance. The variation in total heavy metal concentrations and dissolved fraction in raw municipal wastewater points to a need for site-specific assessments of appropriate technologies for improved heavy metal removal. Studies examining the heavy metal speciation beyond dissolved/particulate are few. Missing or disparate information on process parameters such as hydraulic retention time, pH and composition of return flows makes it hard to generalize the findings from studies concerning heavy metal removal in primary settlers. Coagulation/flocculation and use of low-cost sorbents were identified as the most promising methods for enhancing heavy metal removal during primary settling. Based on the available data on heavy metal speciation and removal during primary settling, sorption technologies may be most effective for enhancing the removal of Cu and Ni, while coagulation may be efficient for Cd, Cr, Cu, Pb, Zn and Hg removal (but not as efficient for Ni removal).

Keywords: primary settling; resource reuse; sludge; effluent; adsorption; coagulation/flocculation

Citation: Sylwan, I.; Thorin, E. Removal of Heavy Metals during Primary Treatment of Municipal Wastewater and Possibilities of Enhanced Removal: A Review. *Water* **2021**, *13*, 1121. https://doi.org/10.3390/w13081121

Academic Editors: Cristina Palet and Julio Bastos-Arrieta

Received: 30 March 2021
Accepted: 16 April 2021
Published: 19 April 2021

1. Introduction

Heavy metals, referring to a group of high-density elements including metals and metalloids, are a concern due to elevated concentrations in the natural environment as a result of anthropogenic activities [1]. One of the anthropogenic sources is wastewater. Heavy metals include biologically essential elements such as copper (Cu), chromium (Cr), zinc (Zn), nickel (Ni), boron (B), iron (Fe) and molybdenum (Mo), and elements which are not essential such as lead (Pb), mercury (Hg), cadmium (Cd), arsenic (As) and silver (Ag). Most heavy metals are toxic to humans even in low amounts. Heavy metals may induce disease to the gastrointestinal, renal, cardiovascular systems, etc., including conditions such as cancer, liver and kidney disease, melancholy and osteoporosis. Heavy metal contamination can also affect plant metabolism and growth [2].

To achieve environmental sustainability, reuse of nutrients, energy and, in some cases, water has become an important aspect of wastewater management [3]. Wastewater treatment, through physical, chemical and biological processes generates treated water (effluent), solid residues (sludge) and gaseous emissions. Degradation of heavy metals is not possible, which means that the majority of influent heavy metals will distribute between effluent and sludge, though heavy metals such as Hg, As and the semi-metal antimony Sb can form volatile compounds which are transferred to the gaseous phase [4,5]. Safe reuse of nutrients from sludge and water requires that associated toxic substances such as heavy metals are minimized. Use of sludge in agriculture is currently the most common option for reuse of nutrients [6], while the most common option for reuse of water is for irrigation and landscaping [7].

Industrial emissions historically were the main source of heavy metals entering municipal wastewater. However, stringent legislation for industrial emission caused heavy metals to decrease markedly during the 1970s and the early 1980s [4]. Based on the current trends of heavy metals' contents in sludge, it was deduced that in some places, diffuse emissions currently constitute the main contribution of heavy metals [8]. Diffuse emission sources include corrosion of pipes and roofs (containing Cu) and runoff from streets (where asphalt, tires and brake linings may contain Cd, Pb and Zn). Furthermore, the use of amalgam for fixing teeth was estimated to contribute a large part of Hg, and food was calculated to contribute 9% of Cd and 25% of Zn in a municipal wastewater treatment plant (WWTP) in Stockholm, Sweden [8].

In order to continue the reduction in heavy metals in sludge and effluent, there is a need to regulate the use of heavy metals in different products further. Meanwhile, it is also relevant to attempt to optimize wastewater treatment systems for minimization of heavy metals in sludge and effluent [4]. Yoshida et al. [5] modelled the ecotoxicity and toxicity impact on humans from municipal wastewater. They found that the heavy metal Zn contributed the largest share of toxic impact with regards to both effluent release and sludge use in agriculture. However, they also found that conventional treatment of wastewater significantly reduced the toxicity exposure from inorganic constituents (including heavy metals) on freshwater and seawater. According to Hospido et al. [9], terrestrial ecotoxicity due to heavy metal content in sludge together with eutrophication are the most significant factors in terms of environmental impact from municipal wastewater treatment.

Several reviews [4,10,11] considered heavy metal removal during conventional municipal wastewater treatment and discussed the importance of assessing heavy metal contents in end products from municipal wastewater treatment: effluent water and sludge. Conventional treatment here refers to primary settling followed by the activated sludge process (ASP) or biological nutrient removal (BNR). Hargreaves et al. [12] reviewed current heavy metal removal and performance of additional treatment processes for enhanced removal to minimize heavy metal concentrations in effluents (not considering heavy metal elimination from sludge). Possible technologies for enhancing the removal of heavy metals during primary settling were not reviewed previously. Increased heavy metal removal during or directly following primary settling (as a separate treatment step) could potentially decrease the heavy metal concentrations in the treated effluent and in the sludge from subsequent biological treatment.

Environmental background concentrations, together with regulations and standards for heavy metal concentration in freshwater, irrigation water and drinking water, are summarized in Table 1 and compared to typical concentrations in municipal wastewater in Europe. Heavy metal concentrations in WWTP influent are generally higher compared to the concentrations of heavy metals found in natural water (background). In Europe, environmental quality standards (EQS) for freshwater were implemented to preserve/attain good ecological and chemical status in all surface- and groundwater, based on the goals of the Water Framework Directive [13]. EQS for Pb, Cd, Ni and Hg exist on the central European level [14], while priority substances identified on the national level led to an extended list of EQS with stricter limits, as exemplified by Swedish regulation [15]. Raw municipal wastewater generally contains higher concentrations of heavy metals compared to both central and local EQS; however, effluent concentrations (after treatment) may not exceed the EQS. In relation to standards and regulation for drinking water and irrigation water, heavy metal concentrations in municipal wastewater do not always exceed the limits.

Background heavy metal concentrations and limiting values for heavy metals in soil and sludge used in agriculture are summarized in Table 2 and compared to typical concentrations in sludge from municipal WWTPs in Europe. European heavy metal limits for sludge use in agriculture [16] are more stringent compared to US limits. The current European legislation is more than 30 years old. Several member states have implemented more strict limits. Updated legislation was drafted on the central European level but not yet implemented. The drafted proposal includes stricter limits for heavy metals and an addition of limit values for synthetic organic contaminants and indicator organisms [17]. Based on the heavy metal concentrations occurring in sludge from European WWTPs, reuse of sludge is partly limited by current legislation, especially in relation to the stricter regulation in some countries.

This review focuses on heavy metal removal during primary treatment of municipal wastewater, with the purpose of reducing heavy metal contents in effluent and in sludge from subsequent biological treatment (secondary sludge). The review includes: (i) occurrence and speciation of heavy metals in raw municipal wastewater (influent to primary settlers), (ii) mechanisms of heavy metal removal during primary settling, (iii) possible treatment technologies for enhanced heavy metal removal in or directly following primary settling, (iv) calculations to estimate theoretical heavy metal removal if enhanced heavy metal removal was introduced based on current removal of particulate and dissolved heavy metals during primary settlement.

2. Material and Methods

In this study, findings on heavy metal removal during primary treatment of municipal wastewater in previous published scientific publications are summarized. The findings are also related to current and planned regulations for water in different parts of the world. Experimental data collected from previous studies were analyzed by correlations of data with respect to heavy metal concentrations and heavy metal removal during primary settling (Sections 4.4 and 7). Further, the potential for improved removal of dissolved and particulate heavy metals was estimated (further details given in Section 7.1).

3. Occurrence of Heavy Metals in Wastewater and Sludge

Based on current legislation, it is relevant to optimize wastewater treatment to reduce heavy metal concentrations in both sludge and effluent. It should be noted that, while concentrations in sludge may directly limit the possibilities of sludge reuse in agriculture, effluent release is not directly prohibited by concentrations exceeding EQS, since the dilution of effluent discharge is considered [18] Gardner et al. [19] showed that dilution of Zn, Cd, Cu and Ni was required for some of UK's WWTPs to meet the national EQS or so called "predicted no effect concentration". The scope of this review includes the metals listed in Tables 1 and 2 (for which water-related regulations and standards exist), and Ag (for which a limiting value was proposed in Sweden) [20].

Table 1. Heavy metal concentrations in municipal WWTP influents compared to concentrations in natural water (background) and regulation/standards for: freshwater (EU/US); irrigation water (Food and Agriculture Organization of the United Nations (FAO)/Canada); drinking water (World Health Organization (WHO)/EU/US); and toxicity (FAO/WHO).

Metal	Background (µg/L)	Freshwater (µg/L)		Irrigation Water (µg/L)	Canada	Drinking Water (µg/L)	EU [b]	US	Typical Municipal WWTP Influent Concentrations (µg/L)	Toxicity (mg/kg Body Weight)
		EQS in Sweden (EU Central EQS) [a]	Water Quality Criteria in the US	FAO		WHO			Europe (Number of Plants) [c]	FAO/WHO [d]
As	0.13–2.71	7 (-)	-	100	100	10	10	10	2.7–12 (15)	Withdrawn (previous: 0.0021 daily)
Cd	6×10^{-4}–0.61	0.08–0.25 (0.08–0.25)	-	10	5	3	5	5	0.4–75 (18)	0.025 monthly
Cr	0.29–11.46	12 (-)	-	100	Cr^{6+}: 8, Cr^{3+}: 5	50	50	100	8–100 (17)	-
Cu	0.23–2.59	8.2 (-)	0.18–20 [e]	200	200 [f]	2000	2000	1300	10–100 (18)	0.500 daily
Pb	0.007–308	0.34 (1.2–1.3)	0.01–7 [e]	5000	200	10	10	15	2–100 (18)	Withdrawn (previous: 0.025 weekly)
Hg	-	0.05 (0.07 [g])	0.77	-	-	6	1	2	0.7–3.6 (6)	0.004
Ni	0.35–5.06	2.3 (4–8.6)	1–115 [e]	200	200	7	20	-	3–100 (17)	-
Zn	0.27–27	7.8 (-)	2.4–260 [e]	2000	1000 (pH < 6.5)	-	-	-	100–1600 (17)	1 daily
Reference	[1]	[14,15]	[12]	[1]	[1]	[1]	[1]	[1]	[4]	[1]

[a] Annual average values. Interval for inland and surface waters, and for Cd also depending on water hardness. [b] Regulations are similar in e.g., Canada and China, though in Canada Cu and Ni are not regulated, and in China Hg is not regulated but Zn is [1]. [c] From WWTPs in France, Greece, Italy, Poland and the UK. The following deviating values were excluded: At one WWTP, Cu concentration was 250–300 µg and Pb concentration was 600 µg/L. At one WWTP, Ni concentration was 800 µg/L. In one of the plants, As concentration was below the detection limit. In one plant, Hg concentration was below the detection limit. (Rule et al. [21] investigated influent heavy metal concentrations to 30 WWTPs in UK and found similar concentrations as given here; no correlation was found between industrial loads and heavy metal concentrations, with exception for Cr concentrations which were higher in plants with >50% ind. load). [d] Provisional tolerable weekly intake for humans, established by a joint FAO/WHO (World Health Organization) expert committee. Limit values for As and Pb were withdrawn in 2011, because the previous limits were no longer considered to be significantly low to assure health protection. [e] Because bioavailibility of heavy metals varies e.g., due to hardness, criteria is different depending on the hardness of the water. [f] 1000 µg/L for tolerant crops. [g] Maximum allowable concentration.

Table 2. Heavy metal concentrations in municipal WWTP sludge compared to natural soil (background), and limit values in agricultural soils (EU/Canada) and in sludge (EU/US Environmental Protection Agency (EPA)) when sludge is used in agriculture (d.w. stands for dry weight).

Metal	Background (mg/kg)		Agricultural Soils (mg/kg d.w.)		Sludge for Agricultural Use (mg/kg d.w.)				Typical Municipal WWTP Sludge Concentrations (mg/kg d.w.)
	Upper Crust	Surface Soil	EU, Directive 86/278/EEC (Soil pH 6–7)	Canada	EU, Directive 86/278/EEC	EU Countries with More Stringent Limits [a]	EU Countries with Far More Stringent Limits [b]	US, EPA/ Federal Regulation	Europe (Number of Plants) [c]
As	1.8	4.7	-	12	-	20–150 [d]	25 [e]	75	4.2–40.4 (6)
Cd	0.1	0.41	1–3	1.4	20–40	2–10	0.8–2	85	1.0–22 (11)
Cr	35	42	-	64	-	70–1000	75–100	-	15.3–856 (10)
Cu	14	14	50–140	63	1000–1750	70–1000	75–600 [f]	4300	38.9–1200 (11)
Pb	15	25	50–300	70	750–1200	45–900	100–120	840	3.1–330 (11)
Hg	0.07	0.07	1–1.5	6.6	16–25	2–10	0.75–2.5	57	0.9–3.2 (6)
Ni	19	18	30–75	45	300–400	25–200	30–50	420	16.6–621 (10)
Zn	52	62	150–300	200	2500–4000	200–3000	300–800 [f]	7500	501–8900 (11)
Reference	[1]	[1]	[1]	[1]	[16]	[4]	[4]	[4]	[4]

[a] Austria, Belgium, Finland, France etc. [b] Netherlands, Sweden, Denmark. In Sweden, a limiting value for Ag was also proposed [20]. [c] Sludge from WWTPs in Greece, Italy, Slovenia, Spain and Sweden (partially industrial load, which ranged from 6–70%, though not given in all cases). [d] Limit values for as only apply in Austria (20) and Belgium (150). [e] As limit only in Denmark. [f] In Denmark: Cu: 1000, Zn: 4000.

4. Heavy Metal Speciation in Raw Municipal Wastewater

The speciation of heavy metals, i.e., the knowledge of which heavy metal ions and complexes are present in which amounts, is important for the potential removal of heavy metals during primary treatment. Heavy metal speciation is affected by hardness, alkalinity, pH and redox potential of the wastewater [10]. The heavy metal speciation in influent wastewater is of importance for the possible removal of heavy metals through different technologies. For example, in the case of sorption processes, dissolved heavy metals can be sorbed, while heavy metals bound to wastewater particles may only interact with an added sorbent if desorption first occurs or if wastewater particles form agglomerates with the sorbent. Heavy metals associated with particulate matter can be removed through settling or filtration if the particles are large enough.

4.1. Different Definitions Applied with Respect to Heavy Metal Speciation in Municipal Wastewater

A number of studies concerning heavy metal load in raw wastewater and distribution of heavy metals between sludge and effluents in WWPTs examined the total heavy metal concentrations without further specifying the speciation of heavy metals [5,21–32]. However, there is also a number of studies where partitioning of heavy metals between the particulate and dissolved phase was determined (e.g., Choubert et al. [33] and Toumi et al. [34]). The distinction between dissolved and particulate matter in wastewater is generally based on size fractioning of the wastewater, where the fraction which passes through a filter of certain pore size is defined as dissolved. A filter with pore size of 0.45 μm is most commonly used; however, in some cases, other pore sizes were applied, e.g., 0.2 μm [35,36]. What are referred to as "dissolved" heavy metals are analyzed after filtering the wastewater, while the "particulate" heavy metals are calculated as the difference between the total heavy metal concentration (analyzed in unfiltered wastewater) and the dissolved heavy metals.

According to Ziolko et al. [10], heavy metals are likely to occur as organometallic complexes, and should be considered in the following three groups: surface-bonded organic ligands, insoluble matter and soluble organic ligands. Truly dissolved heavy metal ions are considered rare. Heavy metals associated with the dissolved phase (<0.45 μm) may be associated with chelating agents of natural origin (e.g., proteins and nucleic acids) or synthetic origin (e.g., nitrilotriacetic acid and ethylenediamine tetraacetic acid (EDTA) [10]. El Samrani et al. [37] examined the different mineral forms of heavy metals in overflow wastewater from a combined collection system for municipal wastewater and stormwater (also called combined sewer overflow). They found that heavy metals associated with sulfide particles were abundant. These were assumed to be formed within the sewer network. Heavy metals were also encountered in the form of alloys, iron oxihydroxides, carbonates, phosphates and sulphates, and associated with clays. The abundance of heavy metal-bearing sulfides was confirmed by [38].

Further knowledge on the speciation of heavy metals could enhance the understanding of heavy metal removal mechanisms. Some attempts were made to elucidate the heavy metal speciation in municipal wastewater. Lawson et al. [36] performed gel chromatography of synthetic wastewater for separation of heavy metal species. Buzier et al. [39] and Gourlay-Francé et al. [40] analyzed the dissolved heavy metals in two different fractions, inert or labile. Buzier et al. [39] described the labile heavy metals as "inorganic metal fraction with a possible additional contribution of small weak organic complexes", while Gourlay-Francé et al. [40] simply described the labile fraction as "truly dissolved in water". The dissolved inert fraction is associated with dissolved organic matter (DOM), which Buzier et al. [39] described as "strong metal complexes and large weak organic complexes". The analysis of heavy metals by Gourlay-Francé et al. [40] and Buzier et al. [39] was performed subsequent to separation of heavy metals species by diffusion and dissociation into a permeable gel.

The dissolved fraction (<0.45 μm) may also be considered to consist of heavy metals associated with colloids and truly dissolved heavy metals. The distinction between the two is based on molecular weight, where the truly dissolved fraction is the fraction smaller than

1 kDa (1000 Dalton), and colloidal matter is defined by the interval 1 kDa–0.45 μm. The organic matter in the truly dissolved size fraction is primarily composed of humic substances, while the colloidal fraction is primarily composed of non-humic macromolecules, including proteins, polysaccharides and aminosugars (which may be associated with nucleic acids) [41].

4.2. Influence of Return Flows on Heavy Metal Speciation

Influent wastewater to primary settlers is commonly mixed with return flows from sludge treatment (thickening and/or dewatering), in which the heavy metal speciation may differ from the speciation in influent wastewater. These return flows are not always considered when calculating the removal of heavy metals over primary clarification or mass balances over whole WWTPs (e.g., Karvelas et al. [35]). However, return flow of sludge liquors was shown to contribute to significant amounts of heavy metals entering primary settlement, according to Goldstone et al. [42–44]. Most of these heavy metals were associated with solids. According to a mass balance study of a Danish WWTP, the amount of heavy metals from return of centrate from sludge dewatering constituted a significant fraction of the heavy metals entering primary treatment: 8% of Cd, Cr, Pb and Zn; 12% of Cu and 5% of Ni (estimated from graph) [5]. Innaa et al. [45] investigated mass flows of Cu in return flows from consolidation of sludge (co-settled sludge from primary treatment and biological trickling filters) and found that these contributed a significant fraction of the total Cu mass flows entering a primary settler. However, despite the increase of Cu concentrations, this return flow together with return of sludge from a trickling filter increased the removal of soluble Cu during primary settling due to increased suspended solids (SS) concentrations. In one of the plants sampled by Gourlay-Francé et al. [40], return liquors from sludge dewatering considerably increased both particle-bound and dissolved concentrations of Cu, Zn and Ni.

4.3. Fractions of Dissolved Heavy Metals in Raw Wastewater

Dissolved fractions of Cu, Cr, Ni, Pb, Cd, Zn, Hg and Ag in raw wastewater are given in Table 3. The fraction of dissolved heavy metals in raw wastewater varies according to different studies. With exception for Ni, the ranges of the fractions of dissolved heavy metals are narrower when considering the studies known not to include return liquors (in the stream that was sampled). This indicates that return sludge liquors influence the heavy metal speciation; however, this conclusion should subject to a level of caution, since the amount of data is limited. Ni is often associated with the dissolved fraction to a higher degree than other heavy metals.

Table 3. Fraction of dissolved heavy metals (%, concentration of dissolved heavy metals in relation to total heavy metal concentrations) in influent wastewater.

Cu	Cr	Ni	Pb	Cd	Zn	Hg	Ag	As	Reference
20	22	78	22	86	-	-	-	-	[39] [a,b]
14	86	59	6	100	12	40	-	67	[46] [c]
24	24	57	18	23	25	-	15	-	[33] [d,b]
26	-	57	17	22	23	26	-	-	[47] [e]
8	59	63	13	70	34	8/58	-	-	[42,43] [a,f,b]
10	-	33	-	-	18	-	-	-	[40] [a,g,b]
14	-	56	-	-	27	-	-	-	
39	-	67	23	-	66	-	-	-	[41] [a]
12	9	79	6	14	8	-	-	-	[35] [h,b]
4	9	70	-	6	75	-	-	-	[29] [i]
55	7	67	5	8	24	9	-	-	[48] [c]
16	21	-	70	36	12	-	-	-	[34] [a]
4–55	7–86	57–79	5–70	6–100	8–75	8–58	-	-	Range, all (11 studies)
8–39	21–59	33–78	13–70	36–86	12–66	8–58	-	-	Range for samples known to not include return liquors (5 studies)

[a] Sampled before addition of return sludge liquors/directly when wastewater entered the plant. [b] The data were estimated from graph. [c] The exact location of raw wastewater sampling is not clear. The sample might thus include return of liquor from sludge processing. [d] Average of ~20 samples, sampled at nine different treatment plants. The data were estimated from graph. Details on sampling points for raw wastewater were not given in the reference, and therefore it is not known whether the sampled flow included return liquors. [e] Average from 16 WWTPs in the UK. Most samples of influent wastewater were taken before addition of return liquors. [f] A second sampling campaign performed ~1.5 yrs. later showed 58% dissolved Hg in influent wastewater. [g] Sampling was made at two different treatment plants. [h] Sampled after addition of return liquors from sludge digestion and after grit removal, but before return of another supernatant stream. In contrast to the other studies, the dissolved fraction was analyzed after filtering through a 0.2 µm filter (instead of 0.45 µm). [i] Sampled after return of sludge from biological treatment and after grit removal. For Ni and Zn, there was a suspicion that samples of dissolved heavy metals were contaminated to some extent, and it can be noted that compared to data from the other studies, these values are in the upper range.

4.4. Correlation between Total Concentrations in Raw Wastewater and the Fraction of Dissolved Heavy Metals

It is hypothesized that high total concentrations of heavy metals in wastewater coincide with high SS concentrations, and thus large complexation or sorption capacity of the solids in the wastewater, which could cause a lower fraction of dissolved heavy metals in raw wastewater. Nielsen and Hrudey (1983) demonstrated this trend; increased SS concentration decreased the concentration of dissolved heavy metals (Ag, Cd, Cr, Cu and Tl). To investigate this trend further, the correlations between total metal concentrations and fractions of dissolved heavy metals are investigated based on literature data, and are shown in Figure 1. With respect to Cu, Ni, Pb and Zn, the correlation is very weak or non-existent (based on $\overline{R^2}$; 0.02 or less, as given in Figure 1). For Cr and Cd, some correlation exists ($\overline{R^2}$ of 0.24 and 0.11). The correlation with respect to Hg appears to be stronger ($\overline{R^2}$ of 0.97); however, the negative correlation was broken when one of the references [47] was excluded, which indicates that the correlation is uncertain.

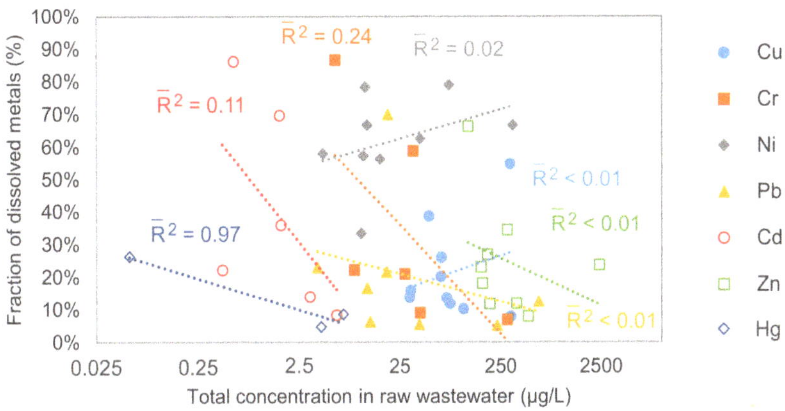

Figure 1. Fraction of dissolved heavy metals in raw municipal wastewater at different total heavy metal concentrations [34,35,39–44,46–48] ([40] includes data from two different WWTPs). (Data from [29,33] (included in Table 3) were not included, since average concentrations of total heavy metals in raw wastewater were not given/not easily read from paper.) (Data for As and Ag not available from multiple sites.) The dotted lines represent the linear regression between the percentage removal and the concentration of the different heavy metals in log form. The adjusted coefficient of determination ($\overline{R^2}$) is given; this was calculated as follows: $1-(1-R^2)(n-1)(n-p-1)$, where R^2 is the coefficient of determination, n is the number of data points for the respective metal and p $(=1)$ is the number of explanatory variables, excluding the constant.

4.5. Fractions of Colloidal/Truly Dissolved or Inert/Labile Heavy Metals

Hargreaves et al. [41] found that while around two-thirds of Ni and Zn in raw wastewater was associated with the dissolved fraction (<0.45 µm), only 18% and 9%, respectively, was found in the truly dissolved fraction. The truly dissolved fractions of Cu and Pb constituted similar shares of the total influent heavy metal concentrations (around 10% and 5%, respectively). In measurements at two different WWTPs in Paris with mixed domestic and industrial influents, Gourlay-Francé et al. [40] found no or very small amounts of Cu and Zn to be labile, while around half of the Ni was labile at one of the sites. Cr and Pb were present in the particulate fraction, but since the dissolved heavy metals were below the detection level, the fraction of labile Cr and Pb was not given. Buzier et al. [39] found around half the total influent concentration of Ni to be in the labile fraction. According to Buzier et al. [39], the fraction of labile heavy metals should constitute less than half the dissolved fraction of Cd, Cr and Pb; however, they did not quantify the labile fraction of these heavy metals, because the concentrations were below the detection level. Labile Cu constituted around a quarter of the dissolved heavy metals in influent wastewater.

Based on findings of Hargreaves et al. [41], the distribution of heavy metals between the particulate, colloidal and truly dissolved fraction was similar in the effluent from primary settling, as in raw wastewater. They therefore concluded that enhanced removal of colloidal matter could enable increased heavy metal removal. Chen et al. [49] found that particle-bound heavy metals in primary settler overflow were mainly associated with particle sizes of 44 µm and smaller, which also indicates that increased removal of small particles could enhance removal of heavy metals. The organic matter in raw municipal wastewater is to a high extent found in the truly dissolved and colloidal phase; around 40–60% was found in these phases, according to a review by Modin et al. [50].

5. Heavy Metal Removal Mechanisms

During primary settling, heavy metals are removed by removal of solids with which the heavy metals are associated. Sorption/desorption of heavy metals in the primary settler together with the speciation or partitioning of heavy metals in the influent wastewater determine the removal capacity [10,51]. Heavy metal removal was shown to correlate positively with SS removal [52]. SS removal was measured frequently in the published literature (measured by [35,41,47]; measured but not reported by [39]; not measured by [46]; in [5], total solids was measured in influent and primary sludge). Buzier et al. [39] found a positive correlation between particulate concentrations of Cu and Cr, and suspended solids concentrations, and proposed that "their removal is likely to be strongly linked to the suspended solids removal". Based on Gardner et al. [47], the SS removal in 16 WWTPs in the UK was in the range ~45–63%; however, the correlation between heavy metal removal and SS removal in each individual plant was not reported.

Removal through volatilization might occur to some extent for certain metals Yoshida et al. [5] reported probable volatilization of the heavy metal Hg (6%) and the semi-metal Sb (44%) during digestion of sludge.

Further factors of importance for removal of heavy metals include wastewater pH and hydraulic retention time (HRT) of the primary settler [51]. Increased HRT was shown to increase heavy metal removal [52]. Information on pH and HRT is not always available in the published literature (with respect to heavy metal removal during primary settling), which makes it hard to generalize the findings fully (pH in influent and effluent from primary settling was reported by [5,41], but not by [35,39,46,47]; HRT was not reported by [5,35,39,41,46,47]).

For optimization of the removal of both heavy metals and organic micropollutants during treatment of municipal wastewater, it was suggested that primary settlers should be optimized for heavy metal removal [53]. This is because removal of organic micropollutants in ASP is improved at longer HRT and solids retention time (SRT), while removal of heavy metals may be negatively influenced [53,54].

Internal return flows and recirculation in WWTPs likely influence the heavy metal removal. Return flows from sludge dewatering (often to the primary settler) can affect the heavy metal removal [10,54], and both increased and decreased removal were reported. Increased removal was hypothesized to be caused by increased concentration of SS (as mentioned above, removal of heavy metals was shown to correlate positively with SS removal), influence of residues of ferric chloride coagulants in the return liquors or changes in redox potential [10]. On the other hand, if the return liquors contain large amounts of heavy metals associated with fine particles (which do not settle easily), the total removal of heavy metals in the primary settler might decrease [10]. Ferric chloride residues could contribute to increased heavy metal removal by increasing the removal of particles in the primary treatment, or, hypothetically, the iron could replace heavy metal ions in EDTA complexes (so that the heavy metals could instead bind to particles) [10]. The exact influence of return flows on redox potential in the primary settler has yet to be examined, to our knowledge. Since the redox potential is expected to increase during municipal wastewater treatment [55], return flows will have higher redox potential than influent flows and could hypothetically have some influence on the redox potential in the primary settler, particularly if the return flows are large in volume. The wastewater redox potential influences the oxidation state of metals (depending on the standard reduction potential of each metal species), and of other compounds in the wastewater (which may cause formation of various metal compounds). Xiao et al. [56] investigated the removal of Cu, Pb, Zn and Cr in constructed wetlands treating municipal wastewater. They found that an increase of redox potential occurred during treatment (−163 mv in the inflow and 184 mV in the outflow). They suggested that the reduction of sulphate (SO_4^{2-}) to sulfide (S^{2-}) (occurring under reducing conditions/low redox values), followed by formation of heavy metal sulfides, played an important role for the distribution of Cu, Pb, and Zn during treatment. In the case of Cr, increases in redox potential may cause conversion of Cr(III) into Cr(VI), which is a more mobile species [56].

Biodegradation is not directly relevant when it comes to heavy metals, though heavy metals could be physically entrapped by biomass (if insoluble), adsorbed to bacterial walls and extracellular polymers or actively taken up by bacteria (if in dissolved state), which means that biological treatment has some effect on heavy metals [10]. In plant layouts where sludge from ASP/BNR is returned to the primary settler (and not directly to the head of the ASP/BNR), adsorption to or uptake by biomass influences heavy metal removal capacity in the primary settler. According to [25], heavy metals could attach to lipids, proteins and polysaccharides at cell surfaces by sorption to carboxyl, hydroxyl, phosphate and sulfonate groups. Extracellular polymeric substances (EPS) were stated to have a high importance in the entrapment/adsorption of micropollutants by bacterial cells [10], and sorption to EPS was shown to be pH-dependent [57] with different sorption capacity for different heavy metals [58].

Process chemicals may contain trace heavy metals which cause significant increases in heavy metal concentrations in wastewater. Buzier et al. [39] found that addition of $FeCl_3$ during tertiary treatment added significant amounts of Cr, Co, Cu and Ni (which may also enter the primary settler with the return flows).

Partitioning Constants and Modeling of Heavy Metal Speciation and Removal

The tendency for heavy metals to occur in the particulate and dissolved phase can be described by the solid/liquid or water/sludge partitioning constant, often referred to as K_P or K_D [40,54]. Gourlay-Francé et al. [40] found that the solid/liquid partitioning constant of heavy metals varied along the treatment process, and no relation was found between K_D and SS or dissolved organic carbon (DOC) concentrations. Similarly, in the model TOXCHEM (described by Parker et al. [59]), the partitioning constant is assumed to take different values in primary clarification compared to in ASP/BNR. According to Wang et al. [60], dissolved organic matter (measured as chemical oxygen demand (COD) in supernatant after sample centrifugation) significantly affects heavy metal partitioning in influent wastewater, but this effect could be disregarded at neutral or low pH (<8). A model for the prediction of heavy metal partitioning between dissolved and particulate fractions was suggested, where heavy metal partitioning is dependent on pH and SS concentration. Measurements by Katsoyiannis and Samara [61] revealed a negative correlation between log K_D and DOC concentrations, indicating that a higher DOC concentration could result in higher amount of heavy metals in the dissolved fraction and thus a lower heavy metal removal in primary settling. Findings by Gourlay-France et al. [40] were contradictory to the findings by Wang et al. [60] and Katsoyiannis and Samara [61]. They calculated a constant for partitioning between labile and inert fractions of dissolved heavy metals and called this partitioning constant K_{DOC}. They found that K_{DOC} was highly variable between different sites in the treatment plant, and thus hypothesized that sorption to DOC may not control heavy metal speciation, or that equilibrium may not be reached due to fast degradation of organic matter. However, another interpretation of their findings could be that varying composition and complexing capacity of DOC in different treatment steps caused the variability in K_{DOC}. Dionisi et al. [54] modelled removal of Cd and Pb in a sequencing batch reactor (a type of ASP), and calibrated K_P based on experimental data. They did not include the organic matter concentration as a variable when they calibrated the K_P value. However, they found that the concentration of biodegradable COD in the influent wastewater together with the specific rate of endogenous metabolism of the biomass had an indirect effect on the heavy metal removal capacity, since they determined the amount of biomass in the process.

When modelling heavy metal removal in full-scale treatment plants, an "aqueous compartment" and a "solid phase" has generally been considered, where heavy metals in the "aqueous compartment" are dissolved, and heavy metals in the "solid phase" are attached to sludge [51]. In models, the removal of heavy metals in the "solid phase" depends upon the removal of SS achieved. The dissolved heavy metals will end up in the effluent if they are not sorbed during the treatment process. Cecchini et al. [46] found that the concentration of dissolved heavy metals in effluents from primary settling followed by ASP was similar to the concentrations in raw wastewater. This indicates that dissolved heavy metals were relatively unaffected during the process, or that sorption and desorption balanced out over the treatment process as a whole. In either case, for this treatment plant, the heavy metals which were not removed during primary settling chiefly ended up in the effluent. Barret et al. [62] suggested a "three compartment-model" for sorption of micropollutants (not heavy metals) to sludge; sorption to particles (analyzed in centrifugation cake) associated with dissolved and colloidal matter (analyzed in centrifugation supernatant filtered at 1.2 μm) and freely dissolved (not measured experimentally). A three-compartment model might increase understanding of the heavy metal fate during physical separation and requires that the full-scale physical model incorporates dynamics of colloidal particles.

Cloutier et al. [63] modelled the removal of particulate heavy metals during primary settling under the assumption that removal of particle-bound heavy metals is relative to the removal of SS in the settler (removal efficiency thus being the same for all heavy metals). The model also assumes that dissolved heavy metals are removed only by mass flow of dissolved heavy metals with sludge, i.e., the concentration of dissolved heavy metals in primary settler overflow is the same as in primary settler influent. These model assumptions do not agree with the data according to, e.g., Karvelas et al. [35]. They found an SS removal of 51%, while the removal of different particle-bound heavy metals was between 22–81%, which indicates that sorption/desorption may occur during primary settling.

6. Technologies for Enhanced Heavy Metal Removal

Enhanced heavy metal removal during primary settling may target dissolved or particle-bound heavy metals. Technologies used for heavy metal removal from wastewaters include chemical precipitation, ion-exchange, adsorption, coagulation/flocculation, electrodialysis, membrane filtration, photocatalysis and electro-chemical treatment technologies [64–66]. These were mainly applied to wastewaters of industrial origin from production of, e.g., paper, pesticides, leather (tanneries), metal objects (plating) and ore (mining) [65]. Some features of these heavy metal removal technologies are summarized in Table 4. Due to large amounts of fibers in municipal wastewater, technologies which involve filtration, e.g., membrane filtration, were presumed not to be relevant for treatment during or directly following primary settling (as fouling may be an issue).

By introducing enhanced removal targeting both dissolved and particle-bound heavy metals, the heavy metal removal during primary settling could theoretically approach 100%. However, in addition to the heavy metal removal capacity of the different technologies, the following conditions are also of relevance for the enhanced removal of heavy metals from municipal wastewater during primary settling:

- Should be suitable for wastewater with high solids/organic matter concentration.
- Low impact in terms of energy use and sludge production.
- Low removal of phosphorus (P) and organic matter.

Table 4. Features of heavy metal removal technologies.

Method	Heavy Metal Removal Capacity	Efficiency at High Solids/ Organic Matter Concentration	Selectivity for Heavy Metals	Cost	Operation	By-Products	Ref.
Adsorption	Wide range, maximum sorption capacity depends on the type of sorbent.	Not directly affected by solids concentration, though the interaction of heavy metals and dissolved organic matter may influence heavy metal sorption.	Efficiency for different heavy metals depends on sorbent properties.	Low-cost (if using low-cost sorbents). However, many sorbents have so far only been tested in lab-scale.	Simple operation (wide pH-range). Separation of sorbent may be a challenge.	Less sludge production. Waste sorbents (if not regenerated).	[60,64–68]
Electrochemical (EC) treatment technologies	Efficient heavy metal removal.	Electroflotation separates heavy metals and organic matter. Some electrochemical methods simultaneously decompose organic matter.	Heavy metal removal or deposition possible, but not selective for heavy metals (different EC treatment generates agglomeration/ floccula-tion/precipitation etc.)	Low chemical usage. Initial investment cost is high. High electricity costs—proportional to the volume of water. Frequent replacement of electrodes.	Easy to operate. Corrosion may be a problem. Quick adaptation to fluctuating flow and pollutant concentration.	Less sludge production. No secondary pollution.	[64–66,69]
Coagulation/ flocculation [a]	Cannot remove heavy metals completely.	Should work well when the heavy metals are associated with colloids. Will simultaneously remove larger amounts of solids.	Not selective for heavy metals, but enhanced flocculants (e.g., with chitosan) have shown increased heavy metal removal efficiencies, as has e.g., flocculation after binding heavy metals to humic acids.	Cost effective, though chemical consumption is high.	Simple operation, requires coagulant/flocculant chemicals.	Increased sludge volume, though good dewaterability.	[64,65]

Table 4. *Cont.*

Method	Heavy Metal Removal Capacity	Efficiency at High Solids/ Organic Matter Concentration	Selectivity for Heavy Metals	Cost	Operation	By-Products	Ref.
Ion-exchange	Synthetic resins can remove nearly all heavy metals from solution.	Easily fouled by organics and other solids in the wastewater.	Non-selective/Removes only limited heavy metal ions. More effective for heavy metals with high ionic charge.	Operational cost is high, which makes it expensive especially at low heavy metal concentrations.	Sensitive to pH.	Resin regeneration may cause secondary pollution.	[64–66]
Chemical precipitation [a]	Not effective at low heavy metal concentration.	Can give high removal of COD.	Complexing agents inhibit hydroxide precipitation. For hydroxide precipitation, optimal pH differs depending on heavy metal while sulfide precipitation gives high degree of heavy metal removal over a broad pH range.	Low capital cost, additional cost for sludge disposal which can make it less economically attractive.	Simple operation, requires large amount of precipitation chemicals. Metal sulfides may form colloidal precipitates that cause settling problems.	Large volumes of low density sludge. Sludge from sulfide precipitation has better thickening and dewatering properties compared to hydroxide precipitation. Hydrogen sulfide formation may be an issue.	[64–66]
Photocatalysis	May be operated at trace concentrations (less than ppm).	Guo et al. [70] investigated Cd^{2+} and Pb^{2+} removal using Gamma irradiation technique (which works in a similar principle as photocatalysis) and found decreased removal at increasing organic matter concentration.	Organic pollutants and heavy metals are removed simultaneously, oxidation and reduction mechanisms are non-selective. Cationic heavy metals may be reduced to less toxic forms and/or deposited in the process.	-	Long retention time.	Less harmful byproducts.	[65,66,70]

[a] Subsequent sedimentation required.

The removal of P and organic matter when enhanced heavy metal removal is introduced is of relevance because (i) the reuse of secondary sludge is more attractive if the organic matter and P content is high, and (ii) certain amounts of COD and P are required for the "treatability" of primary settler effluent in ASP/BNR. Treatability requirements are as follows: for conversion of 1 mg of nitrogen in APS/BNR, 2.9 mg of COD is theoretically required, while to fulfill criteria of less than 10 mg nitrogen in the effluent in full scale BNR, a required ratio of 4.7–8.7 was reported [71]; a general rule to determine the P requirement for heterotrophic growth of biomass is to multiply the COD concentration in primary effluent with 0.005 [71].

Among the treatment technologies described in Table 4, some may have limited potential in relation to the additional requirements. Precipitation may not be efficient when the heavy metal concentrations are low due to large consumption of chemicals and large production of sludge. The ion exchange process is not appropriate for primary treatment of municipal wastewater since it is prone to fouling. Electrochemical treatment methods suffer from high electricity requirements when wastewater volumes are large and may be more appropriate for decentralized wastewater treatment operations. With respect to photocatalysis, there is limited experience with full-scale operation for treatment of municipal wastewater. Loeb et al. [72] pointed out that this may be due to the complexity of large-scale photocatalysis systems. The most attractive methods for enhanced heavy metal removal in this application are thus coagulation/flocculation and adsorption. Coagulation/flocculation could mainly increase the removal of particle-bound heavy metals and heavy metals associated with colloidal matter. However, removal of dissolved heavy metals could be enhanced by use of modified flocculation agents. Addition of a sorbent during or directly following primary treatment could enhance mainly the removal of dissolved heavy metals. Coagulation/flocculation was previously investigated by, e.g., Hargreaves et al. [73,74], while addition of low-cost solid sorbents during primary treatment of municipal wastewater was not previously examined, to our knowledge. Both sorption and coagulation/flocculation are considered easy to operate and relatively cost-effective technologies.

Though not examined in detail here, it is also of importance to consider how enhanced heavy metal removal during primary treatment could affect the removal of heavy metals during secondary treatment and fate of heavy metals during sludge treatment. High removal of BOD during primary treatment may decrease the complexation of heavy metals during secondary treatment [10], which affects the effluent quality. Sludge dewatering may influence both the sludge quality and the amounts of heavy metals entering primary settling (through return liquors, as previously mentioned). Data according to Yoshida et al. [5] indicated that P-loss from sludge during dewatering was greater than loss of heavy metals. In the plant which they examined, sludge dewatering thus had a negative impact on the sludge quality, while heavy metal loss from sludge contributed significantly to the total influent heavy metal load (5–12%). In the case of enhanced heavy metal removal during primary treatment, the primary sludge will contain larger amounts of heavy metals, while secondary sludge will likely contain smaller amounts of heavy metals (depending on the partitioning of heavy metals between secondary sludge and effluent), which may affect the heavy metal losses/fate during sludge dewatering.

6.1. Adsorption Using Low-Cost Sorbents

Low-cost sorbents can be produced from many different materials. The lower cost of such sorbents compared to traditional sorbents is associated with availability—being abundant in nature or produced as a byproduct or waste from other activities/industries. A low-cost sorbent should also require no or limited preprocessing before use, though preprocessing that significantly increases the sorption capacity may be motivated [75]. Low-cost sorbents include:

- Natural materials such as zeolite and clay [75].

- Waste materials such as agricultural residues/plant residues and industrial by-products [76,77] including sewage sludge in wet or dried state [78]. Such waste materials were also referred to as biosorbents when applied as sorbents. Biosorbents are living or dead microorganism biomass (e.g., bacteria, microalgal and fungal biomass) [79,80], or derived from different lignocellulosic materials such as bark, husks, shells, etc. [81].
- Biochars. Production of low-cost biochar use of various organic wastes/residues were investigated, such as saw dust, rice husks, municipal waste, manure, sewage sludge, etc. [82].

Relative sorption capacity (% removal) can generally be expected to increase at low initial metal concentration. Removal of Cd, Cr, Cu, Pb, Zn, Hg and Ag may approach 100% when the initial heavy metal concentration is low [83]. Low concentrations of metals thus facilitate the use of low-cost sorbents with no or little preprocessing (the sorption capacity does not need to be maximized to achieve high removal percentage). Possible challenges of low-cost sorbents are to achieve stable sorption performance (since the properties of waste materials may vary over time) and to separate the sorbent from solution (some sorbents form small, low-density particles) [79,80].

The concurrent removal of COD and P by addition of low-cost sorbents is expected to be relatively small. Utilization of biochars for P removal has generally considered modification of the biochar with metal salts to enhance phosphate sorption capacity ($FeCl_3$, $MgCl_2$) [84]. Peng et al. [85] found that phosphate was not sorbed but released from unmodified sludge-derived biochar. Alkaline sorbents are generally considered appropriate for heavy metal sorption since this is promoted by increased pH, and sorption of phosphate is less efficient at high pH [85].

6.2. *Coagulation/Flocculation*

Coagulation/flocculation is widely used in WWTPs in association with primary settling (for increasing solids and P removal) and in simultaneous precipitation (to enhance P removal in ASP); it is also used as a tertiary treatment step (for P removal) [86]. Coagulation refers to charge neutralization of colloids followed by agglomeration, while flocculation occurs when small particles are joined together through physical bonds—both resulting in formation of larger particles which can be removed through sedimentation. Aluminum, iron and calcium salts are commonly-used coagulants, while polyaluminium chloride, polyacrylamide and polyferric sulphate are commonly-used flocculants [64,86].

Johnson et al. [87] showed that removal of heavy metals (Cr, Cu, Pb, Ni, and Zn) in primary treatment could be more than doubled by adding a coagulant ($FeCl_3$) and an anionic polymer (SS removal increased from 68% to 82%, indicating that the removal of colloidal and dissolved metals also increased). More than 90% removal of Pb, Cu and Zn was shown using commercial aluminum salt coagulants for treatment of combined sewer overflow [88]. One of the possible mechanisms for removal of heavy metals was suggested to be sorption to aluminum oxyhydroxide mineral phases. The concurrent removal of SS was very high. The COD and P removal was not examined, but typical COD removal was reported to be around 60%, which may lead to an insufficient C/N ratio for ASP/BNR. Chemically-enhanced microsieving (as a replacement for primary settling) resulted in a C/N ratio and P concentrations lower than those required for ASP/BNR (at >80% SS removal) [71]. In this context, complete removal of solids (particle-bound heavy metals) is generally not an option when conventional primary settling and ASP/BNR is applied. The typical SS removal during conventional primary settling is ~60% [89].

Increased heavy metal removal in coagulation/flocculation systems is possible through use of polymeric supports, which were impregnated with nanoparticles having high heavy metal affinity. Such flocculation agents include chelating polymers and polymers impregnated with metal hydroxides (e.g., Fe(III), Mn(IV)) [90]. The following are examples of flocculation agents with improved heavy metal removal capacity:

- Polymers in the form of EPS are naturally produced by bacteria. These contain functional groups such as carboxyl, hydroxyl, amino and phosphate groups, which are involved in the removal of heavy metals [91]. Polymers generated by bacteria have also been referred to as bioflocculants. Bioflocculants, which are currently under development, showed a high capacity for heavy metal removal, e.g., for Cd, Cu, Hg and Zn [91]. Liu et al. [92] found that EPS extracted from activated sludge could remove 37–99% of heavy metals (removal increased in the order: Ni < Co < Cd < Cr < Cu < Zn) at a heavy metal concentration range of 10–100 mg/L.
- Heavy metal sorption capacity of polymers may be increased through impregnation of polymers with chelating groups [90]. One such example is the application of the natural material Chitosan, which was shown to remove Cu, Cr and Ni efficiently in lab scale flocculation experiments [93].
- Hargreaves et al. [73] investigated heavy metal removal from municipal wastewater effluent (after treatment in a trickling filter) by addition of ferric chloride (FeCl3), polyethyleneimine (synthetic polymer), chitosan and floculan (biopolymers). Floculan (a tannin-containing, modified plant-based material) performed removal of 77% Cu, 68% Pb and 42% Zn, while efficient Ni removal was not achieved. $FeCl_3$ (which is commonly used in conventional municipal wastewater treatment) achieved similar removal of Cu, Pb and Zn. The drawback of $FeCl_3$ application is that it contains Ni, which may lead to increased Ni concentrations in treated wastewater. It also removes considerable amounts of P (while P removal by floculan was negligible). COD removal of both $FeCl_3$ and floculan was less than 50%. Floculan is a commercially available product, though a drawback is that it has a higher cost compared to $FeCl_3$.

7. Current and Potential Heavy Metal Separation during Primary Settlement

Literature data on the potential removal of total heavy metal concentrations in primary settlers are summarized in Table 5. The removal efficiency is in many cases reported to be highest for Pb, while the lowest removal is often found for Ni, although the variation is large.

In most cases, the sampling of influent was made before addition of return sludge liquors, or it is unclear exactly at what point the influent was sampled, so the removal during primary settling according to Table 5 may not be fully representative of the actual removal capacity in primary settlers. Considering data from Goldstone et al. [42–44], who sampled the influent both before and after addition of return sludge liquors, it is evident that the calculated removal efficiency is larger for all heavy metals if the influent concentrations after addition of return sludge liquors are considered. This is because the return sludge liquors generally have a higher heavy metal concentration than the influent (according to Brown et al. [23], 10–300 times higher) and thus add heavy metals to the primary settler influent.

Brown et al. [23] found that the percentage of heavy metal removal during primary settling was proportional to the total heavy metal concentration in the influent. Collected data from six studies (Figure 2) show that there was a general trend (though not very strong; $\overline{R^2} \leq 0.21$) that increased fractions of particulate heavy metals in raw wastewater can increase the total heavy metal removal during primary settling. This trend could not be shown for Cr and Pb. Since these heavy metals were mainly found in the particulate fraction in all underlying studies, the considerably large variation in removal efficiency might have had other causes than the heavy metal speciation (such as the settler loading/HRT). It should be noted that the total removal efficiencies are uncertain since the influent sampling was often made before addition of return liquors, or the exact sampling point is not known for the underlying studies.

Table 5. Removal of heavy metals during primary settling (%) (data for As not available).

Cu	Cr	Ni	Pb	Cd	Zn	Hg	Ag	Removal Based on Concentration/Mass Flow/Not Specified (C/M/n.s.)	Sampling Made Prior To/After Return Liquors/Not Specified (P/A/n.s.)	Reference
9	2,4	2.5	-	-	14	-	-	M	-	[22] [a]
12–70	17–36	-	28–67	0–25	22–68	13–54	-	C	n.s.	[23] [b]
44	59	29	74	33	-	-	-	C	P	[39] [c]
23	17	2	54	14	19	-	-	M	n.s.	[26]
59	60	67	71	67	73	75	-	n.s.	A	[42–44]
39	55	43	50	41	64	62	-	n.s.	P	[42–44]
22	-	22	23	-	20	52	-	C	P	[41,94]
29	39	24	23	32	22	-	-	M	P [d]	[35] [c]
70	73	23	73	72	74	-	-	C	P	[27,28]
60	68	50	-	39	44	-	-	M	P [d]	[29]
32	55	15	61	50	53	57	-	C	n.s.	[48]
61	47	22	65	68	60	61	43	M	A	[5] [c, e]
31	-	8	31	33	39	54	-	C	P [f]	[47] [c]
12–70	17–73	2–67	23–74	0–72	20–74	13–75	-			Range [g]

[a] Data from pilot plant experiments at influent heavy metal concentrations >10 mg/L. Removal was calculated based on amount of heavy metals in primary sludge (not in primary settler effluent). No return liquors were added in the pilot plant. [b] Range of four different treatment plants. [c] The data were estimated from graph. [d] Influent wastewater was sampled after addition of return sludge liquors and after grit removal, but before return of another supernatant stream. [e] Removal was calculated based on distribution of heavy metals between effluent, primary sludge and secondary sludge, i.e., mass flows. This is comparable to sampling the influent after addition of return sludge liquors. [f] Average from 16 WWTPs. Most samples of influent wastewater were taken before addition of return sludge liquors. [g] Data from Barth et al. [22] were not included, since the study was conducted with metal spiking of influent.

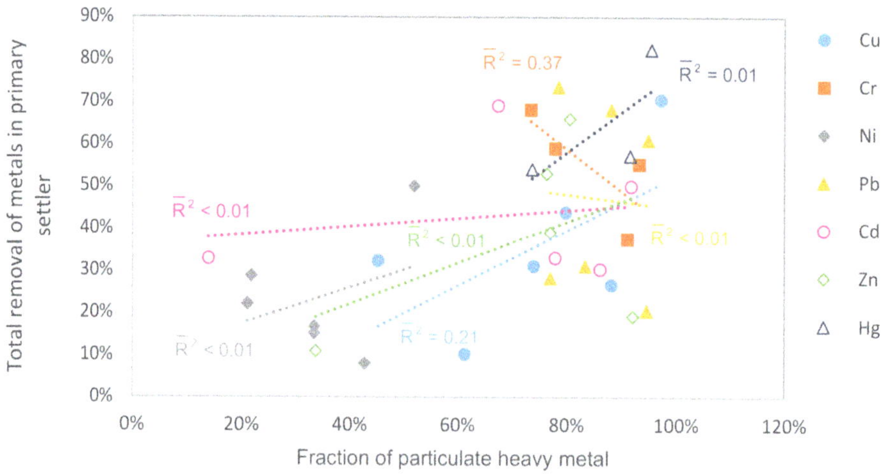

Figure 2. Total removal of heavy metals in primary settler at different fractions of particulate heavy metals in raw wastewater. [35,39,41–44,47,48]. (Data for As and Ag not available.) The dotted lines represent the linear regression between the fraction of particulate heavy metals in raw wastewater and total removal of heavy metals in the primary settler, and strength of correlation is indicated with $\overline{R^2}$ values.

The total removal of heavy metals in the primary settler varied greatly between studies and seemed to increase somewhat at higher fractions of particulate heavy metals in raw wastewater. Nevertheless, the primary settler effluent concentrations correlated with the concentrations in raw wastewater; see Figure 3. A similar correlation was found by Santos and Judd [95] for the removal of heavy metals in ASP and membrane bioreactors.

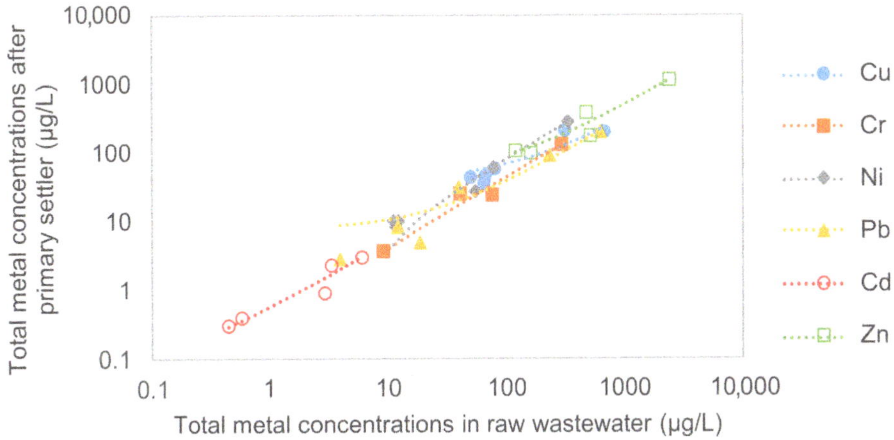

Figure 3. Total heavy metal concentrations after primary settling compared to total heavy metal concentrations in raw wastewater. [35,39,41–44,47,48]. (Sufficient data for As, Ag and Hg not available.) The dotted lines represent the linear regression between the total heavy metals concentrations in raw wastewater and total heavy metal concentrations after primary settling (R^2 values: 0.75, 0.99, >0.99, 0.99, 0.88 and 0.97 for Cu, Cr, Ni, Pb, Cd and Zn, respectively; linear relations between concentration in raw wastewater (x) and concentration after primary settler (y): $y_{Cu} = 0.29x_{Cu} + 39$, $y_{Cr} = 0.45x_{Cr} - 0.42$, $y_{Ni} = 0.86x_{Ni} - 4.9$, $y_{Pb} = 0.31x_{Pb} + 7.5$, $y_{Cd} = 0.50x_{Cd} + 0.067$, and $y_{Zn} = 0.45x_{Zn} + 47$; the regression for Cu is not reliable, because the R2 value is not as high as for the other metals; the regression coefficient can be taken as an estimation of the fraction of each heavy metal remaining after primary settling; the value of the intercept can be interpreted as the average amount of metals added through return flows, and negative values are therefore unrealistic).

7.1. Potential Improvement of Heavy Metal Removal

In Figure 4, removal efficiencies for different heavy metals during primary settling are given based on dissolved and particulate fractions in raw and settled wastewater. Calculations of the total removal were made under the idealized assumption that 100% of heavy metals which are "dissolved" and associated with particles, respectively, could be removed.

Based on these idealized calculations, some indication is given on the potential of enhancing removal of dissolved heavy metals (i.e., applying sorption technologies), or enhancing removal of particle bound metals (i.e., applying coagulation/flocculation).

Enhanced removal of dissolved heavy metals seems to enable improved removal mainly of Cu and Ni. For Pb, Cd and Zn, some of the studies indicate a possible improvement of 50% or more. Sorption using low-cost sorbents may be an appropriate technology for addressing enhanced removal of these metals. It must be stressed the sorption capacity of the applied sorbent for each metal needs to be considered, and that some of the dissolved heavy metals may be unavailable for sorption (as mentioned in previous discussion on the fractioning between colloidal and truly dissolved/labile heavy metals). In the case of Cr and Hg, the possible improvement (in total removal by enhancing removal of dissolved metals) seems to be low.

Enhanced removal techniques associated with particles seem to enable improved removal of all heavy metals, though the possible improvement of Ni removal seems to be low in some of the cases. This indicates that coagulation/flocculation may be an appropriate technology for addressing enhanced removal of these metals.

Figure 4. Current removal during primary settling (removal dissolved, particulate, according to reference and calculated removal) and potential for improved removal of dissolved (sorption) and particulate heavy metals (coagulation/flocculation) for: (**a**) Cu; (**b**) Cr; (**c**) Ni; (**d**) Pb; (**e**) Cd; (**f**) Zn; (**g**) Hg. (Data for As and Ag not available.) The removals according to the references (current dissolved and particulate removal) ([35,39,41–43,47,48]) are given compared to the calculated removal, which were based on the concentrations given (not on mass flows). Negative values indicate that the concentrations after primary settling were higher than the concentrations before primary settling. Comments regarding sampling points: In [39,41], the influent was sampled before addition of return sludge liquors (i.e., the removal achieved by primary settling might be somewhat underestimated). In [42,43], the influent was sampled after addition of return sludge liquors. In [35], the influent was sampled after return of liquors from sludge digestion and after grit removal but before addition of another supernatant stream. In [48], it is not clear if sampling was performed before or after addition of return sludge liquors. In [47], average values from 16 WWTPs in the UK were given, and most samples of influent wastewater were taken before return flows from sludge treatment. Comment on the "Removal according to reference" compared to "Calculated removal": In [39,41,47,48], the removal given in the references was calculated based on the concentrations in the liquid phase. For [39,47,48], the removal calculated here is thus the same as the removal calculated in the references. The calculated removal calculated for [41] differs from the removal given in the reference because of rounding error (rounded off concentrations were given in the reference). In [35], the removal according to the reference was calculated based on mass flows and thus differs somewhat to the removal calculated here. In [42,43], it is unclear if the removal in the reference was calculated based on mass flow or concentration.

Furthermore, emerging wastewater treatment technologies with high solids removal efficiency may be beneficial in terms of residual heavy metals in effluents. A treatment technology which is increasingly implemented is membrane bioreactors, which is a compact technology with high effluent quality. Another wastewater treatment technology of interest is mainstream anammox/AB-system (based on two successive biological treatment stages, A and B, and using no primary clarifier). Mainstream anammox is a technology which is still under development. The advantages of such systems are low external energy requirements and compact construction [96]. With regards to potential heavy metal removal, the solids concentration and removal in the A-stage are typically high which may facilitate removal of both particle-bound and dissolved heavy metals.

8. Conclusions and Outlook

The removal of heavy metals during primary settling is determined by the speciation of heavy metals in raw municipal wastewater and the sorption/desorption of heavy metals occurring in the settler. Volatilization is relevant for some metals, e.g., heavy metal Hg (and semi-metal Sb). Liquors from sludge-thickening and dewatering are often returned to the primary settler. These may increase the heavy metal removal in the primary settler if the concentration of SS is increased or if containing residue of ferric chloride coagulant. Furthermore, changes in redox potential may affect the heavy metal speciation and thus removal. Decreased heavy metal removal may be a result if return liquors contain large amounts of heavy metals associated with fine particles. Heavy metals contained in process chemicals used in secondary or tertiary treatment may contribute to the total heavy metal load entering the primary settler via return liquors. Return sludge from ASP/BNR is sometimes recirculated to the primary settler head, and biomass uptake/sorption may in that case support heavy metal removal.

Mechanisms of heavy metal removal during primary settling are not fully understood. Missing or disparate information on process parameters such as HRT, pH and composition of return flows makes it hard to generalize the findings from different studies.

Knowledge concerning the speciation of heavy metals is of importance to design appropriate removal technologies. Research is not consistent in considering total and dissolved concentrations (<0.45 or 0.2 µm). Studies which attempted to elucidate heavy metal speciation further applied different methods for pre-processing wastewater before analysis of heavy metals, resulting in different definitions of heavy metal speciation. One definition is particulate, dissolved and truly dissolved and the other definition is particulate, inert and labile. The different ways of defining heavy metal speciation in wastewater make it difficult to compare the results of different studies. However, additional studies including such detailed information regarding heavy metal speciation in influents, effluents and sludge could shed further light on the mechanisms of heavy metal removal in WWTPs.

The variation of total heavy metal concentrations and dissolved fraction in raw municipal wastewater points to a need for site-specific assessment of appropriate technologies for improved heavy metal removal.

Modelling may be helpful to increase the understanding of heavy metal removal during primary settling. So far, full-scale modelling of heavy metal removal during primary settling was made under the assumption that the removal of heavy metals is determined by the distribution of heavy metals between particulate and dissolved fractions in raw wastewater and did not consider the possible sorption/desorption which may occur during the process. Knowledge on the speciation of heavy metals and distribution between particulate and colloidal/truly dissolved or inert/labile may be implemented in models. However, in standard WWTP modelling applied at many WWTPs around the world (the "activated sludge model" [51]), the dynamics of colloidal particles are not described. No literature model for heavy metal speciation/removal from municipal wastewater has considered parameters such as hardness, alkalinity and redox potential of the wastewater.

Coagulation/flocculation and use of low-cost sorbents were identified as the most promising methods for enhancing heavy metal removal during or directly following primary settling. In the context of decreased heavy metal concentrations in both effluent and sludge from ASP/BNR, the treatment technology should not only enable high heavy metal removal but also: be suitable in wastewater of high solids concentration; have low impact in terms of, e.g., energy use and sludge production; the concurrent removal of P and organic matter should be low. In the case of coagulation/flocculation, the concurrent removal of organic matter and P needs to be controlled to assure that the C/N ratio and P concentrations are sufficient for subsequent ASP/BNR. Moreover, the possible complexation of heavy metals during ASP/BNR will likely decrease if the MLSS is decreased, which is of importance for the removal from the liquid phase/effluent. Application of low-cost sorbents during primary settling of municipal wastewater was not examined previously, to our knowledge. Based on the heavy metal speciation (Cd, Cr, Cu, Ni, Pb, Zn and Hg) and current removal capacity of primary settlers, it is proposed that sorption technologies may be most effective for enhancing the removal of Cu and Ni, and in some cases of Pb, Cd and Zn. Furthermore, the studies concerning heavy metal speciation and removal reviewed here indicate that the fraction of dissolved Cr and Cd may likely be smaller at higher total influent heavy metal concentrations. Sorption technologies may thus be less efficient in the case when Cr and Cd concentrations are higher. Coagulation/flocculation may be efficient for increasing removal for all the heavy metals studied, with possibly less impact on Ni removal.

Author Contributions: Conceptualization, I.S. and E.T.; formal analysis, I.S.; data curation, I.S.; writing—original draft preparation, I.S.; writing—review and editing, E.T.; supervision, E.T.; project administration, E.T.; funding acquisition, E.T. All authors have read and agreed to the published version of the manuscript.

Funding: This research was funded by VA-kluster Mälardalen, Eskilstuna Energi och Miljö, Vafab Miljö and Mälarenergi.

Institutional Review Board Statement: Not applicable.

Informed Consent Statement: Not applicable.

Data Availability Statement: No new data were created or analyzed in this study. Data sharing is not applicable to this article.

Conflicts of Interest: The authors declare that they have no conflicts of interest. The funders had no role in the design of the study; in the collection, analyses or interpretation of data; in the writing of the manuscript; or in the decision to publish the results.

References

1. Vareda, J.P.; Valente, A.J.; Durães, L. Assessment of heavy metal pollution from anthropogenic activities and remediation strategies: A review. *J. Environ. Manag.* **2019**, *246*, 101–118. [CrossRef] [PubMed]
2. Vardhan, K.H.; Kumar, P.S.; Panda, R.C. A review on heavy metal pollution, toxicity and remedial measures: Current trends and future perspectives. *J. Mol. Liq.* **2019**, *290*, 111197. [CrossRef]
3. Libralato, G.; Ghirardini, A.V.; Avezzù, F. To centralise or to decentralise: An overview of the most recent trends in wastewater treatment management. *J. Environ. Manag.* **2012**, *94*, 61–68. [CrossRef] [PubMed]
4. Cantinho, P.; Matos, M.; Trancoso, M.A.; Dos Santos, M.M.C. Behaviour and fate of metals in urban wastewater treatment plants: A review. *Int. J. Environ. Sci. Technol.* **2016**, *13*, 359–386. [CrossRef]
5. Yoshida, H.; Christensen, T.; Guildal, T.; Scheutz, C. A comprehensive substance flow analysis of a municipal wastewater and sludge treatment plant. *Chemosphere* **2015**, *138*, 874–882. [CrossRef] [PubMed]
6. Milieu; WRC; RPA. *Environmental, Economic and Social Impacts of the Use of Sewage Sludge on Land Draft Summary*; Report 1: Assessment of Existing Knowledge; European Comission/Milieu Ltd.: Brussels, Belgium, 2010.
7. Tao, W.; Sauba, K.; Fattah, K.P.; Smith, J.R. Designing constructed wetlands for reclamation of pretreated wastewater and stormwater. *Rev. Environ. Sci. Bio/Technol.* **2017**, *16*, 37–57. [CrossRef]
8. Lindqvist, A.; Sörme, L.; Söderberg, H. Capacity to Influence Sources of Heavy Metals to Wastewater Treatment Sludge. *Environ. Manag.* **2003**, *31*, 421–428. [CrossRef]

9. Hospido, A.; Moreira, M.T.; Fernández-Couto, M.; Feijoo, G. Environmental performance of a municipal wastewater treatment plant. *Int. J. Life Cycle Assess.* **2004**, *9*, 261–271. [CrossRef]
10. Ziolko, D.; Martin, O.V.; Scrimshaw, M.D.; Lester, J.N. An Evaluation of Metal Removal During Wastewater Treatment: The Potential to Achieve More Stringent Final Effluent Standards. *Crit. Rev. Environ. Sci. Technol.* **2011**, *41*, 733–769. [CrossRef]
11. Santos-Echeandía, J. The fate and transport of trace metals through sewage treatment plant processes. In *Sewage Treatment: Uses, Processes and Impact*; Stephens, A., Fuller, M., Eds.; Nova Science Publication Inc.: New York, NY, USA, 2009; pp. 1–52. Available online: http://www.scopus.com/inward/record.url?eid=2-s2.0-84892042536&partnerID=40&md5=97193d874990933d2a3949b2d4f5dd64 (accessed on 1 September 2020).
12. Hargreaves, A.J.; Constantino, C.; Dotro, G.; Cartmell, E.; Campo, P. Fate and removal of metals in municipal wastewater treatment: A review. *Environ. Technol. Rev.* **2018**, *7*, 1–18. [CrossRef]
13. European Comission. Directive 2000/60/EC of the European Parliament and of the Council of 23 October 2000 Establishing a Framework for Community Action in the Field of Water Policy. *Off. J. Eur. Union* **2000**, *327*, 1–73.
14. European Comission. Directive 2008/105/EC of the European Parliament and of the Council of 16 December 2008 on Environmental Quality Standards in the Field of Water Policy. *Off. J. Eur. Union* **2008**, *348*, 84–97.
15. Irmer, U.; Rau, F.; Arle, J.; Claussen, U.; Mohaupt, V. *Ecological Environmental Quality Standards of "River Basin Specific Pollutants" in Surface Waters—Update and Development Analysis of a European Comparison between Member States*; WFD CIS Working Group A Ecological Status (ECOSTAT); Publications Office of the European Union: Luxembourg, 2014.
16. European Comission. Council Directive 86/278/EEC of 12 June 1986 on the Protection of the Environment, and in Particular of the Soil, When Sewage Sludge is Used in Agriculture. *Off. J. Eur. Union* **1986**, *181*, 6–12.
17. Hudcová, H.; Vymazal, J.; Rozkošný, M. Present restrictions of sewage sludge application in agriculture within the European Union. *Soil Water Res.* **2019**, *14*, 104–120. [CrossRef]
18. Clara, M.; Windhofer, G.; Weilgony, P.; Gans, O.; Denner, M.; Chovanec, A.; Zessner, M. Identification of relevant micropollutants in Austrian municipal wastewater and their behaviour during wastewater treatment. *Chemosphere* **2012**, *87*, 1265–1272. [CrossRef] [PubMed]
19. Gardner, M.; Comber, S.; Scrimshaw, M.D.; Cartmell, E.; Lester, J.; Ellor, B. The significance of hazardous chemicals in wastewater treatment works effluents. *Sci. Total Environ.* **2012**, *437*, 363–372. [CrossRef] [PubMed]
20. Mattsson, A.; Davidsson, F. A strategy for reducing pollutants at source in order to obtain sustainable agricultural recycling of wastewater sludge. *Water Sci. Technol.* **2012**, *66*, 1879–1884. [CrossRef]
21. Rule, K.; Comber, S.; Ross, D.; Thornton, A.; Makropoulos, C.; Rautiu, R. Survey of priority substances entering thirty English wastewater treatment works. *Water Environ. J.* **2006**, *20*, 177–184. [CrossRef]
22. Barth, M.; Ettinger, J.; Salotto, B.; McDermott, G. Summary report on the effects of heavy metals on the biological treatment processes. *Water Pollut.* **1965**, *37*, 86–96.
23. Brown, H.G.; Hensley, C.P.; McKinney, G.L.; Robinson, J.L. Efficiency of Heavy Metals Removal in Municipal Sewage Treatment Plants. *Environ. Lett.* **1973**, *5*, 103–114. [CrossRef]
24. Carletti, G.; Fatone, F.; Bolzonella, D.; Cecchi, F. Occurrence and fate of heavy metals in large wastewater treatment plants treating municipal and industrial wastewaters. *Water Sci. Technol.* **2008**, *57*, 1329–1336. [CrossRef] [PubMed]
25. Chipasa, K.B. Accumulation and fate of selected heavy metals in a biological wastewater treatment system. *Waste Manag.* **2003**, *23*, 135–143. [CrossRef]
26. Davis, J.A., III; Jacknow, J. Heavy metals in wastewater in three urban areas. *J. Water Pollut. Control. Fed.* **1975**, *47*, 2292–2297.
27. Stoveland, S.; Astruc, M.; Lester, J.N.; Perry, R. The balance of heavy metals through a sewage treatment works III. Chromium, Nickel and Zinc. *Sci. Total Environ.* **1979**, *12*, 25–34. [CrossRef]
28. Lester, J.; Harrison, R.; Perry, R. The balance of heavy metals through a sewage treatment works I. Lead, cadmium and copper. *Sci. Total Environ.* **1979**, *12*, 13–23. [CrossRef]
29. Nielsen, J.; Hrudey, S. Metal loadings and removal at a municipal activated sludge plant. *Water Res.* **1983**, *17*, 1041–1052. [CrossRef]
30. Oliveira, A.D.S.; Bocio, A.; Trevilato, T.M.B.; Takayanagui, A.M.M.; Domingo, J.L.; Segura-Muñoz, S.I. Heavy metals in untreated/treated urban effluent and sludge from a biological wastewater treatment plant. *Environ. Sci. Pollut. Res.* **2007**, *14*, 483–489. [CrossRef]
31. Sorme, L.; Lagerkvist, R. Sources of heavy metals in urban wastewater in Stockholm. *Sci. Total Environ.* **2002**, *298*, 131–145. [CrossRef]
32. Mitsuo, C.; Kiyoshi, M.; Hiroshi, S.; Tadahiro, M. The amount of heavy metals derived from domestic sources in Japan. *Water Air Soil Pollut.* **1991**, *57–58*, 829–837.
33. Choubert, J.-M.; Pomiès, M.; Ruel, S.M.; Coquery, M. Influent concentrations and removal performances of metals through municipal wastewater treatment processes. *Water Sci. Technol.* **2011**, *63*, 1967–1973. [CrossRef] [PubMed]
34. Toumi, A.; Nejmeddine, A.; Belkoura, M. The Fate of Heavy Metals (Zn, Cu, Pb, Cd and Cr) in an Integrated Wastewater Treatment Plant: Two Phase Anaerobic Reactor (RAP)—High Rate Algal Pond (HRAP). *Environ. Technol.* **2003**, *24*, 153–159. [CrossRef] [PubMed]
35. Karvelas, M.; Katsoyiannis, A.; Samara, C. Occurrence and fate of heavy metals in the wastewater treatment process. *Chemosphere* **2003**, *53*, 1201–1210. [CrossRef]

36. Lawson, P.S.; Sterritt, R.M.; Lester, J.N. The speciation of metals in sewage and activated sludge effluent. *Water Air Soil Pollut.* **1984**, *21*, 387–402. [CrossRef]
37. El Samrani, A.G.; Lartiges, B.S.; Ghanbaja, J.; Yvon, J.; Kohler, A. Trace element carriers in combined sewer during dry and wet weather: An electron microscope investigation. *Water Res.* **2004**, *38*, 2063–2076. [CrossRef] [PubMed]
38. Houhou, J.; Lartiges, B.; Montarges-Pelletier, E.; Sieliechi, J.; Ghanbaja, J.; Kohler, A. Sources, nature, and fate of heavy metal-bearing particles in the sewer system. *Sci. Total Environ.* **2009**, *407*, 6052–6062. [CrossRef]
39. Buzier, R.; Tusseau-Vuillemin, M.-H.; Meriadec, C.M.D.; Rousselot, O.; Mouchel, J.-M. Trace metal speciation and fluxes within a major French wastewater treatment plant: Impact of the successive treatments stages. *Chemosphere* **2006**, *65*, 2419–2426. [CrossRef]
40. Gourlay-Francé, C.; Bressy, A.; Uher, E.; Lorgeoux, C. Labile, dissolved and particulate PAHs and trace metals in wastewater: Passive sampling, occurrence, partitioning in treatment plants. *Water Sci. Technol.* **2011**, *63*, 1327–1333. [CrossRef]
41. Hargreaves, A.J.; Vale, P.; Whelan, J.; Constantino, C.; Dotro, G.; Campo, P.; Cartmell, E. Distribution of trace metals (Cu, Pb, Ni, Zn) between particulate, colloidal and truly dissolved fractions in wastewater treatment. *Chemosphere* **2017**, *175*, 239–246. [CrossRef]
42. Goldstone, M.E.; Kirk, P.W.W.; Lester, J.N. The behaviour of heavy metals during wastewater treatment I. Cadmiun, Chromium and Copper. *Sci. Total Environ.* **1990**, *95*, 233–252. [CrossRef]
43. Goldstone, M.; Kirk, P.; Lester, J. The behaviour of heavy metals during wastewater treatment II. Lead, nickel and zinc. *Sci. Total Environ.* **1990**, *95*, 253–270. [CrossRef]
44. Goldstone, M.; Atkinson, C.; Kirk, P.; Lester, J. The behaviour of heavy metals during wastewater treatment III. Mercury and arsenic. *Sci. Total Environ.* **1990**, *95*, 271–294. [CrossRef]
45. Inna, D.; Lester, J.; Scrimshaw, M.; Cartmell, E. Speciation and fate of copper in sewage treatment works with and without tertiary treatment: The effect of return flows. *Environ. Technol.* **2013**, *35*, 1–9. [CrossRef] [PubMed]
46. Cecchini, G.; Cirello, P.; Eramo, B. Partitioning Dynamics and Fate of Metals in an Urban Wastewater Treatment Plant. *Environ. Eng. Manag. J.* **2015**, *14*, 1511–1520. [CrossRef]
47. Gardner, M.; Jones, V.; Comber, S.; Scrimshaw, M.D.; Coello-Garcia, T.; Cartmell, E.; Lester, J.; Ellor, B. Performance of UK wastewater treatment works with respect to trace contaminants. *Sci. Total Environ.* **2013**, *456–457*, 359–369. [CrossRef] [PubMed]
48. Oliver, B.G.; Cosgrove, E.G. The efficiency of heavy metal removal by a conventional activated sludge treatment plant. *Water Res.* **1974**, *8*, 869–874. [CrossRef]
49. Chen, K.Y.; Young, C.S.; Jan, T.K.; Rohatgi, N. Trace metals in wastewater effluents. *J. Water Pollut. Control. Fed.* **1974**, *46*, 2663–2675.
50. Modin, O.; Persson, F.; Wilén, B.-M.; Hermansson, M. Nonoxidative removal of organics in the activated sludge process. *Crit. Rev. Environ. Sci. Technol.* **2016**, *46*, 1–38. [CrossRef]
51. Pomiès, M.; Choubert, J.-M.; Wisniewski, C.; Coquery, M. Modelling of micropollutant removal in biological wastewater treatments: A review. *Sci. Total Environ.* **2013**, *443*, 733–748. [CrossRef]
52. Kempton, S.; Sterritt, R.; Lester, J. Heavy metal removal in primary sedimentation I. The influence of metal solubility. *Sci. Total Environ.* **1987**, *63*, 231–246. [CrossRef]
53. Petrie, B.; McAdam, E.J.; Lester, J.N.; Cartmell, E. Assessing potential modifications to the activated sludge process to improve simultaneous removal of a diverse range of micropollutants. *Water Res.* **2014**, *62*, 180–192. [CrossRef]
54. Dionisi, D.; Bornoroni, L.; Mainelli, S.; Majone, M.; Pagnanelli, F.; Papini, M.P. Theoretical and Experimental Analysis of the Role of Sludge Age on the Removal of Adsorbed Micropollutants in Activated Sludge Processes. *Ind. Eng. Chem. Res.* **2008**, *47*, 6775–6782. [CrossRef]
55. Ren, Y.; Wang, J.; Li, H.; Zhang, J.; Qi, P.; Hu, Z. Nitrous oxide and methane emissions from different treatment processes in full-scale municipal wastewater treatment plants. *Environ. Technol.* **2013**, *34*, 2917–2927. [CrossRef]
56. Xiao, H.; Zhang, S.; Zhai, J.; He, Q.; Mels, A.; Ning, K.; Liu, J. Retention and distribution of Cu, Pb, Cr, and Zn in a full-scale hybrid constructed wetland receiving municipal sewage. *Water Sci. Technol.* **2013**, *67*, 2257–2264. [CrossRef]
57. Comte, S.; Guibaud, G.; Baudu, M. Biosorption properties of extracellular polymeric substances (EPS) towards Cd, Cu and Pb for different pH values. *J. Hazard. Mater.* **2008**, *151*, 185–193. [CrossRef]
58. Guibaud, G.; van Hullebusch, E.; Bordas, F. Lead and cadmium biosorption by extracellular polymeric substances (EPS) extracted from activated sludges: pH-sorption edge tests and mathematical equilibrium modelling. *Chemosphere* **2006**, *64*, 1955–1962. [CrossRef]
59. Parker, W.J.; Monteith, H.D.; Bell, J.P.; Melcer, H.; Mac Berthouex, P. Comprehensive fate model for metals in municipal wastewater. *J. Environ. Eng.* **1994**, *20*, 1266–1283. [CrossRef]
60. Wang, J.; Huang, C.; Allen, H.E. Predicting metals partitioning in wastewater treatment plant influents. *Water Res.* **2006**, *40*, 1333–1340. [CrossRef]
61. Katsoyiannis, A.; Samara, C. The fate of dissolved organic carbon (DOC) in the wastewater treatment process and its importance in the removal of wastewater contaminants. *Environ. Sci. Pollut. Res.* **2007**, *14*, 284–292. [CrossRef]
62. Barret, M.; Patureau, D.; Latrille, E.; Carrère, H. A three-compartment model for micropollutants sorption in sludge: Methodological approach and insights. *Water Res.* **2010**, *44*, 616–624. [CrossRef]
63. Cloutier, F.; Jalby, G.; Lessard, P.; Vanrolleghem, P.A. Modélisation dynamique du comportement des métaux lourds des stations d'épuration. *J. Water Sci.* **2009**, *22*, 461–471. [CrossRef]

64. Fu, F.; Wang, Q. Removal of heavy metal ions from wastewaters: A review. *J. Environ. Manag.* **2011**, *92*, 407–418. [CrossRef] [PubMed]
65. Carolin, C.F.; Kumar, P.S.; Saravanan, A.; Joshiba, G.J.; Naushad, M. Efficient techniques for the removal of toxic heavy metals from aquatic environment: A review. *J. Environ. Chem. Eng.* **2017**, *5*, 2782–2799. [CrossRef]
66. Barakat, M. New trends in removing heavy metals from industrial wastewater. *Arab. J. Chem.* **2011**, *4*, 361–377. [CrossRef]
67. Tong, Y.; McNamara, P.J.; Mayer, B.K. Adsorption of organic micropollutants onto biochar: A review of relevant kinetics, mechanisms and equilibrium. *Environ. Sci. Water Res. Technol.* **2019**, *5*, 821–838. [CrossRef]
68. Wei, D.; Ngo, H.H.; Guo, W.; Xu, W.; Du, B.; Khan, M.S.; Wei, Q. Biosorption performance evaluation of heavy metal onto aerobic granular sludge-derived biochar in the presence of effluent organic matter via batch and fluorescence approaches. *Bioresour. Technol.* **2018**, *249*, 410–416. [CrossRef]
69. Muddemann, T.; Haupt, D.; Sievers, M.; Kunz, U. Electrochemical Reactors for Wastewater Treatment. *ChemBioEng Rev.* **2019**, *6*, 142–156. [CrossRef]
70. Guo, Z.; Tang, D.; Liu, X.; Zheng, Z. Gamma irradiation-induced Cd2+ and Pb2+ removal from different kinds of water. *Radiat. Phys. Chem.* **2008**, *77*, 1021–1026. [CrossRef]
71. Väänänen, J. Microsieving in Municipal Wastewater Treatment: Chemically Enhanced Primary and Tertiary Treatment. Ph.D. Thesis, Lund University, Lund, Sweden, 2017.
72. Loeb, S.K.; Alvarez, P.J.J.; Brame, J.A.; Cates, E.L.; Choi, W.; Crittenden, J.; Dionysiou, D.D.; Li, Q.; Li-Puma, G.; Quan, X.; et al. The Technology Horizon for Photocatalytic Water Treatment: Sunrise or Sunset? *Environ. Sci. Technol.* **2019**, *53*, 2937–2947. [CrossRef] [PubMed]
73. Hargreaves, A.J.; Vale, P.; Whelan, J.; Alibardi, L.; Constantino, C.; Dotro, G.; Cartmell, E.; Campo, P. Coagulation–flocculation process with metal salts, synthetic polymers and biopolymers for the removal of trace metals (Cu, Pb, Ni, Zn) from municipal wastewater. *Clean Technol. Environ. Policy* **2018**, *20*, 393–402. [CrossRef]
74. Hargreaves, A.J.; Vale, P.; Whelan, J.; Alibardi, L.; Constantino, C.; Dotro, G.; Cartmell, E.; Campo, P. Impacts of coagulation-flocculation treatment on the size distribution and bioavailability of trace metals (Cu, Pb, Ni, Zn) in municipal wastewater. *Water Res.* **2018**, *128*, 120–128. [CrossRef] [PubMed]
75. Bailey, S.E.; Olin, T.J.; Bricka, R.; Adrian, D. A review of potentially low-cost sorbents for heavy metals. *Water Res.* **1999**, *33*, 2469–2479. [CrossRef]
76. Kurniawan, T.A.; Chan, G.Y.; Lo, W.-H.; Babel, S. Comparisons of low-cost adsorbents for treating wastewaters laden with heavy metals. *Sci. Total Environ.* **2006**, *366*, 409–426. [CrossRef]
77. Wan Ngah, W.S.; Hanafiah, M.A.K.M. Removal of heavy metal ions from wastewater by chemically modified plant wastes as adsorbents: A review. *Bioresour. Technol.* **2008**, *99*, 3935–3948. [CrossRef]
78. Smith, K.; Fowler, G.; Pullket, S.; Graham, N. Sewage sludge-based adsorbents: A review of their production, properties and use in water treatment applications. *Water Res.* **2009**, *43*, 2569–2594. [CrossRef]
79. Qin, H.; Hu, T.; Zhai, Y.; Lu, N.; Aliyeva, J. The improved methods of heavy metals removal by biosorbents: A review. *Environ. Pollut.* **2020**, *258*, 113777. [CrossRef]
80. Niazi, N.; Murtaza, B.; Bibi, I.; Shahid, M.; White, J.; Nawaz, M.; Bashir, S.; Shakoor, M.; Choppala, G.; Wang, H. Removal and Recovery of Metals by Biosorbents and Biochars Derived from Biowastes. *Environ. Mat. Waste* **2016**, 149–177. [CrossRef]
81. Salman, M.; Athar, M.; Farooq, U. Biosorption of heavy metals from aqueous solutions using indigenous and modified lignocellulosic materials. *Rev. Environ. Sci. Bio/Technol.* **2015**, *14*, 211–228. [CrossRef]
82. Mohan, D.; Sarswat, A.; Ok, Y.S.; Pittman, C.U. Organic and inorganic contaminants removal from water with biochar, a renewable, low cost and sustainable adsorbent—A critical review. *Bioresour. Technol.* **2014**, *160*, 191–202. [CrossRef]
83. Chen, T.; Zhou, Z.; Han, R.; Meng, R.; Wang, H.; Lu, W. Adsorption of cadmium by biochar derived from municipal sewage sludge: Impact factors and adsorption mechanism. *Chemosphere* **2015**, *134*, 286–293. [CrossRef]
84. Yin, Q.; Zhang, B.; Wang, R.; Zhao, Z. Biochar as an adsorbent for inorganic nitrogen and phosphorus removal from water: A review. *Environ. Sci. Pollut. Res.* **2017**, *24*, 26297–26309. [CrossRef]
85. Peng, G.; Jiang, S.; Wang, Y.; Zhang, Q.; Cao, Y.; Sun, Y.; Zhang, W.; Wang, L. Synthesis of Mn/Al double oxygen biochar from dewatered sludge for enhancing phosphate removal. *J. Clean. Prod.* **2020**, *251*, 119725. [CrossRef]
86. La Cour Jansen, J.; Arvin, E.; Henze, M.; Harremoës, P. *Wastewater Treatment—Biological and Chemical Processes*, 4th ed.; Polyteknisk Forlag: Copenhagen, Denmark, 2019; p. 372.
87. Johnson, P.D.; Girinathannair, P.; Ohlinger, K.N.; Ritchie, S.; Teuber, L.; Kirby, J. Enhanced removal of heavy metals in primary treatment using coagulation and flocculation. *Water Environ. Res.* **2008**, *80*, 472–479. [CrossRef]
88. El Samrani, A.G.; Lartiges, B.S.; Villiéras, F. Chemical coagulation of combined sewer overflow: Heavy metal removal and treatment optimization. *Water Res.* **2008**, *42*, 951–960. [CrossRef] [PubMed]
89. Metcalf, E. *Wastewater Engineering: Treatment and Resource Recovery*, 5th ed.; McGraw-Hill: New York, NY, USA, 2014.
90. Pan, B.; Pan, B.; Zhang, W.; Lv, L.; Zhang, Q.; Zheng, S. Development of polymeric and polymer-based hybrid adsorbents for pollutants removal from waters. *Chem. Eng. J.* **2009**, *151*, 19–29. [CrossRef]
91. Vimala, R.; Escaline, J.L.; Sivaramakrishnan, S. Characterization of self-assembled bioflocculant from the microbial consortium and its applications. *J. Environ. Manag.* **2020**, *258*, 110000. [CrossRef] [PubMed]

92. Liu, Y.; Lam, M.C.; Fang, H.H. Adsorption of heavy metals by EPS of activated sludge. *Water Sci. Technol.* **2001**, *43*, 59–66. [CrossRef] [PubMed]
93. Sun, Y.; Shah, K.J.; Sun, W.; Zheng, H. Performance evaluation of chitosan-based flocculants with good pH resistance and high heavy metals removal capacity. *Sep. Purif. Technol.* **2019**, *215*, 208–216. [CrossRef]
94. Hargreaves, A.J.; Vale, P.; Whelan, J.; Constantino, C.; Dotro, G.; Cartmell, E. Mercury and antimony in wastewater: Fate and treatment. *Water Air Soil Pollut.* **2016**, *227*, 89. [CrossRef] [PubMed]
95. Santos, A.; Judd, S. The fate of metals in wastewater treated by the activated sludge process and membrane bioreactors: A brief review. *J. Environ. Monit.* **2010**, *12*, 110–118. [CrossRef]
96. Hoekstra, M.; Geilvoet, S.P.; Hendrickx, T.L.G.; Kip, C.S.V.E.T.; Kleerebezem, R.; Van Loosdrecht, M.C.M. Towards mainstream anammox: Lessons learned from pilot-scale research at WWTP Dokhaven. *Environ. Technol.* **2019**, *40*, 1721–1733. [CrossRef]

water

MDPI

Article

Removal of Cobalt (II) from Waters Contaminated by the Biomass of *Eichhornia crassipes*

Ismael Acosta-Rodríguez [1], Adriana Rodríguez-Pérez [2], Nancy Cecilia Pacheco-Castillo [1], Erika Enríquez-Domínguez [1], Juan Fernando Cárdenas-González [2,*] and Víctor-Manuel Martínez-Juárez [3]

1 Laboratorio de Micología Experimental, Facultad de Ciencias Químicas, Universidad Autónoma de San Luís Potosí, 78320 San Luís Potosí, Mexico; iacosta@uaslp.mx (I.A.-R.); nancy.cecilia.pc@gmail.com (N.C.P.-C.); erika.dominguez@uaslp.mx (E.E.-D.)

2 Unidad Académica Multidisciplinaria Zona Media, Universidad Autónoma de San Luís Potosí, 79617 Ríoverde, Mexico; sarai.rodriguez@uaslp.mx

3 Area Académica de Medicina Veterinaria y Zootecnia, Universidad Autónoma del Estado de Hidalgo, 43600 Hidalgo, Mexico; victormj@uaeh.edu.mx

* Correspondence: juan.cardenas@uaslp.mx

Abstract: Due to the increase in contamination of aquatic niches by different heavy metals, different technologies have been studied to eliminate these pollutants from contaminated aquatic sources. So the objective of this work was to determine the removal of cobalt (II) in aqueous solution by the biomass of the aquatic lily or water hyacinth (*Eichhornia crassipes*) which, is one of the main weeds present in fresh water, due to its rapid reproduction, growth, and high competitiveness, by the colorimetric method of the methyl isobutyl ketone. The removal was evaluated at different pHs (4.0–8.0) for 28 h. The effect of temperature in the range from 20 °C to 50 °C and the removal at different initial concentrations of cobalt (II) of 100 to 500 mg/L was also studied. The highest bioadsorption (100 mg/L) was at 28 h, at pH 5.0 and 28 °C, with a removal capacity of 73.1%, which is like some reports in the literature. Regarding the temperature, the highest removal was at 50 °C, at 28 h, with a removal of 89%. At the metal and biomass concentrations analyzed, its removal was 82% with 400–500 mg/L, and 100% with 5 g of natural biomass at 20 h. In addition, this completely removes the metal in situ (100 mg/L in contaminated water, at 7 days of incubation, with 10 g of natural biomass in 100 mL). So, the natural biomass can be used to remove it from industrial wastewater, even if in vivo, only eliminate 17.3% in 4 weeks.

Keywords: contaminants; heavy metals; removal; cobalt; water hyacinth; biomass

Citation: Acosta-Rodríguez, I.; Rodríguez-Pérez, A.; Pacheco-Castillo, N.C.; Enríquez-Domínguez, E.; Cárdenas-González, J.F.; Martínez-Juárez, V.-M. Removal of Cobalt (II) from Waters Contaminated by the Biomass of *Eichhornia crassipes*. *Water* **2021**, *13*, 1725. https://doi.org/10.3390/w13131725

Academic Editors: Cristina Palet and Julio Bastos-Arrieta

Received: 19 May 2021
Accepted: 18 June 2021
Published: 22 June 2021

1. Introduction

The great industrial growth has produced a progressive increase in wastewater discharges from the same and, therefore, a deterioration in water quality. Pollutants pose a danger to both human and environmental health. Some pollutants are of organic origin, such as hydrocarbons and pesticides, and inorganic, such as heavy metals and dye, which play a fundamental role due to their importance and potential danger [1,2]. Although heavy metals are present in water naturally, these are usually found at levels less than 1 mg/L, despite their low concentration levels, they have important chemical and biological implications in systems natural aquatic [3]. When they are above certain concentrations, as occurs in the case of some discharges, they become harmful both for the environment and for living beings, being able to affect the normal metabolism of an organism, either because they bind to non-specific biomolecules or because they cause oxidative damage due to its ability to catalyze oxidation-reduction reactions. This can lead to a deactivation of essential enzymatic reactions, damage to cell structures or DNA (mutagenesis). In humans, short exposures to high levels of certain metals can cause symptoms of acute toxicity, while long exposures to lower levels can lead to allergies or even cancer [4]. Some metals that are of great toxicological and exotoxicological importance are: mercury, chromium, lead, cadmium, nickel, cobalt, and zinc, which, once released into the environment, accumulate and concentrate in the soil and sediments, where they can remain for hundreds of years affecting ecosystems. Therefore, it is more feasible to control the problem from the source of emission before they reach the environment [5]. Currently, the most used methods to remove heavy metals in wastewater are chemical precipitation, filtration, coagulation, solvent extraction, electrolysis, membrane separation, ion exchange, and sorption techniques [6]. Some of these processes can become expensive, given the high operating costs and requirements energetic, as is the case of polymeric resins, which are quite efficient, but not very economical. For example, although ion exchange useful for effective water purification, resins are expensive due to that its synthesis is complex and fossil resources are used as the basis for its preparation. On the other hand, electrochemical treatments, and methods of membrane separation systems are relatively expensive, either because they require sophisticated equipment and because of the high energy expenditure associated with their functioning [6]. The process known as sorption refers to the capacity of certain materials to bind to anions and cations (as is the case with metals). During the sorption, metals bind to active centers found on the surface of materials through different mechanisms [1], and these generally include physical adsorption, ion exchange adsorption, chelation, complex formation, microprecipitation, visible-light-responsive photocatalysts [7], or absorption [8]. The sorption with biological sorbents is also known as biosorption and, as indicated, it can be used in remediation processes with low cost and high efficiency, allowing the minimization of chemical waste or biological [9]. It happens when metal cations are joined by electrostatic interactions to the anionic sites found in biosorbents. These sites, that serve as active centers for biosorption, are in the carboxyl, hydroxyl, amino, or sulfonic groups, which are part of the structure of most naturally occurring polymers [6,10]. In Mexico, agribusiness is one of the most important activities due to its growth in recent years, and it is the one that generates the most by-products that are not used [11], among which are: coffee bagasse, agave, maguey, sugar cane, straws from different crops, organic residues of fruits and vegetables [12]. In this regard, the use of different plant products and other materials like nanocomposite, glutathione-magnetite [13], and natural locust bean gum-based hydrogels [14], with the ability to accumulate and/or bioadsorb heavy metals and other pollutants has been reported.

On the other hand, the water lily or water hyacinth (*Eichhornia crassipes*) is an aquatic macrophyte plant, which is used as an ornamental species for its showy flowers in ponds and aquariums, it is native from South America mainly in the Amazon and Silver basins, it came to Mexico at the end of the 19th century, where it spread rapidly until it became a plague spreading throughout the country, which is considered one of the worst weeds in the world [15]. The only places where it has not been recorded are the states of Baja California Sur, Tlaxcala, and Zacatecas [16]. The water lily has been reported almost always alive, to accumulate and remove different pollutants, among which are heavy metals, for example: the accumulation of copper and cobalt [17], the decrease of arsenic from waters of a town in Peru [18], the accumulation of silver [19], removal of cadmium from wastewater [20], phytoremediation of cyanided water [21], the phytoextraction of arsenic, cadmium, and copper in an artificial wetland with different plants [22], the removal of different heavy metals by the biomass of different aquatic plants (*E. crassipes, Potamogeton lucens* and *Salvinia herzegoi*) [23,24], the phytoremediation of an effluent contaminated with mercury in a wetland in Antioquia (Colombia) [25], the elimination of mercury, cadmium, and arsenic from wastewater [26], and the phytoremediation of methylene blue by the same plant [27]. Therefore, the objective of this work was to evaluate the adsorbent capacity of the biomass of the water lily (*E. crassipes*) in the removal of cobalt (II) in an aqueous solution.

2. Materials and Methods

2.1. Bioadsorbent Used

Initially, the water lily plant was obtained from the San José Dam, San Luis Potosí, S.L.P. Mexico (Figure 1a), located to the west of the city, it is the best known dam in San Luis Potosi, it has an area of 344 hectares, inaugurated on 3 September 1903, with the capacity to hold up to 10 million cubic meters of water retained by a curtain of 32 m high and 100 m long, built almost entirely from a quarry. It is the most prominent dam in the state, and it is a protected natural area in the form of an urban park. To obtain the biomass, the plant was washed 24 h with EDTA at 10% (p/v), and later 1 week in tridesionized water with constant agitation, with water changes every 12 h, and boiled for 60 min to remove dust and adhered organic components, and washed again under the same conditions (Figure 1b). It was dried at 80 °C for 24 h in a bacteriological oven, ground in a blender, and stored in amber flasks until use (Figure 1c).

(a) (b) (c)

Figure 1. Natural biomass of water lily utilized: (**a**) water lily growth in San José Dam. San Luis Potosí, S.L.P. México (**b**) Water lily collected and washed (**c**) Stored and ground biomass.

2.2. Cobalt Solutions

We worked with 100 mL of a 100 mg/L solution of cobalt (II) obtained by diluting 1.0 g/L of standard solution prepared in tridesionized water from $CoCl_2$ (Analit Brand, St. Petersburg, Russia). The pH of the dilution to be analyzed was adjusted to the desired value (4, 5, 6, 7, and 8), with 1 M H_2SO_4 and/or 1 M NaOH, before adding it to the biomass.

2.3. Determination of the Concentration of Divalent Cobalt

Cobalt (II) concentration in aqueous solution was determined by UV spectroscopy in a double-beam UV-Visible spectrophotometer Shimadzu UV-2101PC, using methyl isobutyl ketone [28]. In a separation funnel, to 500 µL of the test solution (100 mg/L of cobalt II), add 0.5 mL of 20%/(p/v) ammonium thiocyanate solution, then add 5 mg of fluoride sodium (to eliminate contaminating ions in the sample), finally, 4 mL of methyl isobutyl ketone are added, stirred gently for 1 min and left to rest for 10 min observing the development of a blue color, and the absorbance of the sample is read at 622 nm. Samples were taken at different times, and the natural biomass was removed by centrifugation (3000 rpm/5 min), and the supernatant was analyzed to determine the concentration of the metal in solution. The instrument was calibrated to zero with a blank of the reagents used without the addition of the metal, determining the cobalt concentration with a calibration curve of 0–100 mg/L of the metal. All experiments were performed 3 times and in duplicate.

3. Results

3.1. *Effect of Incubation Time and pH*

Initially, the bioadsorption of 100 mg/L of cobalt (II) was analyzed, at different incubation times at the following pH values: 4.0, 5.0, 6.0, 7.0, and 8.0 finding that at pH 5.0, 73.1% of the metal was removed at 28 h (Figure 2).

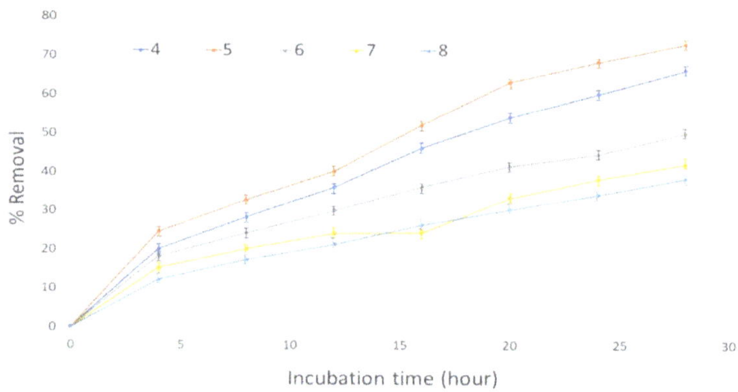

Figure 2. Effect of pH and incubation time on the removal of cobalt (II) by *E. crassipes*. 100 mg/L cobalt (II). 28 °C. 100 rpm. 1 g of natural biomass.

3.2. *Effect of Temperature*

In relation to the temperature, the highest removal was observed at 50 °C, since at 28 h 88.9% of the metal in solution is removed, while at 20 °C, 58.3% is removed at the same time (Figure 3).

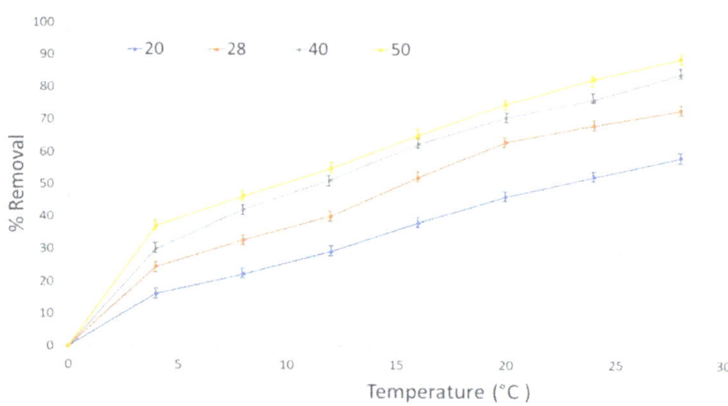

Figure 3. Effect of temperature on the removal of cobalt (II) by *E. crassipes*. 100 mg/L cobalt (II). pH 5.0. 100 rpm. 1 g of natural biomass. 28 h of incubation.

3.3. Effect of the Concentration of Cobalt (II)

Regarding the effect of different concentrations of cobalt (II) in solution, on its removal, at a pH of 5.0 ± 0.2, with 1 g of *E. crassipes* biomass, at 28 °C, and 100 rpm, it was found that, at a higher concentration of the metal, the removal is greater, up to 400–500 mg/L of the metal, it removal 82% of it at 28 h of incubation (Figure 4).

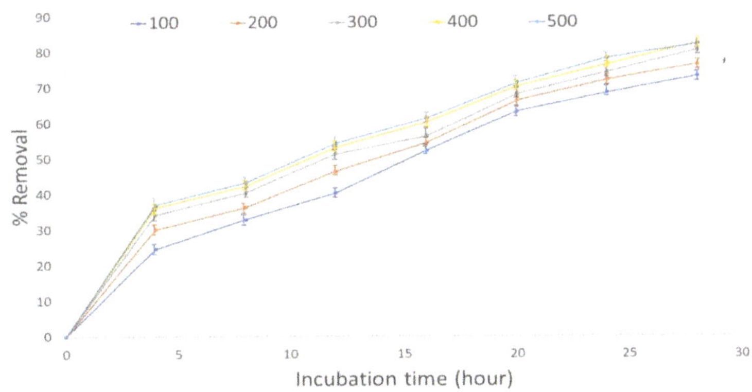

Figure 4. Effect of concentration of cobalt (II) (mg/L) on the removal of cobalt (II) by *E. crassipes*. pH 5.0. 28 °C. 100 rpm. 1 g of natural biomass. 28 h of incubation.

3.4. Effect of Initial Biomass Concentration

In Figure 5, the effect of biomass concentration on metal removal is observed. If the concentration of this is increased, the elimination of the metal in solution is increased, because with 5 g of biomass, 100% is removed after 20 h, while with 1 g of biomass, 63.2% is removed in the same incubation time.

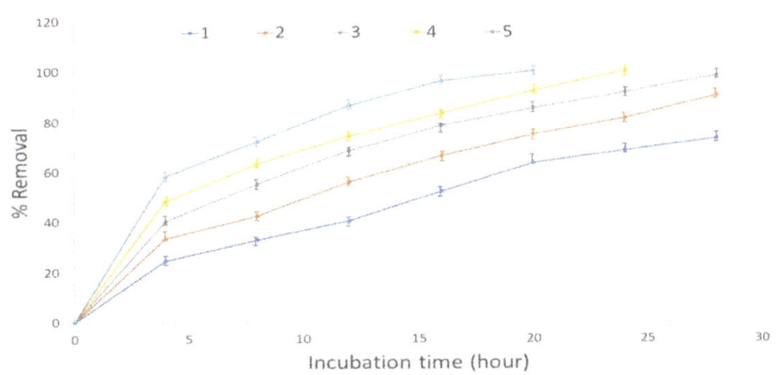

Figure 5. Effect of *E. crassipes* biomass (g/L) on the removal of 100 cobalt (II). pH 5.0. 28 °C. 100 rpm.

3.5. Removal of Co (II) in Industrial Wastes

A cobalt (II) bioremediation test was carried out from water contaminated with 100 mg/L of the metal (adjusted), obtained from an industrial effluent lagoon (Tenorio Tank) located to the east of the capital city of San Luis Potosí, S.L.P. Mexico, with 5 and 10 g of natural biomass, observing that this removal 90.1% and 100% of the metal, respectively, at 7 days of incubation (Figure 6).

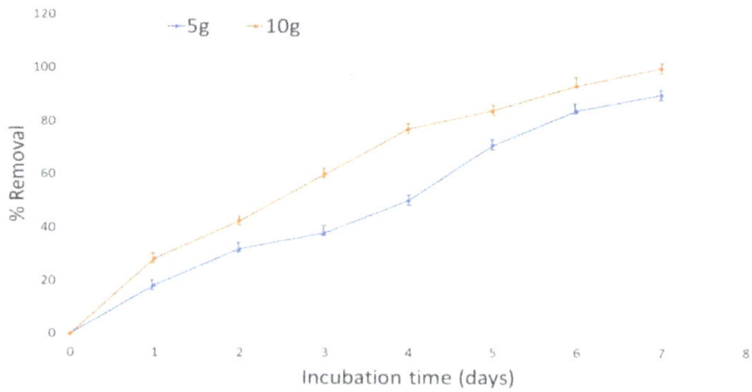

Figure 6. Cobalt (II) bioremediation by *E. crassipes* biomass from contaminated water (100 mL) with 100 mg of cobalt (II)/L. pH 5.0 (adjusted). 28 °C. 100 rpm. 5 and 10 g of biomass.

3.6. Removal of Cobalt (II) In Vivo

Finally, an in vivo experiment was carried out with water lily plants (lily A and lily B), incubating them in the presence and absence of a 100 mg/L solution of cobalt (II) (adjusted) and pH 6.8, in a final volume of 250 mL, from a lagoon of industrial effluents, taking 5 mL samples every week for a month, determining the concentration of the metal in the supernatant, finding that at 4 weeks of incubation, 17.3% of the metal under study was removed (Figure 7), which is less than that reported in the previous experiment, and also, the plant that was incubated in the presence of the metal, grew less and slowly lost its green color (data not shown).

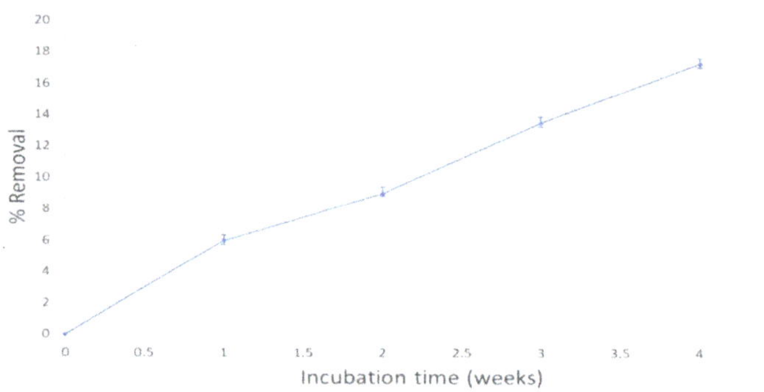

Figure 7. In vivo removal of 100 mg/L cobalt (II) (adjusted) from water contaminated (250 mL) by *E. crassipes* live. pH 6.8. 28 °C. Static conditions.

4. Discussion

Cobalt has an irreplaceable functionality in modern technology, it is an element that occurs naturally in the earth's crust. It is a very small part of our environment. It is a component of vitamin B12, which helps in the production of red blood cells. Animals and humans need very small amounts to stay healthy. Metal poisoning can occur when the human is exposed to large amounts of this element, an excessive metal exposure can result in a range of symptoms/conditions in humans including goitre and reduced thyroid activity [29]. There are three basic ways that cobalt can cause poisoning: ingest it in excess, inhale it in large quantities into the lungs, or by constant contact with the skin. EPA has not established a Reference Concentration (RfC) or a Reference Dose (RfD) for cobalt.

The California Environmental Protection Agency 3 (CalEPA) has established a chronic reference exposure level of 0.000005 milligrams per cubic meter (mg/m^3) for cobalt based on respiratory effects in rats and mice [30,31]. Therefore, it is very important to try to eliminate it from the different environmental niches, using technologies like natural biomasses, that are efficient, cheap, and that do not pollute the environment anymore [32,33].

In this work, we analyze the application of the water lily plant biomass for the removal of cobalt (II) in solution. Initially, the bioadsorption of the metal was analyzed, at different incubation times and pH values finding that at 28 h and pH 5.0, the removal was 73.1%. In this regard, the literature reports an optimal incubation time of 24 h with removal of the same metal of 93%, 77.5%, and 70.4%, for the fungus *Paecilomyces* sp., *Penicillium* sp. and *Aspergillus niger*, respectively [33], 20 min with 20 mg/L using *Citrus lemon* leaves powder [34], 180 and 120 min for the removal of 25–27 mg/L (76%), and 24.3 mg/L (24.3%), with red marine algae *Callithamnion corymbosum* biomass [35], and XAD-4 resin modified with *Anoxybacillus kestanbolienis* [36], a time of 60 min for the removal of 2.557 mg/g, with *Luffa cylindrica* as biosorbent [37], 120 min for elimination of 57.34% of 100 mg/L with palm kernel mesocarp fiber modified [38].

With respect to pH, the optimum was 5.0 (73.1% of removal at 28 h) and has been reported a pH optimum of 2.0, 4.4, 5.0, 6.0, and 4.5 for *C. lemon* leaves powder and *C. corymbosum* biomass [34,35], XAD-4 resin modified with *A. kestanbolienis* [35], *L. cylindrica,* [37], and palm kernel mesocarp fiber modified [38] as biosorbents, respectively. This phenomenon can be explained based on the less competition between positively charged H^+ and Co^{+2} ions for the similar functional group. As the pH rises, more ligands are exposed and the number of negatively charged groups on the adsorbent matrix probably increases, improving the removal of the cationic species [39]. Also, it was found that the higher temperature, the bioadsorption of the metal is greater. In this regard, our results are similar for three fungal species the literature reports an optimal incubation time of 24 h with a maximum adsorption of 100%, 97.1%, and 94.1%, respectively, at 50 °C for *Paecilomyces* sp., *Penicillium* sp. and *A. niger*, respectively [33], for *P. catenlannulatus,* as the uptake of cobalt (II) increase with increasing temperature from 20–40 °C [40], *Cocos nucifera* L. leaf powder [41], and *F. benghalensis* leaf powder [42], but are different for *C. corymbosum* biomass, bark of Eucalyptus, and spent green tea leaves, for the removal of the same heavy metal, which the optimum temperature was 22 °C [9,35,43], and for the removal of cobalt (II) with *L. cylindrica* as biosorbent, where a range of temperature of 10 °C to 40 °C, not affect the removal [37]. On the other hand, enhancement of the adsorption capacity of the fungal biomasses, at higher temperatures may be attributed to the activation of the adsorbing surface, the accelerated diffusivity of metal with the increasing temperature and the increase in the mobility of metal ions [39].

Regarding the effect of different metal concentrations in solution, it was found that, at a higher concentration of this, the removal is greater. This may be due to the increased number of competing for the functional groups of the surface of the biomass ions [23], and this is similar for bark of Eucalyptus, with a range of 20–200 mg/L of metal concentration [43], for the Cyanobacterium *Spirulina platensis,* because the cobalt (II) biosorption, increased proportionately with increasing its concentration, attaining a maximum value of 181 mg cobalt (II)/g with 600 mg/L initial concentration of the metal [44], for leaves of *Tectona grandis* (teak) tree were collected from the farm lands in Vellore District, India [45], and for the fungi *Paecilomyces* sp., *Penicillium* sp. and *A. niger*, respectively, in which if increase the metal concentration of 50–300 mg/L, increase the percentage of removal [33]. However, it is different for the algae *H. valentiae*, for which a higher initial concentration of cobalt (II), decreases the percentage removal [39], too for the calcium alginate of seaweed *Macrocistis piryfera*, [46], rice straw natural and activated rice straw [47].

On the other hand, if the biomass concentration is increased, the elimination of the metal in solution is increased, with more biosorption sites of the same, because the amount of added biosorbent determines the number of binding sites available for metal biosorption. These results are similar for *C. lemon* leaves powder, in which the adsorption capacity of cobalt (II) was increased with an increase in an adsorbent dose up to 0.1–0.5 g, however, the percent removal decreases at higher concentrations, and the maximum adsorption occurs at 60 ppm [34], as well as with 50 to 200 mg of XAD-4 resin modified with *A. kestanbolienis* [36], with 0.5–1.0 g for palm kernel mesocarp fiber modified [38], bark of Eucalyptus, with a range of 0.02–0.6 g/L of biomass concentration [43], and spent green tea leaves, with 0.04–0.25 mg/L of biomass [9], but are different for *C. corymbosum* biomass, in which the removal is not increased if the biomass concentration increases [35], and for *L. cylindrica* and palm kernel mesocarp fiber modified as biosorbent, in which if increase the biomass concentration of 10–100 mg/L, diminish the removal capacity [37,38].

It has been suggested that the water lily or water hyacinth has some advantages to treat industrial and domestic wastewater, sewage effluents, and sludge ponds, due to its high absorption capacity for different organic and inorganic pollutants, it can tolerate an extremely polluted environment and has a great capacity for biomass production [23,24]. Therefore, a cobalt (II) bioremediation test was carried out from water contaminated from an industrial effluent lagoon (Tenorio Tank), observing that this removal 90.1% and 100% of the metal, with 5 and 10 g of biomass, respectively at 7 days of incubation. which coincides with some reports in the literature with different biomasses and microorganisms in the removal of the same metal such as: the fungi *Paecilomyces* sp., *Penicillium* sp. and *A. niger* [33], for different water samples with XAD-4 resin modified with *A. kestanbolienis* [36], the adsorption of cobalt ions was evaluated using sediment samples from water bodies [48], the restoration of different soils and water by different biosorbents [49], the accumulation of copper and cobalt by *E. crassipes* [17], the bioremediation of waters contaminated with different heavy metals with the biomasses of the green bitter orange peel (*Citrus aurantium*), kumquat (*Citrus japonica*) and fine palomino grape seed (*Vitis vinifera* "Palomino") [50].

An in vivo experiment was also carried out with water lily plants, incubating them in presence and absence from water contaminated with the metal, from a lagoon of industrial effluents, finding that at 4 weeks of incubation, 17.3% of the metal under study is removed. In this regard, it has been reported that the water hyacinth can survive in presence of 60 μM of copper and cobalt during 60 days of incubation [17], the survival of water hyacinth in presence of a mixture of heavy metal concentrations of up to 3 mg/L, while at high concentrations of some metals (100 mg Cd/L), they caused a rapid discoloration of the plants [51], the removal of 0.02 mg/L of cadmium (II) from acid mining drainage with *E. crassipes*, for 14 days [52], a removal efficiency of 83.57% from simulated water with 2 mg/L of cadmium (II) using *E. crassipes* during 11 days [20], the removal of arsenic, cadmium, and Cupper from artificial wetlands for 60 days at a pH of 6.5, with different plants [22], the removal of 71% of 0.325 mg/L of mercury of wastewater from the mining industry for 7 months [25], and the phytoremediation of Nickel and Lead from soil and waters contaminated, using the same plant (water lily) [53].

5. Conclusions

In this study, cobalt (II) removal by the water lily biomass was analyzed. The performance of the biosorbent was examined as a function of the operating conditions, in particular: incubation time, pH, temperature, initial metal ion, and biomass concentration, as well as the phytoremediation of water contaminated in vivo and in vitro. The experimental evidence shows a strong effect of the experimental conditions. Maximum biosorption capacity values showed that the biosorbents used are very effective in the removal of cobalt (II) from aquatic systems in the conditions analyzed. Where the ease of production and economical parameters are concerned, it was observed that *E. crassipes* biomass, is a very promising biomaterial for removal or recovery of the heavy metal studied.

Author Contributions: Conceptualization, I.A.-R., A.R.-P. and N.C.P.-C.; methodology, I.A.-R., E.E.-D., A.R.-P. and J.F.C.-G.; software, J.F.C.-G.; validation, I.A.-R. and V.-M.M.-J.; formal analysis, I.A.-R. and A.R.-P.; investigation, I.A.-R., J.F.C.-G., A.R.-P., and N.C.P.-C.; resources, E.E.-D.; data curation, J.F.C.-G. and V.-M.M.-J.; writing—original draft preparation, I.A.-R., N.C.P.-C. and E.E.-D.; writing—review and editing, I.A.-R., N.C.P.-C., J.F.C.-G., and V.-M.M.-J.; visualization, A.R.-P.; supervision, I.A.-R. and E.E.-D.; project administration, I.A.-R. and J.F.C.-G.; funding acquisition, I.A.-R. All authors have read and agreed to the published version of the manuscript.

Funding: This research received no external funding.

Institutional Review Board Statement: Not applicable.

Informed Consent Statement: Not applicable.

Data Availability Statement: No new data were created or analyzed in this study. Data sharing is not applicable to this article.

Acknowledgments: This research work was carried out thanks in part to the support of the Directorate of the Faculty of Chemical Sciences of the Autonomous University of San Luis Potosi and the Directorate of the Center for Research and Extension of the Middle Zone, El Balandrán.

Conflicts of Interest: The authors declare no conflict of interest.

References

1. Díaz de Alba, M.I.; Gestoso Rojas, J.A.; Granado Castro, M.D.; Galindo Riaño, M.D.; Casanueva Marenco, M.J. Potencial uso de residuos agroalimentarios en la eliminación de metales contaminantes para la mejora del medioambiente. In Proceedings of the XLI Jornadas de Viticultura y Enología de la tierra de Barros. I Congreso Agroalimentario de Extremadura, Almendralejo, Spain, 7–10 May 2019; pp. 367–385. Available online: https://dialnet.unirioja.es/servlet/articulo?codigo=7617885 (accessed on 21 June 2021).
2. Kapahy, M.; Sachdeva, S. Bioremediation options for heavy metals pollution. *J. Health Poll.* **2019**, *9*, 1–20. [CrossRef]
3. Pandey, S.; Ramontja, J. Natural bentonite clay and it composites for dye removal: Current state and future potential. *Am. J. Chem. Appl.* **2016**, *3*, 8–19.
4. Vacar, C.L.; Covaci, E.; Chakraborty, S.; Li, B.; Weindorf, D.C.; Frentiu, T.; Parve, M.; Podar, D. Heavy metals-resistant filamentous fungi as potential mercury bioremediators. *J. Fung.* **2021**, *7*, 386. [CrossRef] [PubMed]
5. Kharangate-Lad, A.; D'Souza, N.C. Current Approaches in Bioremediation of Toxic Contaminants by Application of Microbial Cells; Biosurfactants and Bioemulsifiers of Microbial Origin. In *Rhizobiont in Bioremediation of Hazardous Waste*; Kumar, V., Prasad, R., Kumar, M., Eds.; Springer: Singapore, 2021. [CrossRef]
6. Muhammad, G.; Mehmood, A.; Shahid, M.; Ashraf, R.S.; Altaf, M.; Hussain, M.A.; Raza, M.A. Biochemical Methods for Water Purification. In *Methods for Bioremediation of Water and Wastewater Pollution. Environmental Chemistry for a Sustainable World*; Ahamed, M.I., Lichtfouse, E., Asiri, A.M., Eds.; Springer: Cham, Switzerland, 2020; Volume 51. [CrossRef]
7. Pandey, S.; Mandari, K.K.; Kim, J.W.; Kang, M.; Fosso-Kankeau, E. *Recent Advancement in Visible-Light-Responsive Photocatalysts in Heterogeneous Photocatalytic Water Treatment Technology*; Fosso-Kankeu, E., Pandey, S., Ray, S.S., Eds.; Scrivener Publishing LLC: Salem, MA, USA, 2020; pp. 167–196. [CrossRef]
8. Pandey, S. A comprehensive review on recent developments in bentonite-based materials used as adsorbents for wastewater treatment. *J. Mol. Liq.* **2017**, *241*, 1091–1113. [CrossRef]
9. Imessaoudene, D.; Bensacia, N.; Chenoufi, F. Removal of Cobalt (II) from aqueous solution by spent green teleaves. *J. Radioan. Nucl. Chem.* **2020**, *324*, 1245–1253. [CrossRef]
10. Pandey, S.; Do, J.Y.; Kim, J.W.; Kang, M. Fast and highly efficient catalytic degradation of dyes using κ-carrageenan stabilized silver nanoparticles nanocatalyst. *Carbo. Polym.* **2020**, *230*, 115971. [CrossRef] [PubMed]
11. Valdez, V.I.; Acevedo, B.J.; Hernández, S.C. Distribution and potential of bioenergy resources from agricultural activities in Mexico. *Ren. Sust. Ener. Rev.* **2010**, *14*, 2147–2153. [CrossRef]
12. Sánchez-Silva, J.M.; González-Estrada, R.R.; Blancas-Benitez, F.J.; Fonseca-Cantabrana, A. Utilización de subproductos agroindustriales para la bioadsorción de metales pesados. *TIP Rev. Esp. Cien. Quím. Biol.* **2020**, *23*, 1–18. [CrossRef]
13. Sharma, M.; Chaudhar, K.; Kumar, M.; Yadab, P.; Sachdev, K.; Janud, C.; Gupta, J. Highly efficient, economic, and recyclable glutathione decorated magnetically separable nanocomposite for uranium (VI) adsorption from aqueous solution. *Mat. Tod. Chem.* **2020**, *18*, 100379. [CrossRef]
14. Pandey, S.; Doa, J.Y.; Kimb, A.; Hang, M. Fast and highly efficient removal of dye from aqueous solution locust gum based hydrogels as adsorbent. *Int. J. Biol. Macrom.* **2020**, *143*, 60–75. [CrossRef]
15. Tovar-Jiménez, X.; Favela-Torres, E.; Volke-Sepulveda, T.L.; Escalante-Espinosa, E.; Diaz-Ramírez, I.J.; Córdova-López, J.A.; Tellez-Jurado, A. Influence of the geographical area and morphological part of the water hyacinth on its chemical composition. *Ing. Agríc. Biosist.* **2019**, *11*, 39–52. [CrossRef]

16. Bonilla-Barbosa, J.R.; Santamaría, B. Plantas acuáticas exóticas y traslocadas invasoras. In *Especies Acuáticas Invasoras en México*, 1st ed.; Mendoza, R., Koleff, P., Eds.; Comisión Nacional para el Conocimiento y Uso de la Biodiversidad: México City, Mexico, 2014; pp. 223–247. [CrossRef]
17. Leenu, S.; Sheela, D. Impact of Co and Cu (heavy metals) on *Eichhornia crassipes* (Mart) solms, with special reference to bioaccumulation and phytoremediation. *Asian J. Sci. Technol.* **2016**, *7*, 2996–2998.
18. Rojas Adrianzen, L.P.; Suyon Diaz, E.P. Eficiencia de Fitoremediación con Jacinto de Agua (*Eichhornia crassipes*) Para Disminuir Concentraciones de Arsénico en Aguas del Centro Poblado Cruz del Medano-Morrope-2019. Bachelor's Thesis, Escuela Profesional de Ingeniería Ambiental, Facultad de Ciencias e Ingeniería, Universidad de Lambayeque, Chiclayo, Perú, 2020. Available online: https://repositorio.udl.edu.pe/handle/UDL/314 (accessed on 21 June 2021).
19. Gallardo Pazmiño, A.S. Bioensayos estáticos para determinar la acumulación de plata en lirio Acuático (*Eichhornia crassipes*). Master's Thesis, Facultad de Ciencias Ambientales, Universidad Internacional SEK, Quito, Ecuador, 2015. Available online: http://repositorio.uisek.edu.ec/handle/123456789/1090 (accessed on 21 June 2021).
20. Sandoval Vilchez, J.D. Eficiencia del Jacinto de Agua *Eichhornia crassipes* y Lenteja de Agua *Lemna minor* L. en la Remoción de Cadmio en Aguas Residuales. Bachelor's Thesis, Ingeniero Ambiental, Facultad de Ingeniería Geográfica, Ambiental y Ecoturismo, Universidad Nacional Federico Villarreal, Lima, Perú, 2019. Available online: http://repositorio.unfv.edu.pe/handle/UNFV/3256 (accessed on 21 June 2021).
21. Gómez Gutiérrez, C.; Guarín Álvarez, S.L. Evaluar el Efecto del Proceso de Fitorremediación con *Eichhornia crassipes* en un agua Cianurada Mediante Cromatografía de Gases. Bachelor's Thesis, Facultad de Ingeniería y Arquitectura, Universidad Católica de Manizales, Manizales, Colombia, 2020. Available online: http://repositorio.ucm.edu.co (accessed on 21 June 2021).
22. Islas Olvera, J. Fitoextracción de Arsénico, Cadmio y Cobre, Mediante las Especies *Eichhornia crassipes, Miriophyllum aquaticum* y *Wolffia colombiana*. Master's Thesis, División de Ciencias Básicas e Ingeniería, Universidad Autónoma Metropolitana, Unidad Azcapotzalco, México City, México, 2020. Available online: http://hdl.handle.net/11191/6850 (accessed on 21 June 2021).
23. Ali, S.; Abbas, S.; Rizwan, M.; Zaheer, I.E.; Yava, I.; Ünay, A.; Abdel-Daim, M.M.; Bin-Jumah, M.; Hasanuzzaman, M. Kalderis, Application of Floating Aquatic Plants in Phytoremediation of Heavy Metals Polluted Water: A Review. *Sustainability* **2020**, *12*, 1927. [CrossRef]
24. Eid, M.E.; Shaltout, K.S.; Moghanm, F.S.; Youssef, M.S.G.; El-Mohsnawy, E.; Haroun, S.A. Bioaccumulation and translocation of nine heavy metals by *Eichhornia crassipes* in Nile Delta, Egypt: Perspectives for phytoremediation. *Int. J. Phytorem.* **2019**, *21*, 11–18. [CrossRef]
25. Domínguez, M.C.; Gómez, S.; Ardila, A.N. Fitorremediación de mercurio presente en aguas residuales provenientes de la industria minera. *UG Ciencia* **2016**, *22*, 227–237. [CrossRef]
26. Mi, N.; Idrees, I.; Danish, P.; Ahmad, S.; Ali, Q.; Malik, A. Potential of water hyacinth (*Eichhornia crassipes* L.) for phytoremediation of heavy metals from wastewater. *Biol. Clin. Sci. Res. J.* **2020**, *2020*, 1–6.
27. Lara González, L. Fitorremediación para la extracción del colorante "azul de metileno" mediante el uso de *Eichhornia crassipes*. Bachelor's Thesis, Facultad de Planeación Urbana y Regional, Universidad Autónoma del Estado de México, Toluca de Lerdo, México, 2017. Available online: http://hdl.handle.net/20.500.11799/67471 (accessed on 21 June 2021).
28. Charlot, G. *Colorimetric Determination of Elements*, 12th ed.; Elsevier Publishing Company: Amsterdam, The Netherlands; London, UK; New York, NY, USA, 1964.
29. Ziwa, G.; Crane, R.; Hudson-Edwards, K.A. Geochemistry, Mineralogy and Microbiology of Cobalt in Mining-Affected Environments. *Minerals* **2021**, *11*, 22. [CrossRef]
30. Agency of Toxic substances and Disease Registry (ATSDR). *Toxicological Profile for Public Health Service*; U.S. Departament of Health and Human Services: Atlanta, GA, USA, 1992.
31. Drinking Water Regulations and Contaminants United States Environmental Protection Agency (EPA). 2018. Available online: https://www.epa.gov/sdwa/drinking-water-regulations-and-contaminants (accessed on 21 June 2021).
32. Dusengemungu, L.; Kasali, G.; Gwanama, C.; Ouma, K.O. Recent advances in biosorption of copper and cobalt by filamentous fungi. *Front. Microbiol.* **2020**, *11*, 582016. [CrossRef] [PubMed]
33. Cárdenas González, J.F.; Rodríguez Pérez, A.S.; Vargas Morales, J.M.; Martínez Juárez, V.M.; Acosta Rodríguez, I.; Michel Cuello, C.; Gallegos Fonseca, G.; Escalera Chávez, M.E.; Muñoz Morales, A. Biosorption of Cobalt (II) from aqueous solutions by fungal biomasses. *Bioinor. Chem. Appl.* **2019**, *2019*, 8757149. [CrossRef]
34. Ullah, A.; Ali, I.; Rasheed, S.; Shah, M.; Munir, T.; Nawaz, S. Adsorptive Removal of Cobalt (II) and Chromium (III) from Aqueous Solution onto *Citrus Limon* Leaves Powder. *Lett. Appl. Nano Biosci.* **2021**, *10*, 2446–2458. [CrossRef]
35. Lucaci, A.R.; Bulgariu, D.; Ahmad, I.; Bulgariu, L. Equilibrium and Kinetics Studies of Metal Ions Biosorption on Alginate Extracted from Marine Red Algae Biomass (*Callithamnion corymbosum* sp.). *Polymers* **2020**, *12*, 1988. [CrossRef]
36. Ozdemir, S.; Kilinc, E.; Fatih, S. A novel biosorbent for preconcentrations of Co(II) and Hg(II) in real samples. *Sci. Rep.* **2020**, *10*, 445. [CrossRef] [PubMed]
37. Bouzida, A.; Djedid, M.; Ad, C.; Benalia, M.; Hafez, B.; Elmsellem, H. Biosorption of CO (II) ions from aqueous solutions using selected local *Luffa Cylindrica*: Adsorption and characterization studies. *Mor. J. Chem.* **2021**, *9*, 156–167. [CrossRef]
38. Fang, Y.H. Biosorption of Cobalt (II) ions using palm kernel mesocarp fiber. Bachelor's Thesis, Faculty of Engineering and Green Technology, University Tunku Abdul Ramhan, Petaling Jaya, Malaysia, 2019. Available online: http://eprints.utar.edu.my/id/eprint/3913 (accessed on 21 June 2021).

39. Vafajoo, L.; Cheraghi, R.; Dabbagh, R.; McKay, G. Removal of cobalt (II) ions from aqueous solutions utilizing the pre-treated 2-*Hypnea Valentiae* algae: Equilibrium, thermodynamic, and dynamic studies. *Chem. Eng. J.* **2018**, *31*, 39–47. [CrossRef]
40. Li, F.; Gao, Z.; Li, X.; Fang, L. The effect of environmental factors on the uptake of 60Co by Paecilomyces catenlannulatus. *J. Radioanal. Nucl. Chem.* **2014**, *299*, 1281–1286. [CrossRef]
41. Hymavathi, D.; Prabhakar, G. Optimization, equilibrium, and kinetic studies of adsorptive removal of cobalt(II) from aqueous solutions using *Cocos nucifera* L. *Chem. Eng. Commun.* **2017**, *204*, 1094–1104. [CrossRef]
42. Hymavathi, D.; Prabhakar, G. Studies on the removal of cobalt(II) from aqueous solutions by adsorption with *Ficus benghalensis* leaf powder through response Surface methodology. *Chem. Eng. Commun.* **2017**, *204*, 1401–1411. [CrossRef]
43. Essaadaoui, Y.; Lebkiri, A.; EL Houssein, R.; Kadiri, L.; Ouass, A. Adsorption of cobalt from aqueous solutions onto Bark of Eucalyptus. *Med. J. Chem.* **2018**, *7*, 145–155. [CrossRef]
44. Vannela, R.; Verma, S.K. Co^{2+}, Cu^{2+}, and Zn^{2+} Accumulation by Cyanobacterium *Spirulina platensis*. *Biotechnol. Prog.* **2006**, *22*, 1282–1293. [CrossRef]
45. Vilvanathan, S.; Shanthakumar, S. Column Adsorption Studies on Nickel and Cobalt Removal from Aqueous Solution Using Native and Biochar Form of *Tectona grandis*. *Environ. Prog. Sustain. Ener.* **2017**, *36*, 1030–1038. [CrossRef]
46. Rabanal Atalaya, M.M. Estudio de la biosorción de Co (II) por el alga marina *Macrocistis piryfera*. Bachelor's Thesis, Profesional Químico, Universidad Nacional Mayor de San Marcos, Lima, Perú, 2006. Available online: https://hdl.handle.net/20.500.12672/3282 (accessed on 21 June 2021).
47. Swelam, A.A.; Awad, M.B.; Salem, A.S.M.A.; El-Feky, A.S. Biosorption of Cobalt(II) Ions from Aqueous Solution using Rice Straw and its Modification. *J. Sci. Eng. Res.* **2017**, *4*, 121–129.
48. Montes de Oca, R.; Solache-Ríos, M.; Jimenez-Reyes, M.; García-Sánchez, J.J.; Almazán-Sánchez, P.T. Adsorption of cobalt by using inorganic components of sediment samples from water bodies. *Int. J. Sed. Res.* **2021**, *36*, 524–531. [CrossRef]
49. Rana, A.; Sindhu, M.; Kumar, A.; Kumar Dhaka, R.; Chahar, M.; Singh, S.; Nain, L. Restoration of heavy metals-contaminated soil and water through biosorbents: A review of current understanding and future challenge. *Phys. Plant.* **2021**, 1–24. [CrossRef]
50. Sánchez Ponce. L. Aplicación de Residuos Vegetales en el Tratamiento de aguas Contaminadas por Metales. Bachelor's Thesis, Grado en Química, Facultad de Ciencias, Universidad de Cadiz, Cadiz, Spain, 2018. Available online: http://hdl.handle.net/10498/23560 (accessed on 21 June 2021).
51. Soltan, M.E.; Rashed, M.N. Laboratory study on the survival of water hyacinth under several conditions of heavy metal concentrations. *Adv. Environ. Res.* **2003**, *7*, 321–334. [CrossRef]
52. Palihakkara, C.R.; Dassanayake, S.; Jayawardena, C.; Senanayake, I.P. Floating Wetland Treatment of Acid Mine Drainage using *Eichhornia crassipes* (Water Hyacinth). *J. Health Poll.* **2018**, *8*, 14–19. [CrossRef] [PubMed]
53. Hussain, S.T.; Mahmood, T.; Malik, S.A. Phytoremediation technologies for Ni^{++} by water Hyacinth. *Afr. J. Biotechnol.* **2010**, *9*, 8648–8660. [CrossRef]

water

MDPI

Article

Highly Efficient Removal of Cu(II) Ions from Acidic Aqueous Solution Using ZnO Nanoparticles as Nano-Adsorbents

Eduardo Leiva [1,2,*], Camila Tapia [1,3] and Carolina Rodríguez [1]

[1] Departamento de Química Inorgánica, Facultad de Química y de Farmacia, Pontificia Universidad Católica de Chile, Avenida Vicuña Mackenna 4860, Macul, Santiago 7820436, Chile; camilatapiape@gmail.com (C.T.); cnrodriguez@uc.cl (C.R.)
[2] Departamento de Ingeniería Hidráulica y Ambiental, Pontificia Universidad Católica de Chile, Avenida Vicuña Mackenna 4860, Macul, Santiago 7820436, Chile
[3] Departamento de Química, Facultad de Ciencias, Universidad de Chile, Ñuñoa 7800003, Chile
* Correspondence: ealeiva@uc.cl; Tel.: +56-2-2354-7224; Fax: +56-2-2354-5876

Abstract: Water pollution by heavy metals has significant effects on aquatic ecosystems. Copper is one of the heavy metals that can cause environmental pollution and toxic effects in natural waters. This encourages the development of better technological alternatives for the removal of this pollutant. This work explores the application of ZnO nanoparticles (ZnO-NPs) for the removal of Cu(II) ions from acidic waters. ZnO NPs were characterized and adsorption experiments were performed under different acidic pHs to evaluate the removal of Cu(II) ions with ZnO NPs. The ZnO NPs were chemically stable under acidic conditions. The adsorption capacity of ZnO NPs for Cu(II) was up to 47.5 and 40.2 $mg \cdot g^{-1}$ at pH 4.8 and pH 4.0, respectively. The results revealed that qmax (47.5 $mg \cdot g^{-1}$) and maximum removal efficiency of Cu(II) (98.4%) are achieved at pH = 4.8. In addition, the surface roughness of ZnO NPs decreases approximately 70% after adsorption of Cu(II) at pH 4. The Cu(II) adsorption behavior was more adequately explained by Temkin isotherm model. Additionally, adsorption kinetics were efficiently explained with the pseudo-second-order kinetic model. These results show that ZnO NPs can be an efficient alternative for the removal of Cu(II) from acidic waters and the adsorption process was more efficient under pH = 4.8. This study provides new information about the potential application of ZnO NPs as an effective adsorbent for the remediation and treatment of acidic waters contaminated with Cu(II).

Keywords: adsorption; copper; zinc-oxide nanoparticles; nanomaterials; acid mine drainage

Citation: Leiva, E.; Tapia, C.; Rodríguez, C. Highly Efficient Removal of Cu(II) Ions from Acidic Aqueous Solution Using ZnO Nanoparticles as Nano-Adsorbents. *Water* **2021**, *13*, 2960. https://doi.org/10.3390/w13212960

Academic Editors: Cristina Palet and Julio Bastos-Arrieta

Received: 5 September 2021
Accepted: 14 October 2021
Published: 20 October 2021

1. Introduction

Surface and groundwater contamination by heavy metals is a growing concern. One of the major sources of heavy metals is the contamination derived from acid mine drainage (AMD) release [1–3]. AMD is characterized by low pH and high concentrations of sulfate and dissolved metals and metalloids, causing severe damage to aquatic ecosystems [1,2]. Among the metals that are commonly released are iron, copper, lead, zinc, silver, arsenic, aluminum, manganese, antimony, selenium, among others [4]. Heavy metals and metalloids affect the quality of surface and groundwater resources, mainly because they are non-biodegradable, toxic at low concentrations, and easy to accumulate in the tissues of various living organisms [5,6]. They can cause serious harm to human health from cancer to nervous system problems [7–9]. Thus, the study and development of sustainable technologies to remove these pollutants have gained attention in recent years.

One of the metals that are most widely present in surface waters is copper, mainly because of its multiple industrial applications [10–12]. In addition, many mining operations release acid runoff with high Cu concentrations, affecting unique ecosystems [13,14]. In native systems, Cu is usually found as a divalent cation (Cu(II)) and is quite mobile at low pH values, which makes the treatment of copper-enriched AMD more and more relevant [15]. Copper is quite toxic even at low concentrations and its antimicrobial effects can cause substantial damage to the biodiversity of both microorganisms and higher organisms [16]. Thus, polluted waters with copper must be treated to reduce their impact on the environment [16]. The World Health Organization (WHO) establishes that the maximum limit of the concentrations of copper ions in drinking water should not exceed 2 $mg \cdot L^{-1}$. In recent years, significant efforts have been made to develop technologies for the treatment and removal of copper from polluted waters, in order to respond to the environmental effects of copper and the projected scarcity of water resources. In this context, technologies based on adsorption, ion exchange, photocatalysis, filtration and reverse osmosis methods have been extensively studied [16–27]. However, adsorption methods are the most cost-effective and have been the subject of various studies. The costs associated with adsorption technologies for water treatment vary from USD 10 to 200 per million liters [28,29] while the other conventional technologies such as ion exchange, reverse osmosis, ultrafiltration and electrodialysis have costs ranging from USD 15 to 450 per million liters [29] (Appendix A). Despite this, in recent years improvements in the removal process based on the use of nano-adsorbents have emerged as an alternative that can improve the adsorption capacity because of characteristic properties of nanomaterials such as higher surface area and greater potential for functionalization.

Various materials have been studied for the removal of Cu from wastewaters. The use of adsorbent materials based on carbon, such as activated carbon and graphene, has shown to be a suitable alternative for the removal of Cu, with adsorption capacities over 30 and 40 mg/g, respectively [30,31]. Other adsorbents studied at low pH have presented variable efficiency, with adsorption capacities of 20.97 and 31.7 mg/g for waste slurry and pecan shells activated carbon, respectively [32,33]. In addition, some nanomaterials that have shown high efficiency to remove copper from polluted waters are graphene oxides [34], biopolymers-based adsorbents [35,36], organic polymers [37], magnetic nanoparticles [38,39] carbon nanotubes [40–42], silica-based nanomaterials [43], metallic NPs [44]. Among the metallic nanoparticles that have been studied for the removal and adsorption of heavy metals, Zinc oxide (ZnO) has shown excellent results because of its catalytic properties. ZnO nanoparticles (NPs) have a high potential to be used as nano-adsorbents since they have a large specific surface area and various functional groups that favor their interaction and removal of heavy metals from aqueous solutions [45]. It has other advantages, such as high resistance to chemical and optical corrosion, high chemical stabilization, biocompatibility, environmentally friendly and is non-toxic in nature [46,47]. On one hand, ZnO NPs have antibacterial properties that inhibit the growth of both Gram-positive and Gram-negative bacteria, which may be optimal as an antifouling mechanism in sorbent media [48,49]. ZnO NPs are cheap compared to carbon-based nanomaterials, can be produced on a large scale, have a good photocatalytic performance and have a high removal efficiency of various inorganic pollutants such as heavy metals [47,48,50]. Li et al., 2014 [51] reported that ZnO NPs can adsorb heavy metals by the various kinds of hydroxyl groups present on their surface. Several studies have used ZnO for the removal of toxic pollutants, such as Cr (VI), Ni(II), Pb(II) [52–56]. However, few studies have focused on the removal of Cu(II) [57,58]. Wang et al. [59] showed that ZnO particles encapsulated in hollow microspheres are more efficient for the removal of Cu(II), Cd(II) and Pb(II) cations than commercial ZnO particles. Meanwhile, Primo et al., 2020 [60] showed a high removal of Cu(II) ions with ZnO NPs synthesized using the Aloe vera green synthesis route. Despite this, there are very few studies that have focused on the removal of Cu(II) from acidic waters where Cu(II) ions can be found in higher concentrations. Indeed, the adsorption of Cu(II) ions onto ZnO NPs is poorly covered in the literature, and there are practically no studies in acidic waters. Hence, the relevance of pH in the adsorption efficiency of Cu(II) should be studied to evaluate the potential scaling of ZnO nano-adsorbents in acidic water treatment technologies.

In this work, we presented our preliminary results about the use ZnO NPs as nano-adsorbents of Cu(II) ions from AMD waters. Cu(II) removal rates were evaluated in batch adsorption tests under different pHs. Experimental data were fitted using different isotherms models in adsorption experiments. Using ZnO NPs has the advantage of having a good cost-effectiveness ratio compared to other nanomaterials such as graphene, CNTs and magnetic nanoparticles (Appendix B). The study of the adsorption capacity under different pHs will allow exploring potential uses as emerging and sustainable technologies for the removal of Cu(II) from AMD waters. In addition, knowing the removal effectiveness in AMD waters allows determining its chemical and functional stability under more aggressive conditions, which can give it a comparative advantage compared with other commercial adsorbents and can promote effective scaling in real conditions.

2. Materials and Methods

2.1. Materials

Commercially produced ZnO NPs ($\leq$100 nm) were purchased from SigmaAldrich (St. Louis, MO, USA) and in this study were used without further purification. Cooper-enriched synthetic acid wastewater was prepared by adding $NaNO_3$ and $CuSO_4{\cdot}5H_2O$ to deionized (DI) water. Sulfuric acid (H_2SO_4), hydrochloric acid (HCl), and absolute ethanol (C_2H_5OH) were purchased from Merck. All the reagents and solvents used were of analytical reagent grade and all solutions were prepared with deionized (DI) water.

2.2. Preparation of ZnO Nanoadsorbents

ZnO NPs with an average size of 100 nm were used in batch adsorption experiments. The ZnO NPs suspension was obtained through the dispersion into absolute ethanol (0.33 mg$\cdot$mL^{-1}), and then the solution was sonicated for 40 min. Finally, the ZnO NPs were separated from the liquid by centrifugation (4000 rpm, t = 30 min), washed three times with ethanol and kept hermetically sealed until their use in adsorption studies.

2.3. Characterization Techniques

Prior to adsorption experiments, ZnO NPs were characterized using Raman spectroscopy, scanning electron microscope (SEM) coupled with energy-dispersive X-ray spectroscopy (SEM-EDX), field emission scanning electron microscopy (FESEM) and Brunauer–Emmett–Teller analysis.

Raman spectra were recorded using WITec Alpha 300 RA confocal Raman microscope with AFM (WITec GmbH, Ulm, Germany). The scanning electron micrographs were determined by using a scanning electron microscope (SEM) (JSM-IT300LV, JEOL, Tokyo, Japan) coupled with energy-dispersive X-ray spectroscopy (Oxford Instruments, High Wycombe, UK) (SEM-EDX). The scanning electron micrographs (SEM-EDX) were used to confirm the nanoparticle size and elemental mapping on the nanometer scale of ZnO NPs. On the other hand, field emission scanning electron microscopy (FESEM) images of the ZnO NPs were obtained using a scanning electron microscope (Quanta 250 FEG, FEI Co., Hillsboro, OR, USA) equipped with an EDX (XFlash 5010; Bruker AXS Microanalysis, Berlin, Germany). The FESEM images were used to analyze the surface topography and fine morphology of the ZnO NPs.

The specific surface area and pore volume of the ZnO NPs were determined by Brunauer–Emmett–Teller analysis (BET) N2 adsorption–desorption analysis (Micromeritics Instruments Corp., Norcross, GA, USA). Finally, the pH of the point of zero charge (pH_{PZC}) and zeta potential for ZnO NPs was determined by adjusting 0.01 M NaCl solutions to different pH values from 2 to 12 using 0.1 M NaOH and 0.1 M HCl added dropwise. Then, 30 mg of ZnO NPs were added to test tubes with 40 mL of the different pH solutions and shaken at 380 rpm at room temperature for 24 h. The final pH and zeta potential were measured using a pH meter (PHC301, HACH, Loveland, CO, USA) [41,61].

2.4. Batch Adsorption Studies

Equilibrium isotherms for Cu(II) were obtained in batch adsorption studies using ZnO NPs as nano-adsorbents. Experiments were performed Cu(II) concentrations ranging between 3.0 and 24.0 mg·L^{-1}. The Cu(II) solution was prepared by adding $CuSO_4 \cdot 5H_2O$ to deionized (DI) water.

The experiments were performed under two pHs: (1) with the resulting pH of the prepared solutions (Cu(II) solution + ZnO NPs) (pH = 4.8), and (2) with the pH adjusted to 4 by adding 0.01 M HCl (pH = 4.0) drop-wise and with continuous pH measurement until pH 4.0 was reached. For batch experiments, 20 mg of the ZnO NPs were added into 40 mL tubes with variable concentrations of Cu(II). The experiments were conducted in triplicate with shaking (380 rpm) in the dark at room temperature for 20 h for adsorption/desorption equilibrium. After the solutions reached equilibrium, they were centrifuged (6000 rpm, 5 min) to separate the solution from the ZnO NPs. The resulting supernatant was then filtered using 0.22 μm membranes to analyze the residual Cu(II) concentrations. Scanning electron microscopy (SEM) coupled with energy-dispersive X-ray spectroscopy (SEM-EDX) was used to analyze the morphology and elemental composition of the surface of ZnO NPs before and after Cu(II) adsorption in batch experiments.

The sorption capacity at equilibrium C_e (mg·g^{-1} sorbent) was calculated using Equation (1):

$$q_e = \frac{(C_0 - C_e)\cdot V}{m}, \tag{1}$$

where C_0, C_e, V, and m correspond to the initial concentration (mg·L^{-1}), the aqueous-phase equilibrium metal concentration (mg·L^{-1}), the volume of suspension (L), and the mass of the adsorbent (g), respectively.

Similarly, the Cu(II) ions removal efficiency (η) of ZnO particles was calculated using Equation (2):

$$\eta\ (\%) = \left(\frac{C_0 - C_t}{C_0}\right)\cdot 100, \tag{2}$$

where C_0 and C_t were the concentration of metal ions at the initial and time t, respectively.

To study the effect of other ions on adsorption efficiency, multimetallic water was prepared at three different concentrations: (1) 3 mg·L^{-1} of Cu (Added as $CuSO_4 \cdot 5H_2O$), Mn (Added as $MnSO_4 \cdot H_2O$) and Al (Added as $KAl(SO_4)_2 \cdot 12H_2O$); (2) 12 mg·L^{-1} of Cu, Mn and Al; and (3) 22 mg·L^{-1} of Cu, Mn an Al. In this way, it is sought to emulate more realistic wastewater, with the presence of other ions in the solution.

2.5. Adsorption Isotherms

From data obtained experimentally in batch configurations, the Langmuir, Freundlich and Temkin isotherm models were fitted. The Langmuir model assumed that the adsorption surface sites have identical energy and each adsorbate molecule (Cu(II) in this study) is positioned in a single place, forming a monolayer of sorption on the adsorbent surface [62,63]. On the contrary, the Freundlich model describes reversible heterogeneous adsorption without restricting the adsorption process to a single monolayer [64]. For this reason, the Freundlich isotherm predicts that the adsorbate concentration on the adsorbent will increase without saturation according to how to increase the adsorbate concentration in the liquid solution [65].

The sorption capacity q (mg·g^{-1} sorbent) was obtained using Langmuir, Freundlich and Temkin models. Temkin Model was used to evaluate if the adsorption behavior can be better described for another model [66]. For this, Equations (3)–(5) were used for each model, respectively [63].

Langmuir model:

$$q = \frac{q_L K_l C_e}{1 + K_L C_e}, \tag{3}$$

where q_L is the amount of adsorption corresponding to a monolayer coverage, K_L is the Langmuir constant associated with the energy of adsorption, and C_e is the metal concentration at the equilibrium in an aqueous solution (mg·L^{-1}).

Freundlich model:

$$q = K_F C_e^{\frac{1}{n}}, \quad (4)$$

where K_F is the constant related to adsorption capacity, n corresponds to the constant associated with adsorption intensity, and C_e is the metal concentration at the equilibrium in aqueous solution ($mg \cdot L^{-1}$).

Temkin model:

$$q = BInA_tC_e, \quad (5)$$

where B is an abbreviation of $RT(bt)^{-1}$, where R, T and bt represent the gas constant (8.314 $J \cdot mol^{-1} \cdot K^{-1}$), absolute temperature (K) and Temkin isotherm constant, respectively, A_t corresponds to the Temkin isotherm equilibrium binding constant ($L \cdot g^{-1}$) and C_e is the metal concentration at the equilibrium in aqueous solution ($mg \cdot L^{-1}$).

2.6. Surface Roughness Analysis

The surface roughness of the ZnO NPs was analyzed before and after Cu(II) adsorption. Specifically, Gwyddion software was used to examine the surface characteristics of SEM images of ZnO NPs [67]. Several surface roughness representative parameters were selected according to Zhao et al., 2019 [68]. The parameters selected were mean roughness (R_a), the mean square roughness (R_q), the surface skewness (R_{sk}) and the kurtosis coefficient (R_{ku}).

Roughness average (R_a) is the average deviation of all points roughness profile from a mean line over the evaluation length and is used to represent the mean value of the surface roughness of the sample [68]. R_a is calculated using Equation (6):

$$R_a = \frac{1}{N}\sum_{j=1}^{N}|r_j|, \quad (6)$$

where the N are the number of scanning points on the sample.

Root mean square roughness (R_q) is defined as the average of the measured height deviations taken within the evaluation length and measured from the mean line [68]. R_q is used to determine the degree of change in the surface roughness of a sample and is calculated using Equation (7):

$$R_q = \sqrt{\frac{1}{N}\sum_{j=1}^{N} r_j^2}, \quad (7)$$

The surface skewness (R_{sk}) is a measure of the asymmetry of the amplitude distribution function of the sample and quantifies the symmetry of the variation in a profile about its mean line [68]. Specifically, R_{sk} represents the integrity of the surface roughness of a specific sample [68]. Thus, a value of R_{sk} equal to zero shows that the surface height distribution is normal. A negative R_{sk} value shows that the surface height distribution is biased to the left, which represents that there is more area where the sample surface height is above the mean value. On the contrary, a positive R_{sk} value shows that the distribution is biased to the right and therefore there is more area where the sample surface height is below the mean value [68]. R_{sk} is calculated using Equation (8):

$$R_{sk} = \frac{1}{NR_q^3}\sum_{j=1}^{N} r_j^3, \quad (8)$$

Kurtosis (R_{ku}) is related to the uniformity of the amplitude distribution function of the sample [68]. A value of R_{ku} equal to zero shows that the surface height distribution of the sample follows a normal distribution. A negative value shows that the waveform associated with the surface height distribution is flat, so the surface height of the sample is distributed throughout the sample, while a positive value of R_{ku} shows the waveform has a peak, and the surface height of the sample is concentrated at the mean value of one or several peaks [68]. R_{ku} is calculated using Equation (9):

$$R_{ku} = \frac{1}{NR_q^4} \sum_{j=1}^{N} r_j^4, \quad (9)$$

2.7. Kinetic Experiments

The kinetic behavior of Cu(II) adsorption with ZnO NPs was studied using 20 mg of ZnO NPs in 40 mL of Cu(II) solution. Cu(II) concentration used was 25 mg·L^{-1}. The solutions were shaken at 380 rpm at room temperature (22–25 °C) and sampling was performed at 10 min, 30 min, 1 h, 2 h, 4 h, and 24 h. The Cu(II) removal was plotted in function of time. In order to analyze the uptake rates of Cu(II) ions, a kinetic analysis using pseudo-first-order and pseudo-second-order kinetic equations was performed.

The pseudo-first-order kinetic model was represented by Equation (10) according to Ding et al. [69].

$$\log(q_e - q_t) = \log q_e - \frac{k_1}{2.303} t, \quad (10)$$

where k_1 corresponds to first-order adsorption constant (min^{-1}), q_e is the adsorption capacity at equilibrium (mg·g^{-1}), q_t is the adsorption capacity at the time t (mg·g^{-1}).

The pseudo-second-order kinetic model was represented by Equation (11) according to Ho and McKay [70].

$$\frac{t}{q_t} = \frac{1}{k_2 q_e^2} + \frac{t}{q_e}, \quad (11)$$

where k_2 is the second-order adsorption constant (g·mg^{-1}·min^{-1}), q_e is the adsorption capacity at equilibrium (mg·g^{-1}), q_t is the adsorption capacity at the time t (mg·g^{-1}).

The linearized form of these kinetic models was plotted. From the trend line of the experimental data, the kinetic constants and parameters were determined using the slope and intercept values obtained in each case.

The approaching equilibrium factor (R_w) was calculated to characterize the kinetic curve behavior of the pseudo-second-order model using Equation (12) according to Wu et al. [71]

$$R_w = \frac{1}{1 + k_2 q_e t_{ref}}, \quad (12)$$

where k_2 is the second-order adsorption constant (g·mg^{-1}·min^{-1}), q_e is the adsorption capacity at equilibrium (mg·g^{-1}) and t_{ref} is the longest operation time (based on kinetic experiments).

Finally, the Gibbs free energy (ΔG^0) was determined as a thermodynamic parameter that indicates the degree of the spontaneity of an adsorption process where a higher negative value indicates a more energetically favorable adsorption process [72]. The ΔG^0 parameter was determined as follows [73,74].

$$\Delta G^0 = -RT \ln K \quad (13)$$

where R, T and K is the gas constant (8.314 J·mol^{-1}·K^{-1}), absolute temperature (K) and the equilibrium constant (L·g^{-1}).

2.8. Chemical Analyzes

Cu(II) cations concentrations were measured using a UV-vis spectrophotometer (DR3900, Hach, Loveland, CO, USA) and pH (PHC301, Hach, Loveland, CO, USA) was measured to adjust pH of experimental solutions before batch adsorption experiments of second pH (pH adjusted to 4.0) and immediately after sample collection using a multi-meter (Hq40d Multi, Hach, Loveland, CO, USA).

2.9. Quality Assurance/Quality Control (QA/QC)

Quality assurance/quality control (QA/QC) procedures were performed for all the analyzes carried out in this study, to ensure the quality, reproducibility, and accuracy of the obtained results. All equipment was calibrated prior to its use in this study and periodically during its development according to the instrument guidelines. All the chemical reagents used in this study were analytical grade. Additionally, all materials used in the experiments and sampling were neatly cleaned and rinsed with Milli-Q water.

The accuracy and precision of the measurements of Cu(II) were checked and compared against blank samples and synthetic standard samples of known concentration. The batch adsorption isotherms were carried out in triplicate and the analytical measurements of pH and Cu(II) were verified by performing triplicate readings.

3. Results and Discussion

3.1. Adsorbent Characterization

3.1.1. Raman Spectroscopy

Raman spectrum of ZnO NPs is shown in Figure 1. The main peaks are identified in the figure. There was no detectable variation in the peaks frequency of different points analyzed in different areas of the sample. The main peaks at 95 cm^{-1} and 438 cm^{-1} are characteristic of ZnO and correspond to phonon frequencies E_2^{low} and E_2^{high}, respectively [75,76]. The peak at 330 is associated with the process $E_2^{high} - E_2^{low}$ [77]. Therefore, the peaks obtained in the Raman analysis are consistent with the characteristic peaks of ZnO. In summary, the analysis of this spectrum indicates that the NPs used in this study effectively correspond to ZnO NPs.

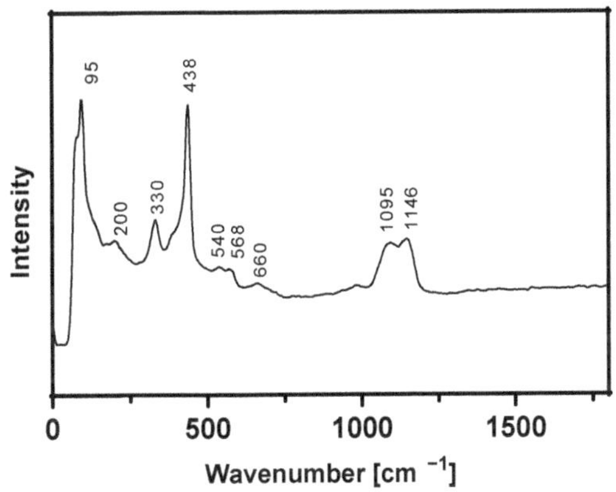

Figure 1. Raman spectrum of the ZnO NPs.

3.1.2. Brunauer–Emmett–Teller (BET) Analysis

Figure 2 shows adsorption/desorption curve of ZnO NPs. A hysteresis loop with typical characteristics of type H3 can be observed in isotherms, typical of type 3 and 5 isotherms. In addition, from the analysis, it can be observed that ZnO NPs showed higher values for BET surface area, reaching a value of 45.58 $m^2 \cdot g^{-1}$. The pore size distribution curve was determined using the Barrett–Joyner–Halenda (BJH) method. The average pore volume (Vp) and pore diameter (Dp) were found to be 30.87 nm and 0.3 $cm^3 \cdot g^{-1}$, respectively. Similar results were observed by Zafar et al. [78] reaching values 0.211 $cm^3 \cdot g^{-1}$ for Vp and 27.44 nm for Dp. These data show that ZnO NPs have a large exposed surface, which can favor surface adsorption processes of heavy metals.

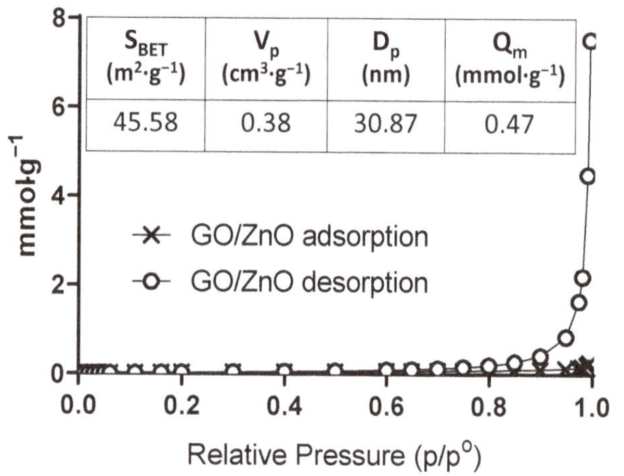

Figure 2. BET adsorption–desorption isotherms for ZnO NPs.

3.1.3. Scanning Electron Microscopy (SEM)

The morphology of ZnO NPs can be observed in Figure 3a,b. The ZnO NPs show small spherical particles with uniform shapes (Arrows). The different sizes of NPs could be a consequence of the formation of interconnected agglomerates between NPs, which could cause a decrease in the available surface area. The field emission scanning electron microscopy (FESEM) (images not shown) confirmed that the ZnO NPs are nano in size (<100 nm).

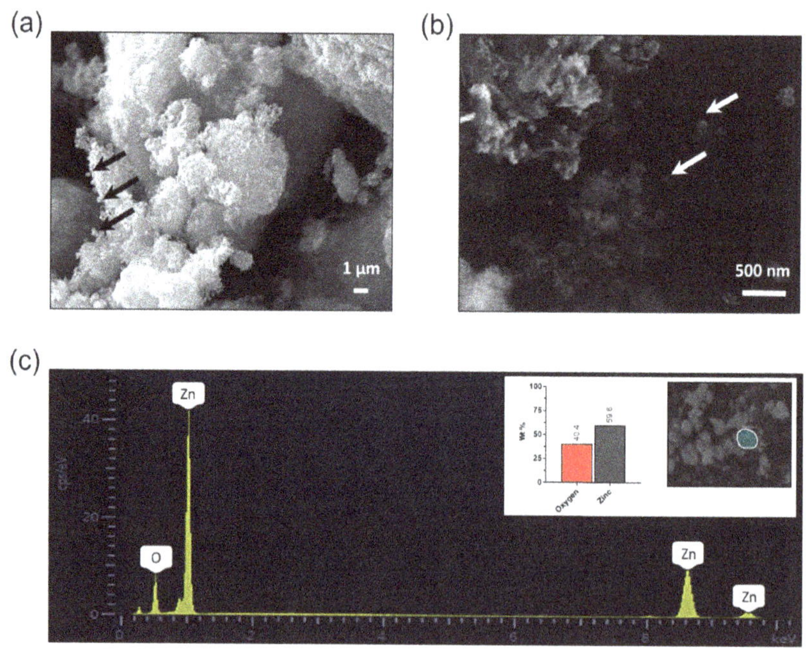

Figure 3. Scanning electron micrographs (**a**) and scanning transmission electron microscopy (STEM) (**b**) of ZnO NPs. Representative EDX spectrum and elemental mapping of ZnO NPs (**c**).

The surface chemical composition of ZnO NPs was determined by EDX analysis. The energy peaks of Zn and O are clear in EDX spectrum of ZnO NPs (Figure 3c). From the elemental analysis, it is observed that ZnO NPs, as expected, are composed of Zinc (59.6% wt) and Oxygen (40.4% wt). Furthermore, the absence of other peaks suggests that the purity of the NPs used in this study was high.

The pHpzc value was obtained for ZnO NPs through the plot of initial pH values and final pH (Figure 4a). A value of 6.21 for ZnO NPs was determined. Other studies have reported higher pHpzc values. For example, Kataria and Garg [79] reported pHpzc values of 6.9. Meanwhile, Chauhan et al. [80] reported pHpzc values of 7.5. The differences observed with other studies may be a consequence of the method of synthesis, the aggregation of the NPs, the presence of impurities, among other factors. The pHpzc corresponds to the pH in which the positive and negative charges are equal on the surface of an adsorbent [81]. Therefore, the surface of ZnO NPs will be positively charged if pH < pHpzc, while it will have a net negative charge if pH > pHpzc. This parameter is important for the heavy metal adsorption process by ZnO NPs, since at values higher than pH = 6.21 the ZnO NPs surface will be negatively charged enhanced electrostatic attraction between heavy metal cations (such as Cu(II)) and ZnO surface. On the contrary, at lower pH values (<pHpzc) the adsorption process will not be as effective as there are higher repulsive forces. Even so, the adsorption process is not determined solely by this parameter, but will also depend on the surface area, the density of pores, the presence of competitors, among other factors. Conversely, the measurement of the zeta potential reveals that the pH where the surface potential of the material is zero, the isoelectric point (IEP) [82] (Figure 4b), is 10.2, a value higher than that obtained at the pH_{PZC}. The values of pH_{PZC} and IEP must be equal if H^+ and OH^- are the only potential determining ions. Therefore, the difference obtained between these two values may be because other specific adsorptions are occurring [83,84].

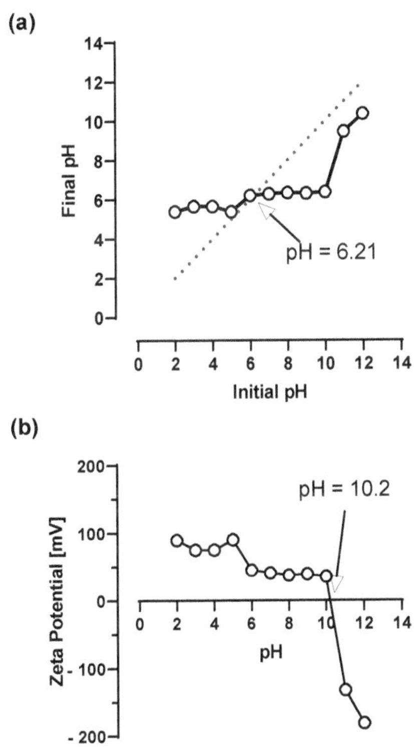

Figure 4. (**a**) Point zero charge (pHpzc) plot final pH vs. initial pHi and (**b**) zeta potential at a function of pH.

3.2. Adsorption Experiments

Adsorption experiments were carried out at two pHs, to study the effect of pH in adsorption capacity at pH values typical of AMD waters in northern Chile. The sorption capacity q_e at different equilibrium Cu(II) concentrations (C_e) are presented graphically in Figure 5. According to experimental data, the maximum sorption capacity for Cu at pH1 (pH = 4.8) was 47.5 mg·g^{-1} (Figure 5a), reaching a maximum removal rate of 98.4% (Figure 5c), while at pH2 (pH = 4.0), the maximum sorption capacity was 40.2 mg·g^{-1} (Figure 5b) and the maximum removal rate was 93.7% (Figure 5d). Furthermore, it is possible to observe that at pH = 4.8, removal rates close to 100% are reached even at low concentrations of the metal ion, while at pH = 4.0, the sorption capacity reaches the maximum values at initial concentrations higher than 8 mg·L^{-1}. Previously, Gu et al. [85] studied the adsorption of Cu and other metals onto ZnO NPs. They reported a maximum sorption capacity of around 16 mg·g^{-1} in a multimetallic solution, where Cu is the second with more affinity with the adsorbent after Cr, which demonstrates the effectiveness of ZnO NPs for copper adsorption.

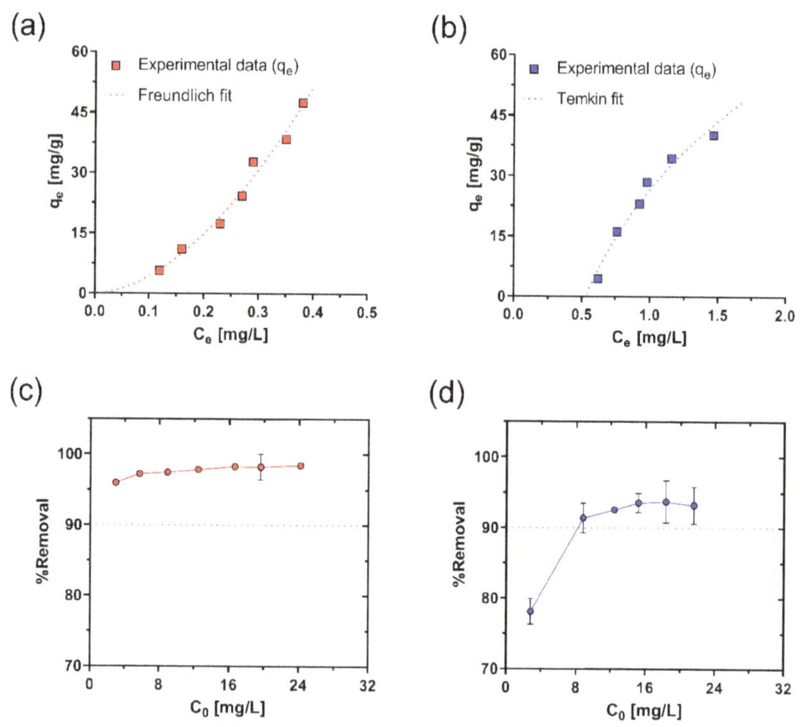

Figure 5. Adsorption isotherms for Cu(II) under (**a**) pH1 (pH = 4.8) with Freundlich isotherm model fit, and (**b**) pH2 (pH = 4.0) with Temkin isotherm model fit using ZnO NPs as nanoadsorbent. Removal percentages for Cu(II) under (**c**) pH1 (pH = 4.8) and (**d**) pH2 (pH = 4.0).

The effect of pH on the removal efficiency of Cu was studied (Appendix C). According to the results obtained, it is possible to observe that in the pH range of 4 to 10, the removal percentage was higher than 90%. At pH 2, the removal efficiency decreases considerably, reaching an average of 29%. Therefore, the removal percentage increases with increasing pH A substantial increase is observed between pH 2 and 5, and then it stabilizes around 99%, with a maximum value of 99.7% at pH 8. These results are consistent with the previous study of Yoshida [86] who observed leaching of ZnO at pH below 3. Although there are no previous studies on the effect of pH on Cu adsorption by ZnO NPs, some studies have reported the effect of pH in the adsorption of different metals using ZnO. Sheela et al. [87] Sheela reported that the removal rates of Zn, Cd and Hg increased when the pH of the solution increased in a range of 4 to 8, and particularly, an increase between 5 and 15% in the removal rates was observed with increasing the pH from 4 to 5. Gu et al. [85] observed that the Cr adsorption capacity in ZnO NPs remained practically constant, with a slight tendency to decrease, between pH 3 and 7. The predominant factor that determines the effect of pH on adsorption corresponds to pH_{PZC}, which could explain the differences between the results reported by the previous studies. In our study, the pH_{PZC} value for ZnO NPs was 6.21, which indicates that at a pH higher than the pH_{PZC}, the adsorbent surface is negatively charged, so the affinity for metal ions such as Cu is improved [88–90]. Therefore, the slight increase in adsorption capacity at higher pH observed in our study can be explained because of the pH_{PZC} value.

The study carried out using multimetallic water with equal concentrations of Cu(II), Mn(II) and Al(III) shows that the removal efficiency of Cu is not affected by the presence of other ions in the solution. In the range of concentrations studied, the removal of Cu remained above 95% both in the tests carried out without pH adjustment and at pH 4.0 (Figure 6). Likewise, it is possible to observe that the ZnO NPs also showed a great affinity for the removal of Al, with removal efficiencies above 94%. Finally, Mn did not show significant adsorption on ZnO. Therefore, for the range of concentrations studied, the effect of the competition is not significant for the sorption capacity of Cu (II).

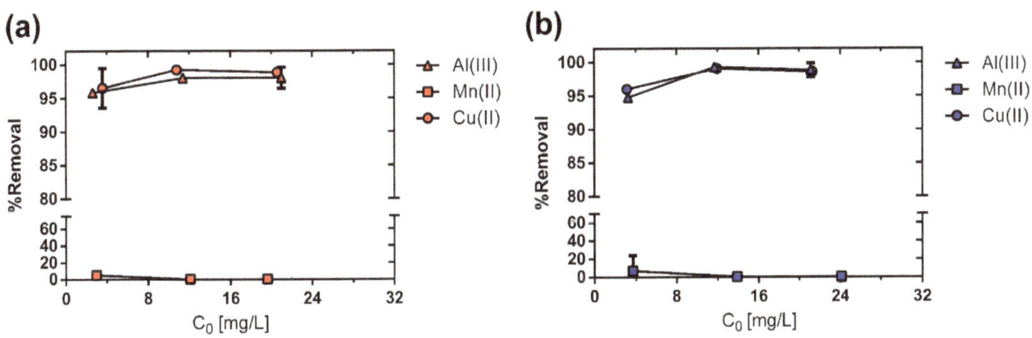

Figure 6. Removal percentages for Cu(II), Mn(II) and Al(III) under (**a**) pH1 (pH = 4.8) and (**b**) pH2 (pH = 4.0) at different initial concentrations.

Because at low pH the ZnO NPs dissolve in the aqueous solution, the desorption process is not feasible by acidifying the adsorbent medium, as has been reported for many other materials [41,91,92]. For the recovery of ZnO, it is possible to consider ultra-sonication methods to promote the desorption process and thus allow the reuse of the material. Previous studies have shown the efficiency of this method for activated carbon, among others [93–96]. There are no previous studies that account for the effectiveness of ultra-sonication for the desorption of ZnO NPs, remaining as a future perspective from this work.

SEM images of ZnO NPs at 500×, 2000×, 4000× and 5000× magnifications before adsorption of Cu(II) ions are shown in Figure 7 a–d. As seen, the ZnO NPs before contacting with Cu(II) solution are shown as clusters of aggregate particles with a rough surface and flat and irregular shapes. It has been reported that the surface energy of the photocatalyst materials, such as ZnO and TiO_2 NPs increases due to its smaller particle size [97]. This may support the agglomeration observed in the ZnO NPs. The specific morphologies observed in the highly porous structure of ZnO NPs support their use as nano-adsorbents for metal cations such as Cu(II) ions. Figure 7e shows the EDX spectra of ZnO NPs before adsorption of Cu(II) ions. As seen in Figure 3c, it is observed that the atomic ratio (Zn/O) of ZnO NPs is near to 1.5:1, which confirms the chemical nature of the ZnO NPs used for the adsorption process of Cu(II) ions.

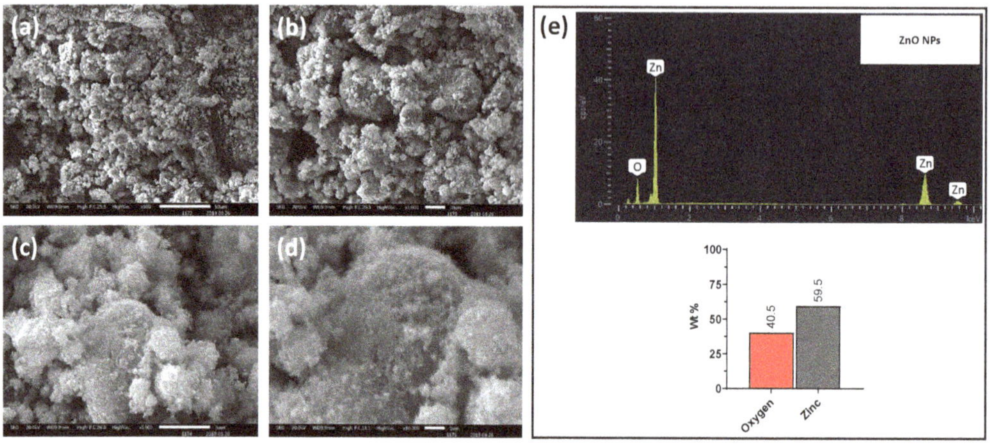

Figure 7. (**a–d**) Scanning electron micrographs before adsorption of Cu(II) ions (pH = 7.0) (**e**) SEM-EDX spectra and elemental mapping (atomic percentage %) of ZnO NPs before adsorption of Cu(II) ions. The magnifications shown in SEM images were 500× (**a**), 2000× (**b**), 4000× (**c**) and 5000× (**d**).

Figure 8 a–d shows SEM images of ZnO NP at magnifications of 500×, 1000×, 5000× and 10,000× after adsorption of Cu (II) ions. As can be seen in the SEM images, the morphology and size of ZnO NPs after reacting with the Cu(II) solution do not change significantly because of the adsorption of Cu(II) ions. The shape of ZnO NPs has no clear change after the adsorption of Cu(II) ions. This is consistent with those reported in other studies that have used ZnO NPs to adsorb metal ions such as Cr(III) [85,98]. Similarly, after adsorption of Cu (II) ions, the ZnO NPs appear to be aggregated in clusters, which occurs because of their higher surface energy. For adsorption processes, it has been observed that microporous materials (pore size < 2 nm) are more selective for adsorption and separation, while macroporous materials (pore size > 50 nm) and mesoporous materials (pore size between 2–50 nm) due to their larger pores allow different adsorbates to penetrate through them [99]. The EDX results (Figure 8e) show the adsorption of Cu(II) ions, which confirms what was seen in the adoration isotherms (Figure 5). The atomic mapping showed 2.85% weight percentage of Cu, 25.8% weight percentage of O and 71.4% weight percentage of Zn. This strongly supports the idea that Cu(II) was adsorbed onto the surface of ZnO NPs, which shows the key role these NPs play in the uptake of Cu (II) ions from AMD waters. Interestingly, the removal of Cu(II) ions by ZnO NPs are not significantly affected by acidic conditions (pH = 4.0). A remarkable aspect is the ability of ZnO NPs to maintain their structure before and after the adsorption of Cu(II) ions at acidic pH. In fact, it can be seen that the morphology of ZnO NPs does not change considerably at pH = 4 (adsorption of Cu(II) ions). Thus, this may be a key factor in promoting and scaling the use of ZnO NPs in acidic water treatment systems and metal cation removal applications. Even so, it is necessary to deepen the changes at the nanostructure level of ZnO NPs, which could be the consequence of the most aggressive acidic conditions.

To complement the adsorption analysis, the experimental data were fitted using Langmuir, Freundlich and Temkin isotherm models. The parameters for these models are summarized in Table 1. The Langmuir isotherm presented an R^2 value of 0.965 for the pH1. However, the values of the parameters q_m and K_L presented negative values, which cannot be because both parameters represent mass properties. With pH2, the coefficient of determination showed a poor fit and, again, the model parameters presented negative values. In this way, it is ruled out that the adsorption process studied is explained by a type of monolayer adsorption [100]. For the case of pH1, the best fit corresponds to the Freundlich model, which describes multilayer adsorption onto a heterogeneous surface of the adsorbent [101–103]. Here, the stronger binding sites are occupied until the adsorption energy decreased [104]. At pH2, the model that best describes the adsorption process corresponds to the Temkin isotherm, in which is assumed that the heat of adsorption of all the molecules in the layer decreases linearly rather than logarithmically as equilibrium adsorption capacity increases [101]. Both settings are presented in Figure 5a,b. Although the best fit model is different in the studied pH values, both have the particularity of describing a type of multilayer adsorption.

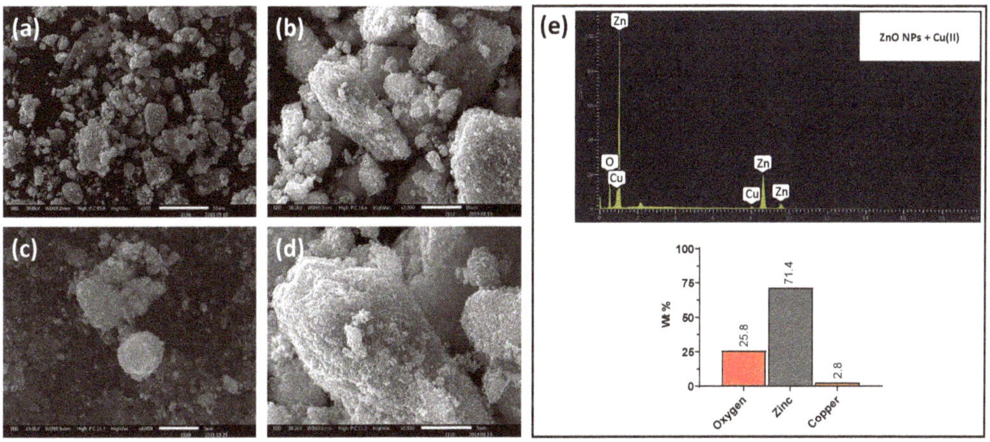

Figure 8. (**a**–**d**) Scanning electron micrographs after adsorption of Cu(II) ions (pH =4.0). (**e**) SEM-EDX spectra and elemental mapping (atomic percentage %) of ZnO NPs after adsorption of Cu(II) ions. The magnifications shown in SEM images were 500× (**a**), 1000× (**b**), 5000× (**c**) and 10,000× (**d**).

Table 1. Parameters for the Langmuir, Freundlich and Temkin isotherm models for Cu(II) adsorption performed under two pHs.

		Langmuir			Freundlich			Temkin		
pH	**Nano-Adsorbent**	q_m **(mg·g^{-1})**	K_L **(L·mg^{-1})**	R^2	K_F **(L·g^{-1})**	n	R^2	K_T **(L·g^{-1})**	B_T **(mg·L^{-1})**	R^2
pH 4.8	ZnO	−18.28	−2.1	0.965	264.85	0.56	0.988	8.69	34.45	0.908
pH 4.0	ZnO	−6.64	−0.75	0.739	22.75	0.42	0.830	1.89	41.98	0.975

At pH values lower than 5, the dominant copper species is its divalent form Cu(II) and in the pH range studied in this work (4.0–4.8), copper speciation should be similar in both pHs [30,105]. Although the differences in the sorption capacity between the two pHs are slight, the differences are mainly explained by the surface charge of the ZnO NPs as a function of the pH_{PZC} value. In this way, the chemical interaction between the functional groups of the NPs and Cu(II) could explain the adsorption mechanism. Although both pH values are below the pH_{PZC}, and yet the removal rates obtained were close to 100%, it is possible to deduce that physical adsorption plays an important role in the adsorption mechanism. This can also be confirmed by the high BET surface area value found for ZnO NPs (45.58 $m^2 \cdot g^{-1}$, Figure 2), which may support the idea that physical adsorption is the dominant mechanism overcoming the repulsions generated by being below the pH_{PZC}.

The thermodynamic parameter ΔG^0 was determined using the equilibrium constant obtained from Temkin and Freundlich fit for the experiments with and without adjusting pH, respectively since these isotherm models presented a better fit for each case. The results are presented in Table 2. According to the values obtained, the adsorption test at pH 4.8 presented a higher negative value than in the case at pH 4, so that at a higher pH the adsorption process is energetically more favorable.

Table 2. Gibbs free energy (ΔG^0) for adsorption processes under two pHs.

pH	K ($L \cdot g^{-1}$)	ΔG^0 ($kJ \cdot mol^{-1}$)
pH 4.8	264.85	−13,829.74
pH 4.0	1.89	−1577.96

3.3. *Surface Analysis: Roughness*

To analyze changes at the nanostructure level of the ZnO NPs before and after reacting with Cu(II) solution, SEM images and their corresponding 3D surface profiles were determined through image Gwyddion software. SEM images are directly related to their real relief and therefore these images can be used to determine changes in the surface roughness of a specific material. Likewise, changes in roughness can be the consequence of chemical reactions on the surfaces of ZnO NPs. In this way, these analyzes can provide quantitative data on changes in surface roughness because of the effect of chemical adsorption conditions or interactions with specific adsorbates. Figure 9 shows 3D images of the ZnO NP before (Figure 9a) and after (Figure 9b) adsorption of Cu(II) ions. It is clearly observed that the surface roughness of ZnO NPs prior to the adsorption of Cu(II) ions is greater than after contacting the Cu(II) solution at pH = 4.0. This can be a consequence of more aggressive pH conditions that wear away the nano-adsorbent surface (ZnO NPs).

(a)

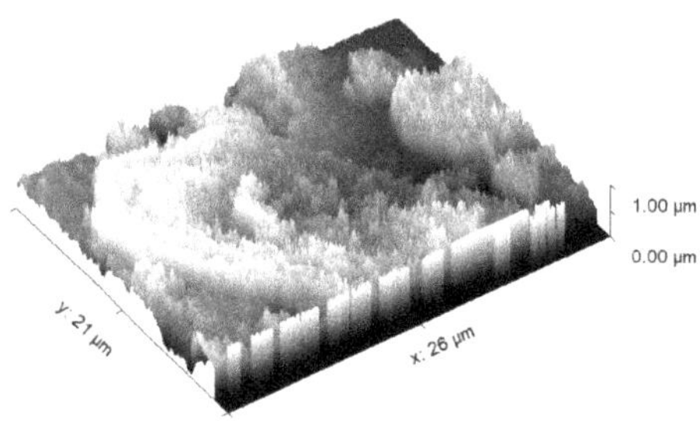

Roughness average (R_a) [nm]	30.0 ± 8.7
Root mean square roughness (R_q) [nm]	37.9 ± 11.1
Surface skewness (R_{sk})	0.21 ± 0.07
Kurtosis (R_{ku})	3.3 ± 0.3

(b)

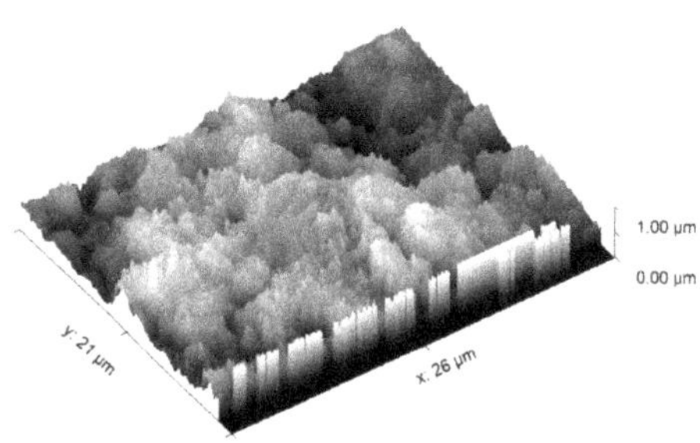

Roughness average (R_a) [nm]	9.3 ± 1.3
Root Mean Square Roughness (R_q) [nm]	11.9 ± 1.7
Surface skewness (R_{sk})	− 0.06 ± 0.05
Kurtosis (R_{ku})	3.2 ± 0.2

Figure 9. 3-D SEM images and surface roughness parameters of ZnO NPs before (**a**) and after (**b**) Cu(II) adsorption. The 3-D SEM images and surface roughness parameters (R_a, R_q, R_{sk} and R_{ku}) were obtained by the Gwyddion software.

For quantitative analysis, the surface roughness parameters R_a, Rq, Rsk and R_{ku} were obtained by the Gwyddion software (Figure 9a,b). The R_a values of ZnO NPs before adsorption of Cu(II) ions shows an average value of 29.997 ± 8.663, while R_a values of ZnO NPs after adsorption of Cu(II) ions shows an average value of 9.348 ± 1.281, evidencing a marked decrease in surface roughness by approximately 70%. This shows that more aggressive acidic conditions (pH = 4.0) can produce changes at the nanostructural level, although these changes are not evident in changes in the morphology, shape, and particle size of the ZnO NPs. The R_q of ZnO NPs before and after adsorption Cu(II) ion were 37.920 ± 11.077 and 11.877 ± 1.690, respectively. This shows that surface roughness variation in ZnO NPs before adsorption of Cu(II) ions is higher than that in ZnO NPs after adsorption of Cu(II) ions, which suggests that acidic solutions enriched in Cu(II) can homogenize the surface of the ZnO NPs. The R_{sk} values were very similar for the samples of ZnO NPs before and after the adsorption of Cu(II) ions. Even so, the R_{sk} values were positive for ZnO NPs before the adsorption of Cu(II) ions, which shows that in these samples there are more troughs than peaks on the surface. On the contrary, R_{sk} values were negative for ZnO NPs after adsorption of Cu(II) ions, showing that in these samples that there are fewer troughs than peaks. Finally, the R_{ku} values were similar for both conditions, before (3.344 ± 0.284) and after (3.232 ± 0.213) the adsorption of Cu(II) ions. This shows that the shape of the pore size distribution is concentrated in both conditions. It has been observed that the joint strength between adhesive and adherent is influenced by the surface roughness of the adherents [106]. Therefore, the surface roughness of the nano-adsorbent (e.g., ZnO NPs) could also be critical for adsorbate–adsorbent interactions.

3.4. Kinetic Studies of the Adsorption

The study of the adsorption kinetics showed that after 240 min of contact time, removal of over 90% of Cu(II) is achieved (Figure 10). The experimental data were fitted with pseudo-first-order and pseudo-second-order kinetic models. Table 3 summarizes the kinetic parameters obtained for both models. According to R^2 and q_e values, it could be determined that the pseudo-second-order presented a better fit. The pseudo-second-order describes a sorption process mainly controlled by the adsorption reaction at the liquid/solid interface at the adsorbent, in contrast to the pseudo-first-order model, which describes a diffusion-controlled process [41,107]. Previous studies about the use of ZnO NPs for the adsorption of different pollutants, such as Zn, Cd, Hg, Cr, among others, have shown that the pseudo-second-order model is the one that best adjusts to the adsorption kinetics [85,87,108,109]. The approaching equilibrium factor (R_w) obtained was 0.024, which is related to a type of kinetic curve largely curved. This value also shows a well-approaching equilibrium level [69].

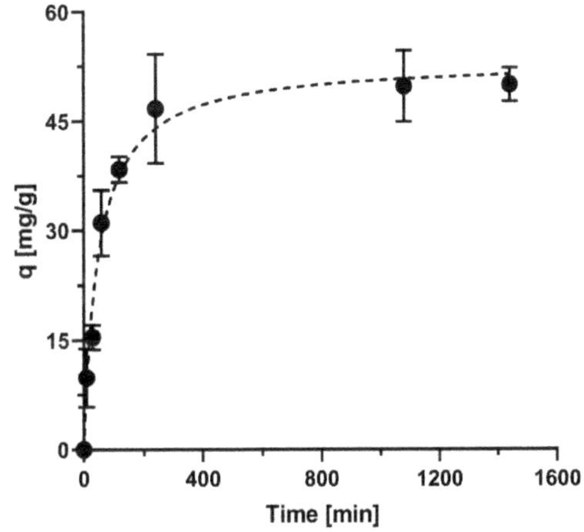

Figure 10. Effect of contact time on removal Cu(II) ions. The kinetic curve for Cu(II) is based on the pseudo-second-order model.

Table 3. Kinetic adsorption parameters for pseudo-first-order and pseudo-second-order models.

Nano-Adsorbent	q_e^{exp} (mg·g^{-1})	Pseudo-First-Order			Pseudo-Second-Order		
		k_1 (1·min^{-1})	q_{e1} (mg·g^{-1})	R^2	k_2 (g·mg^{-1} min^{-1})	q_{e2} (mg·g^{-1})	R^2
ZnO	49.94	0.011	28.94	0.922	0.0054	51.28	0.999

4. Conclusions

The results show that ZnO NPs have a great affinity for Cu(II) ions, getting adsorption capacities even higher than previous studies. The characterization analyzes of the nano-adsorbent allowed us to observe that the ZnO NPs form agglomerations, which could decrease the total surface area and give an underestimated result of this value through the BET analysis. Raman confirmed the presence of ZnO NPs and SEM-EDX and FESEM confirmed that the ZnO NPs are of nano size (<100 nm).

The study of different pHs demonstrates that even though large variations in adsorption capacity do not occur when the pH varies between 4.0 and 4.8, a slight improvement is observed at pH 4.8, which is mainly explained by the value of pH_{PZC} of the nano-adsorbent, corresponding to 6.21. It is expected that, at pH values higher than pH_{PZC}, the adsorption capacity will be even higher. In addition, it is possible to observe that at pH 4.8 the removal rate is higher at low concentrations than at pH 4. The results suggest that the adsorption process occurs by a physical mechanism rather than by chemical adsorption since at the evaluated pHs there are higher repulsions than attraction forces on the surface of ZnO NPs. Likewise, the surface roughness analysis showed a marked decrease in surface roughness of ZnO NPs by approximately 70% after adsorption of Cu(II) ions at pH 4.0, which shows that more acidic conditions can produce changes at the nanostructural level. Although these changes are not clear in the changes in the morphology, shape, particle size of the ZnO NPs and in the removal rates of Cu(II) ions. Adjustments with isotherm models allowed determining that adsorption occurs mainly because of the formation of multi-layers on the surface of the ZnO NPs. The kinetics showed that the pseudo-second-order model better fit the experimental data.

These findings contribute to a better understanding of the adsorption of Cu metal ions onto ZnO NPs and the effect of slight variations in pHs. However, additional efforts are necessary to improve the knowledge of the effect of pH under more extreme conditions and considering multimetallic waters that represent a more realistic scenario in the composition of AMD characteristic waters.

Author Contributions: The manuscript was written by C.R. and E.L., but all the authors contributed to its preparation and review. Experiments were performed by C.T.; Data analyses were carried out by C.R. and C.T. in discussion with E.L. All authors have read and agreed to the published version of the manuscript.

Funding: This research received funding from FIC-R Fondo de Innovación para la Competitividad Gore Coquimbo BIP nos. 30485965-0, CORFO-L2 L2 ISV93456 and FONDECYT Iniciación 11191154 (2019–2022) grants.

Institutional Review Board Statement: Not applicable.

Informed Consent Statement: Not applicable.

Data Availability Statement: The data presented in this study are available on request from the corresponding author.

Acknowledgments: Thanks are also extended to the reviewers for their corrections and suggestions, who contributed significantly to improving the quality of this manuscript.

Conflicts of Interest: The authors declare no conflict of interest.

Appendix A

Table A1. Costs of wastewater treatment technologies.

Technology	Cost per Million Liters	Reference
Adsorption	USD 10 to 200 USD 50 to 150	[28,29]
Ion Exchange	USD 50 to 200	[29]
Reverse Osmosis	USD 200 to 450	[29]
Micro- and ultra-filtration	USD 15 to 400	[29]
Electrodialysis	USD 15 to 400	[29]

Appendix B

Table A2. Costs of nanomaterials for adsorption treatment technologies.

Nanomaterial	Cost per Gram	Reference
ZnO nanoparticles	USD 7.6	[110]
Graphene	USD 632.73	[111]
Carbon nanotubes	USD 263	[112]

Appendix C

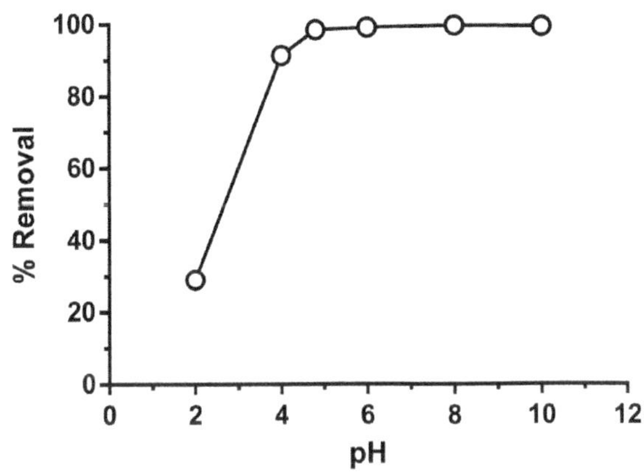

Figure A1. Effect of pH on Cu(II) ions removal. The initial concentration was 24 mg·L^{-1} in each case.

References

1. Blowes, D.W.; Ptacek, C.J.; Jambor, J.L.; Weisener, C.G. The Geochemistry of Acid Mine Drainage. In *Treatise on Geochemistry*; Elsevier Inc.: Amsterdam, The Netherlands, 2003; Volume 9, pp. 149–204, ISBN 9780080548074.
2. Naidu, G.; Ryu, S.; Thiruvenkatachari, R.; Choi, Y.; Jeong, S.; Vigneswaran, S. A critical review on remediation, reuse, and resource recovery from acid mine drainage. *Environ. Pollut.* **2019**, *247*, 1110–1124. [CrossRef] [PubMed]
3. Kefeni, K.K.; Msagati, T.A.M.; Mamba, B.B. Acid mine drainage: Prevention, treatment options, and resource recovery: A review. *J. Clean. Prod.* **2017**, *151*, 475–493. [CrossRef]
4. Akcil, A.; Koldas, S. Acid Mine Drainage (AMD): Causes, treatment and case studies. *J. Clean. Prod.* **2006**, *14*, 1139–1145. [CrossRef]
5. Carolin, C.F.; Kumar, P.S.; Saravanan, A.; Joshiba, G.J.; Naushad, M. Efficient techniques for the removal of toxic heavy metals from aquatic environment: A review. *J. Environ. Chem. Eng.* **2017**, *5*, 2782–2799. [CrossRef]
6. Vardhan, K.H.; Kumar, P.S.; Panda, R.C. A review on heavy metal pollution, toxicity and remedial measures: Current trends and future perspectives. *J. Mol. Liq.* **2019**, *290*, 111197. [CrossRef]
7. Visa, M. Synthesis and characterization of new zeolite materials obtained from fly ash for heavy metals removal in advanced wastewater treatment. *Powder Technol.* **2016**, *294*, 338–347. [CrossRef]

8. Malik, A. Metal bioremediation through growing cells. *Environ. Int.* **2004**, *30*, 261–278. [CrossRef]
9. Ziagova, M.; Dimitriadis, G.; Aslanidou, D.; Papaioannou, X.; Litopoulou Tzannetaki, E.; Liakopoulou-Kyriakides, M. Comparative study of Cd(II) and Cr(VI) biosorption on Staphylococcus xylosus and Pseudomonas sp. in single and binary mixtures. *Bioresour. Technol.* **2007**, *98*, 2859–2865. [CrossRef]
10. Chouyyok, W.; Shin, Y.; Davidson, J.; Samuels, W.D.; Lafemina, N.H.; Rutledge, R.D.; Fryxell, G.E.; Sangvanich, T.; Yantasee, W. Selective removal of copper(II) from natural waters by nanoporous sorbents functionalized with chelating diamines. *Environ. Sci. Technol.* **2010**, *44*, 6390–6395. [CrossRef]
11. Akbari, M.; Hallajisani, A.; Keshtkar, A.R.; Shahbeig, H.; Ali Ghorbanian, S. Equilibrium and kinetic study and modeling of Cu(II) and Co(II) synergistic biosorption from Cu(II)-Co(II) single and binary mixtures on brown algae C. indica. *J. Environ. Chem. Eng.* **2015**, *3*, 140–149. [CrossRef]
12. Ahmad, A.; Rafatullah, M.; Sulaiman, O.; Ibrahim, M.H.; Chii, Y.Y.; Siddique, B.M. Removal of Cu(II) and Pb(II) ions from aqueous solutions by adsorption on sawdust of Meranti wood. *Desalination* **2009**, *247*, 636–646. [CrossRef]
13. Stylianou, M.A.; Inglezakis, V.J.; Moustakas, K.G.; Malamis, S.P.; Loizidou, M.D. Removal of Cu(II) in fixed bed and batch reactors using natural zeolite and exfoliated vermiculite as adsorbents. *Desalination* **2007**, *215*, 133–142. [CrossRef]
14. Salomons, W. Environmental impact of metals derived from mining activities: Processes, predictions, prevention. *J. Geochem. Explor.* **1995**, *52*, 5–23. [CrossRef]
15. Rodríguez, C.; Leiva-Aravena, E.; Serrano, J.; Leiva, E. Occurrence and removal of copper and aluminum in a stream confluence affected by acid mine drainage. *Water* **2018**, *10*, 516. [CrossRef]
16. Al-Saydeh, S.A.; El-Naas, M.H.; Zaidi, S.J. Copper removal from industrial wastewater: A comprehensive review. *J. Ind. Eng. Chem.* **2017**, *56*, 35–44. [CrossRef]
17. Andrejkovičová, S.; Sudagar, A.; Rocha, J.; Patinha, C.; Hajjaji, W.; Da Silva, E.F.; Velosa, A.; Rocha, F. The effect of natural zeolite on microstructure, mechanical and heavy metals adsorption properties of metakaolin based geopolymers. *Appl. Clay Sci.* **2016**, *126*, 141–152. [CrossRef]
18. Satyro, S.; Marotta, R.; Clarizia, L.; Di Somma, I.; Vitiello, G.; Dezotti, M.; Pinto, G.; Dantas, R.F.; Andreozzi, R. Removal of EDDS and copper from waters by TiO2 photocatalysis under simulated UV-solar conditions. *Chem. Eng. J.* **2014**, *251*, 257–268. [CrossRef]
19. Li, X.; Deng, G.; Zhang, Y.; Wang, J. Rapid removal of copper ions from aqueous media by hollow polymer nanoparticles. *Colloids Surf. A Physicochem. Eng. Asp.* **2019**, *568*, 345–355. [CrossRef]
20. Matouq, M.; Jildeh, N.; Qtaishat, M.; Hindiyeh, M.; Al Syouf, M.Q. The adsorption kinetics and modeling for heavy metals removal from wastewater by Moringa pods. *J. Environ. Chem. Eng.* **2015**, *3*, 775–784. [CrossRef]
21. Wang, X.; Yu, S.; Jin, J.; Wang, H.; Alharbi, N.S.; Alsaedi, A.; Hayat, T.; Wang, X. Application of graphene oxides and graphene oxide-based nanomaterials in radionuclide removal from aqueous solutions. *Sci. Bull.* **2016**, *61*, 1583–1593. [CrossRef]
22. Gros, F.; Baup, S.; Aurousseau, M. Copper cementation on zinc and iron mixtures: Part 1: Results on rotating disc electrode. *Hydrometallurgy* **2011**, *106*, 119–126. [CrossRef]
23. Ferrer, O.; Gibert, O.; Cortina, J.L. Reverse osmosis membrane composition, structure and performance modification by bisulphite, iron(III), bromide and chlorite exposure. *Water Res.* **2016**, *103*, 256–263. [CrossRef]
24. Xu, Y.C.; Wang, Z.X.; Cheng, X.Q.; Xiao, Y.C.; Shao, L. Positively charged nanofiltration membranes via economically mussel-substance-simulated co-deposition for textile wastewater treatment. *Chem. Eng. J.* **2016**, *303*, 555–564. [CrossRef]
25. Caprarescu, S.; Corobea, M.C.; Purcar, V.; Spataru, C.I.; Ianchis, R.; Vasilievici, G.; Vuluga, Z. San copolymer membranes with ion exchangers for Cu(II) removal from synthetic wastewater by electrodialysis. *J. Environ. Sci.* **2015**, *35*, 27–37. [CrossRef] [PubMed]
26. Dong, Y.; Liu, J.; Sui, M.; Qu, Y.; Ambuchi, J.J.; Wang, H.; Feng, Y. A combined microbial desalination cell and electrodialysis system for copper-containing wastewater treatment and high-salinity-water desalination. *J. Hazard. Mater.* **2017**, *321*, 307–315. [CrossRef] [PubMed]
27. Kanakaraju, D.; Ravichandar, S.; Lim, Y.C. Combined effects of adsorption and photocatalysis by hybrid TiO_2/ZnO-calcium alginate beads for the removal of copper. *J. Environ. Sci.* **2017**, *55*, 214–223. [CrossRef] [PubMed]
28. Ali, I.; Gupta, V.K. Advances in water treatment by adsorption technology. *Nat. Protoc.* **2007**, *1*, 2661–2667. [CrossRef] [PubMed]
29. Gupta, V.K.; Ali, I.; Saleh, T.A.; Nayak, A.; Agarwal, S. Chemical treatment technologies for waste-water recycling—An overview. *RSC Adv.* **2012**, *2*, 6380–6388. [CrossRef]
30. Demiral, H.; Güngör, C. Adsorption of copper(II) from aqueous solutions on activated carbon prepared from grape bagasse. *J. Clean. Prod.* **2016**, *124*, 103–113. [CrossRef]
31. Yang, S.T.; Chang, Y.; Wang, H.; Liu, G.; Chen, S.; Wang, Y.; Liu, Y.; Cao, A. Folding/aggregation of graphene oxide and its application in Cu2+ removal. *J. Colloid Interface Sci.* **2010**, *351*, 122–127. [CrossRef]
32. Bansode, R.R.; Losso, J.N.; Marshall, W.E.; Rao, R.M.; Portier, R.J. Adsorption of metal ions by pecan shell-based granular activated carbons. *Bioresour. Technol.* **2003**, *89*, 115–119. [CrossRef]
33. Lee, S.M.; Davis, A.P. Removal of cu(II) and cd(II) from aqueous solution by seafood processing waste sludge. *Water Res.* **2001**, *35*, 534–540. [CrossRef]
34. Sitko, R.; Turek, E.; Zawisza, B.; Malicka, E.; Talik, E.; Heimann, J.; Gagor, A.; Feist, B.; Wrzalik, R. Adsorption of divalent metal ions from aqueous solutions using graphene oxide. *Dalton Trans.* **2013**, *42*, 5682–5689. [CrossRef]
35. Abou-Zeid, R.E.; Dacrory, S.; Ali, K.A.; Kamel, S. Novel method of preparation of tricarboxylic cellulose nanofiber for efficient removal of heavy metal ions from aqueous solution. *Int. J. Biol. Macromol.* **2018**, *119*, 207–214. [CrossRef]

36. Mautner, A.; Maples, H.A.; Kobkeatthawin, T.; Kokol, V.; Karim, Z.; Li, K.; Bismarck, A. Phosphorylated nanocellulose papers for copper adsorption from aqueous solutions. *Int. J. Environ. Sci. Technol.* **2016**, *13*, 1861–1872. [CrossRef]
37. Cegłowski, M.; Gierczyk, B.; Frankowski, M.; Popenda, Ł. A new low-cost polymeric adsorbents with polyamine chelating groups for efficient removal of heavy metal ions from water solutions. *React. Funct. Polym.* **2018**, *131*, 64–74. [CrossRef]
38. Karami, S.; Zeynizadeh, B. Reduction of 4-nitrophenol by a disused adsorbent: EDA-functionalized magnetic cellulose nanocomposite after the removal of Cu^{2+}. *Carbohydr. Polym.* **2019**, *211*, 298–307. [CrossRef]
39. Liu, F.; Zhou, K.; Chen, Q.; Wang, A.; Chen, W. Application of magnetic ferrite nanoparticles for removal of Cu(II) from copper-ammonia wastewater. *J. Alloys Compd.* **2019**, *773*, 140–149. [CrossRef]
40. Salam, M.A. Removal of heavy metal ions from aqueous solutions with multi-walled carbon nanotubes: Kinetic and thermodynamic studies. *Int. J. Environ. Sci. Technol.* **2013**, *10*, 677–688. [CrossRef]
41. Rodríguez, C.; Leiva, E. Enhanced heavy metal removal from acid mine drainage wastewater using double-oxidized multiwalled carbon nanotubes. *Molecules* **2019**, *25*, 111. [CrossRef]
42. Rodríguez, C.; Briano, S.; Leiva, E. Increased adsorption of heavy metal ions in multi-walled carbon nanotubes with improved dispersion stability. *Molecules* **2020**, *25*, 3106. [CrossRef]
43. Abuhatab, S.; El-Qanni, A.; Al-Qalaq, H.; Hmoudah, M.; Al-Zerei, W. Effective adsorptive removal of Zn^{2+}, Cu^{2+}, and Cr^{3+} heavy metals from aqueous solutions using silica-based embedded with NiO and MgO nanoparticles. *J. Environ. Manag.* **2020**, *268*, 110713. [CrossRef]
44. Li, S.; Wang, W.; Liang, F.; Zhang, W.X. Heavy metal removal using nanoscale zero-valent iron (nZVI): Theory and application. *J. Hazard. Mater.* **2017**, *322*, 163–171. [CrossRef]
45. Tan, P.; Sun, J.; Hu, Y.; Fang, Z.; Bi, Q.; Chen, Y.; Cheng, J. Adsorption of Cu2+, Cd2+ and Ni2+ from aqueous single metal solutions on graphene oxide membranes. *J. Hazard. Mater.* **2015**, *297*, 251–260. [CrossRef] [PubMed]
46. Salahuddin, N.; El-Kemary, M.; Ibrahim, E.M. Synthesis and characterization of ZnO nanoparticles via precipitation method: Effect of annealing temperature on particle size. *Nanosci. Nanotechnol.* **2015**, *5*, 82–88.
47. Ruszkiewicz, J.A.; Pinkas, A.; Ferrer, B.; Peres, T.V.; Tsatsakis, A.; Aschnera, M. Neurotoxic effect of active ingredients in sunscreen products, a contemporary review. *Toxicol. Rep.* **2017**, *4*, 245–259. [CrossRef] [PubMed]
48. Siddiqi, K.S.; Rahman, A.U.; Tajuddin, H.A. Properties of zinc oxide nanoparticles and their activity against microbes. *Nanoscale Res. Lett.* **2018**, *13*, 1–13. [CrossRef]
49. Sawai, J. Quantitative evaluation of antibacterial activities of metallic oxide powders (ZnO, MgO and CaO) by conductimetric assay. *J. Microbiol. Methods* **2003**, *54*, 177–182. [CrossRef]
50. Chen, C.C.; Liu, P.; Lu, C.H. Synthesis and characterization of nano-sized ZnO powders by direct precipitation method. *Chem. Eng. J.* **2008**, *144*, 509–513. [CrossRef]
51. Li, Y., Lu, X.; Qi, H.; Li, X.; Xiao, X.; Gao, J. Ursolic acid induces apoptosis through mitochondrial intrinsic pathwayand suppression of ERK1/2 MAPK in HeLa cells. *J. Pharmacol. Sci.* **2014**, *125*, 202–210. [CrossRef]
52. Qamar, M.; Gondal, M.A.; Yamani, Z.H. Laser-induced efficient reduction of Cr(VI) catalyzed by ZnO nanoparticles. *J. Hazard. Mater.* **2011**, *187*, 258–263. [CrossRef]
53. Chakrabarti, S.; Chaudhuri, B.; Bhattacharjee, S.; Ray, A.K.; Dutta, B.K. Photo-reduction of hexavalent chromium in aqueous solution in the presence of zinc oxide as semiconductor catalyst. *Chem. Eng. J.* **2009**, *153*, 86–93. [CrossRef]
54. Shirzad-Siboni, M.; Farrokhi, M.; Darvishi Cheshmeh Soltani, R.; Khataee, A.; Tajassosi, S. Photocatalytic reduction of hexavalent chromium over ZnO nanorods immobilized on kaolin. *Ind. Eng. Chem. Res.* **2014**, *53*, 1079–1087. [CrossRef]
55. Siboni, M.S.; Samadi, M.T.; Yang, J.K.; Lee, S.M. Photocatalytic reduction of Cr(VI) and Ni(II) in aqueous solution by synthesized nanoparticle ZnO under ultraviolet light irradiation: A kinetic study. *Environ. Technol.* **2011**, *32*, 1573–1579. [CrossRef]
56. Lee, J.H.; Seo, J.U.; Chung, Y.J.; Lee, J.C.; Park, S. Removal of heavy metal ions from aqueous Pb-EDTA and Cu-EDTA solutions using nanosized ZnO powders by solution-combustion method. *Key Eng. Mater.* **2006**, *317–318*, 837–840. [CrossRef]
57. Lee, J.C.; Kim, H.S.; Lee, J.H.; Park, S. Photocatalytic removal of Cu ions from aqueous Cu-EDTA solution using solution combusted zinc oxide nanopowder. *J. Nanosci. Nanotechnol.* **2008**, *8*, 5284–5287. [CrossRef]
58. Le, A.T.; Pung, S.Y.; Sreekantan, S.; Matsuda, A.; Huynh, D.P. Mechanisms of removal of heavy metal ions by ZnO particles. *Heliyon* **2019**, *5*, e01440. [CrossRef]
59. Wang, X.; Cai, W.; Liu, S.; Wang, G.; Wu, Z.; Zhao, H. ZnO hollow microspheres with exposed porous nanosheets surface: Structurally enhanced adsorption towards heavy metal ions. *Colloids Surf. A Physicochem. Eng. Asp.* **2013**, *422*, 199–205. [CrossRef]
60. De Primo, J.O.; Bittencourt, C.; Acosta, S.; Sierra-Castillo, A.; Colomer, J.-F.; Jaerger, S.; Teixeira, V.C.; Anaissi, F.J. Synthesis of zinc oxide nanoparticles by ecofriendly routes: Adsorbent for copper removal from wastewater. *Front. Chem.* **2020**, *8*, 1100. [CrossRef]
61. Pashai Gatabi, M.; Milani Moghaddam, H.; Ghorbani, M. Point of zero charge of maghemite decorated multiwalled carbon nanotubes fabricated by chemical precipitation method. *J. Mol. Liq.* **2016**, *216*, 117–125. [CrossRef]
62. Vasconcelos, H.L.; Camargo, T.P.; Gonçalves, N.S.; Neves, A.; Laranjeira, M.C.M.; Fávere, V.T. Chitosan crosslinked with a metal complexing agent: Synthesis, characterization and copper(II) ions adsorption. *React. Funct. Polym.* **2008**, *68*, 572–579. [CrossRef]
63. Langmuir, I. The constitution and fundamental properties of solids and liquids. II. Liquids. *J. Am. Chem. Soc.* **1917**, *39*, 1848–1906. [CrossRef]
64. Freundlich, H.M.F. Over the adsorption in solution. *J. Phys. Chem.* **1906**, *57*, 1100–1107.

65. Vaghetti, J.C.P.; Lima, E.C.; Royer, B.; da Cunha, B.M.; Cardoso, N.F.; Brasil, J.L.; Dias, S.L.P. Pecan nutshell as biosorbent to remove Cu(II), Mn(II) and Pb(II) from aqueous solutions. *J. Hazard. Mater.* **2009**, *162*, 270–280. [CrossRef]
66. Tempkin, M.I.; Pyzhev, V. Kinetics of ammonia synthesis on promoted iron catalyst. *Acta Phys. Chim. USSR* **1940**, *12*, 327.
67. Nečas, D.; Klapetek, P. Gwyddion: An open-source software for SPM data analysis. *Open Phys.* **2012**, *10*, 181–188. [CrossRef]
68. Zhao, S.; Li, Y.; Wang, Y.; Ma, Z.; Huang, X. Quantitative study on coal and shale pore structure and surface roughness based on atomic force microscopy and image processing. *Fuel* **2019**, *244*, 78–90. [CrossRef]
69. Ding, P.; Huang, K.L.; Li, G.Y.; Zeng, W.W. Mechanisms and kinetics of chelating reaction between novel chitosan derivatives and Zn(II). *J. Hazard. Mater.* **2007**, *146*, 58–64. [CrossRef]
70. Ho, Y.S.; McKay, G. Comparative sorption kinetic studies of dye and aromatic compounds onto fly ash. *J. Environ. Sci. Heal. Part A Toxic/Hazard. Subst. Environ. Eng.* **1999**, *34*, 1179–1204. [CrossRef]
71. Wu, F.C.; Tseng, R.L.; Huang, S.C.; Juang, R.S. Characteristics of pseudo-second-order kinetic model for liquid-phase adsorption: A mini-review. *Chem. Eng. J.* **2009**, *151*, 1–9. [CrossRef]
72. Jia, C.S.; Zhang, L.H.; Peng, X.L.; Luo, J.X.; Zhao, Y.L.; Liu, J.Y.; Guo, J.J.; Tang, L.D. Prediction of entropy and Gibbs free energy for nitrogen. *Chem. Eng. Sci.* **2019**, *202*, 70–74. [CrossRef]
73. Özcan, A.; Özcan, A.S.; Tunali, S.; Akar, T.; Kiran, I. Determination of the equilibrium, kinetic and thermodynamic parameters of adsorption of copper(II) ions onto seeds of Capsicum annuum. *J. Hazard. Mater.* **2005**, *124*, 200–208. [CrossRef]
74. Aksu, Z. Determination of the equilibrium, kinetic and thermodynamic parameters of the batch biosorption of nickel(II) ions onto Chlorella vulgaris. *Process Biochem.* **2002**, *38*, 89–99. [CrossRef]
75. Arguello, C.A.; Rousseau, D.L.; Porto, S.P.S. First-Order Raman Effect in Wurtzite-Type Crystals. *Phys. Rev.* **1969**, *181*, 1351. [CrossRef]
76. Ristić-Djurović, J.L.; Fernández-Izquierdo, L.; Hadžić, B.; Jiménez-Hernández, L.; Díaz-García, A.M.; Mitrić, J.; Babić, B.; Romčević, M.; Ćirković, S.; Romčević, N. Raman spectroscopy of zinc oxide nanoplatelets modified with ruthenium (II) complexes. *J. Raman Spectrosc.* **2019**, *50*, 1829–1838. [CrossRef]
77. Cuscó, R.; Alarcón-Lladó, E.; Ibanez, J.; Artús, L.; Jiménez, J.; Wang, B.; Callahan, M.J. Temperature dependence of Raman scattering in ZnO. *Phys. Rev. B* **2007**, *75*, 165202. [CrossRef]
78. Zafar, M.N.; Dar, Q.; Nawaz, F.; Zafar, M.N.; Iqbal, M.; Nazar, M.F. Effective adsorptive removal of azo dyes over spherical ZnO nanoparticles. *J. Mater. Res. Technol.* **2019**, *8*, 713–725. [CrossRef]
79. Kataria, N.; Garg, V.K. Removal of Congo red and Brilliant green dyes from aqueous solution using flower shaped ZnO nanoparticles. *J. Environ. Chem. Eng.* **2017**, *5*, 5420–5428. [CrossRef]
80. Chauhan, A.K.; Kataria, N.; Garg, V.K. Green fabrication of ZnO nanoparticles using Eucalyptus spp. leaves extract and their application in wastewater remediation. *Chemosphere* **2020**, *247*, 125803. [CrossRef]
81. Kragović, M.; Stojmenović, M.; Petrović, J.; Loredo, J.; Pašalić, S.; Nedeljković, A.; Ristović, I. Influence of alginate encapsulation on point of zero charge (pH pzc) and thermodynamic properties of the natural and Fe(III)-modified zeolite. *Procedia Manuf.* **2019**, *32*, 286–293. [CrossRef]
82. Zhao, T.; Fang, M.; Tang, Z.; Zhao, X.; Wu, F. Adsorption, aggregation and sedimentation of titanium dioxide nanoparticles and nanotubes in the presence of different sources of humic acids. *Sci. Total Environ.* **2019**, *692*, 660–668. [CrossRef] [PubMed]
83. Alvarez-Silva, M.; Mirnezami, M.; Uribe-Salas, A.; Finch, J.A. Point of zero charge, isoelectric point and aggregation of phyllosilicate minerals. *Can. J. Metall. Mater. Sci.* **2013**, *49*, 405–410. [CrossRef]
84. Suttiponparnit, K.; Jiang, J.; Sahu, M.; Suvachittanont, S.; Charinpanitkul, T.; Biswas, P. Role of surface area, primary particle size, and crystal phase on titanium dioxide nanoparticle dispersion properties. *Nanoscale Res. Lett.* **2010**, *6*, 1–8. [CrossRef]
85. Gu, M.; Hao, L.; Wang, Y.; Li, X.; Chen, Y.; Li, W.; Jiang, L. The selective heavy metal ions adsorption of zinc oxide nanoparticles from dental wastewater. *Chem. Phys.* **2020**, *534*, 110750. [CrossRef]
86. Yoshida, T. Leaching of zinc oxide in acidic solution. *Mater. Trans.* **2003**, *44*, 2489–2493. [CrossRef]
87. Sheela, T.; Nayaka, Y.A.; Viswanatha, R.; Basavanna, S.; Venkatesha, T.G. Kinetics and thermodynamics studies on the adsorption of Zn(II), Cd(II) and Hg(II) from aqueous solution using zinc oxide nanoparticles. *Powder Technol.* **2012**, *217*, 163–170. [CrossRef]
88. Rodríguez, C.; Tapia, C.; Leiva-Aravena, E.; Leiva, E. Graphene oxide–zno nanocomposites for removal of aluminum and copper ions from acid mine drainage wastewater. *Int. J. Environ. Res. Public Health* **2020**, *17*, 1–18.
89. Lee, S.; Zhang, Z.; Wang, X.; Pfefferle, L.D.; Haller, G.L. Characterization of multi-walled carbon nanotubes catalyst supports by point of zero charge. *Catal. Today* **2011**, *164*, 68–73. [CrossRef]
90. Pyrzyńska, K.; Bystrzejewski, M. Comparative study of heavy metal ions sorption onto activated carbon, carbon nanotubes, and carbon-encapsulated magnetic nanoparticles. *Colloids Surf. A Physicochem. Eng. Asp.* **2010**, *362*, 102–109. [CrossRef]
91. Hu, J.; Shipley, H.J. Evaluation of desorption of Pb (II), Cu (II) and Zn (II) from titanium dioxide nanoparticles. *Sci. Total Environ.* **2012**, *431*, 209–220. [CrossRef]
92. Grover, V.A.; Hu, J.; Engates, K.E.; Shipley, H.J. Adsorption and desorption of bivalent metals to hematite nanoparticles. *Environ. Toxicol. Chem.* **2012**, *31*, 86–92. [CrossRef] [PubMed]
93. Lim, J.L.; Okada, M. Regeneration of granular activated carbon using ultrasound. *Ultrason. Sonochem.* **2005**, *12*, 277–282. [CrossRef] [PubMed]
94. Rege, S.U.; Yang, R.T.; Cain, C.A. Desorption by ultrasound: Phenol on activated carbon and polymeric resin. *AIChE J.* **1998**, *44*, 1519–1528. [CrossRef]

95. Breitbach, M.; Bathen, D.; Schmidt-Traub, T. Effect of ultrasound on adsorption and desorption processes. *Ind. Eng. Chem. Res.* **2003**, *42*, 5635–5646. [CrossRef]
96. Korkut, O.; Sayan, E.; Lacin, O.; Bayrak, B. Investigation of adsorption and ultrasound assisted desorption of lead (II) and copper (II) on local bentonite: A modelling study. *Desalination* **2010**, *259*, 243–248. [CrossRef]
97. Yousefi, A.; Allahverdi, A.; Hejazi, P. Effective dispersion of nano-TiO2 powder for enhancement of photocatalytic properties in cement mixes. *Constr. Build. Mater.* **2013**, *41*, 224–230. [CrossRef]
98. Ma, J.; Zuo-Jiang, S.; He, Y.; Sun, Q.; Wang, Y.; Liu, W.; Sun, S.; Chen, K. A facile, versatile approach to hydroxyl-anchored metal oxides with high Cr (VI) adsorption performance in water treatment. *R. Soc. Open Sci.* **2016**, *3*, 160524. [CrossRef]
99. MiarAlipour, S.; Friedmann, D.; Scott, J.; Amal, R. TiO_2/porous adsorbents: Recent advances and novel applications. *J. Hazard. Mater.* **2018**, *341*, 404–423. [CrossRef]
100. Foo, K.Y.; Hameed, B.H. Insights into the modeling of adsorption isotherm systems. *Chem. Eng. J.* **2010**, *156*, 2–10. [CrossRef]
101. Saadi, R.; Saadi, Z.; Fazaeli, R.; Fard, N.E. Monolayer and multilayer adsorption isotherm models for sorption from aqueous media. *Korean J. Chem. Eng.* **2015**, *32*, 787–799. [CrossRef]
102. Amin, M.T.; Alazba, A.A.; Shafiq, M. Nonspontaneous and multilayer adsorption of malachite green dye by Acacia nilotica waste with dominance of physisorption. *Water Sci. Technol.* **2017**, *76*, 1805–1815. [CrossRef]
103. Zhang, X.; Ge, Y.; Zhu, G.; Tang, J.; Xing, X.; Li, N. Effect of acid and hydrothermal treatments on the multilayer adsorption of Cr(VI) and dyes on biomass-derived nano/mesoporous carbon. *J. Mater. Res.* **2019**, *34*, 3020–3029. [CrossRef]
104. Zeldowitsch, J. Adsorption site energy distribution. *Acta Phys. Chim. URSS* **1934**, *1*, 961–973.
105. Cuppett, J.D.; Duncan, S.E.; Dietrich, A.M. Evaluation of copper speciation and water quality factors that affect aqueous copper tasting response. *Chem. Senses* **2006**, *31*, 689–697. [CrossRef]
106. Prolongo, S.G.; Rosario, G.; Urena, A. Study of the effect of substrate roughness on adhesive joints by SEM image analysis. *J. Adhes. Sci. Technol.* **2006**, *20*, 457–470. [CrossRef]
107. Simonin, J.P. On the comparison of pseudo-first order and pseudo-second order rate laws in the modeling of adsorption kinetics. *Chem. Eng. J.* **2016**, *300*, 254–263. [CrossRef]
108. Khezamia, L.; Tahaa, K.K.; Amamic, E.; Ghiloufid, I.; El Mird, L. Removal of cadmium (II) from aqueous solution by zinc oxide nanoparticles: Kinetic and thermodynamic studies. *Desalin. Water Treat.* **2017**, *62*, 346–354. [CrossRef]
109. Zhang, F.; Lan, J.; Yang, Y.; Wei, T.; Tan, R.; Song, W. Adsorption behavior and mechanism of methyl blue on zinc oxide nanoparticles. *J. Nanoparticle Res.* **2013**, *15*, 1–10. [CrossRef]
110. Sigma-Aldrich. Zinc Oxide, Dispersion. Available online: https://www.sigmaaldrich.com/CL/es/product/aldrich/721077?context=product (accessed on 22 September 2021).
111. Sigma-Aldrich. Graphene Powder. Available online: https://www.sigmaaldrich.com/CL/es/product/aldrich/900561?context=product (accessed on 22 September 2021).
112. Sigma-Aldrich. Carbon Nanotube, Multi-Walled. Available online: https://www.sigmaaldrich.com/CL/es/product/aldrich/698849?context=product (accessed on 22 September 2021).

water MDPI

Article

Recovery of Cr(III) from Tannery Effluents by Diafiltration Using Chitosan Modified Membranes

Asmaa Zakmout [1,2], Fatma Sadi [2], Svetlozar Velizarov [1,*], João G. Crespo [1] and Carla A. M. Portugal [1,*]

1 LAQV-REQUIMTE, Department of Chemistry, Campus da Caparica, NOVA School of Science and Technology, Universidade Nova de Lisboa, 2829-516 Caparica, Portugal; zakmout_a@yahoo.fr (A.Z.); jgc@fct.unl.pt (J.G.C.)

2 Laboratory of Electrochemistry, Corrosion Metallurgy and Inorganic Chemistry, Faculty of Chemistry, University of Sciences and Technology, Houari Boumediene, USTHB, B.P. 32 El Alia, Bab Ezzouar 16111, Algiers, Algeria; sadifatma4444@yahoo.fr

* Correspondence: s.velizarov@fct.unl.pt (S.V.); cmp@fct.unl.pt (C.A.M.P.)

Citation: Zakmout, A.; Sadi, F.; Velizarov, S.; Crespo, J.G.; Portugal, C.A.M. Recovery of Cr(III) from Tannery Effluents by Diafiltration Using Chitosan Modified Membranes. *Water* **2021**, *13*, 2598. https://doi.org/10.3390/w13182598

Academic Editor: Margaritis Kostoglou

Received: 11 July 2021
Accepted: 15 September 2021
Published: 21 September 2021

Abstract: The selective recovery of chromium remaining in tannery effluents after the leather tanning process is highly desirable to potentiate its reuse, simultaneously minimizing the ecotoxicity of these effluents. To the best of our knowledge, this work evaluates for the first time the ability of a chitosan-based membrane for selective recovery of chromium from a tannery wastewater by subsequent diafiltration and selective chromium desorption, envisaging their integration after tannery wastewater treatment by reverse osmosis (RO). A polyethersulfone (PES) microfiltration membrane top-coated with a chitosan layer (cs-PES MF022) was used for selective recovery of Cr(III), from concentrate streams obtained by treatment of synthetic and real tannery effluents through reverse osmosis (RO), through a diafiltration process. The diafiltration of the RO concentrates was conducted by an intermittent addition of water acidified to pH 3.6. The prepared cs-PES MF022 membranes were able to retain 97% of the total mass of Cr(III) present in the RO concentrates, from a real tannery effluent, with a selectivity of 4.2 and 5 in reference to NH_4^+ and Cl^-, respectively, 12.9 and 14.6 in reference to K and Na, and >45 in reference to Mg, Ca, and S. Such a high selectivity is explained by the preferential adsorption of Cr(III) onto chitosan, and by the relatively high permeability of cs-PES MF022 membranes to the other ionic species. Proof of concept studies were performed to investigate the desorption of Cr(III) at pH 2 and 5.8. A higher Cr(III) desorption degree was obtained at pH 2, leading to a final solution enriched in Cr(III), which may be re-used in tannery operations, thus improving the process economy and reducing the hazardous impact of the effluents discharged by this industry.

Keywords: tannery effluent; chromium recovery; chitosan membrane; diafiltration; selective adsorption; selective desorption

1. Introduction

Leather tanning requires the use of a large variety of chemicals, such as tannins, sulphates, phenolics, surfactants, and ion salts. In particular, Cr(III) is used in the form of chromium sulphate $Cr_2(OH)_2(SO_4)_2$ for the conversion of collagen from skin into commercial leather. This process results in the production of high wastewater volumes with appreciable chromium content [1] that increases the ecotoxicological impact of the tannery effluents [2]. Wastewater treatment is thus mandatory before discharge this water stream, which must possess physicochemical characteristics that conform with local regulations for a safe discharge into the ecosystems. In addition, the recovery of chromium allowing for its subsequent reuse can reduce the costs associated with the leather tanning process.

Several physicochemical and biological approaches have been used to reduce the organic and inorganic content of tannery wastewaters, including coagulation/flocculation induced by specific chemical agents [3], electrocoagulation [4], solvent extraction [5], adsorption [6], aerobic or anaerobic biological treatment [7], incineration [8], chemical oxidation [9], membrane filtration [10–12], and via process integration, such as by combining coagulation with microfiltration [13] or electrodialysis [14,15]. Among the adsorption-based methods, different adsorbents, such as clays [16,17], activated carbon derived from different biological [18] and non-biological materials [19] with distinct morphologies [20], and chitosan-based compounds [6,21–26] have proved to allow for an efficient capture of metals from target effluents.

Chitosan is an eco-friendly and inexpensive biomaterial obtained from the partial deacetylation of the acetoamine groups of chitin. Chitosan-based flakes/beads [6], nanofibres [22], membranes [12,23], or films [24] produced by chitosan crosslinking have been prepared and tested in terms of their capacity to adsorb metal ionic species. The outstanding capacity of chitosan to adsorb heavy metal ions [22], including copper [25,26] and chromium [23], such as Cr(III), has been mainly ascribed to the ability of amine to act as chelators of these ionic species. However, a potential interaction with hydroxyl groups should also be considered. The chelating efficiency was found to be dependent on the electric charge of the species involved and, thus, dependent on pH [6]. In strongly acidic solutions, Cr(III) is mainly present in the form of cationic species Cr^{3+}, $Cr(OH)_2{}^+$, and $Cr(OH)^{+2}$, whereas the Cr precipitation observed at more alkaline conditions (pH > 6) is explained by a decrease of the solubility of chromium hydroxides at increasing pH values. Thus, due to the prevalence of positively charged amine ($R\text{-}NH_3{}^+$) and hydroxyl ($R\text{-}OH_2{}^+$) groups of chitosan, the adsorption of the positively charged Cr(III) species is unfavoured at lower pH values, but increases with the deprotonation of these chemical groups as the pH increases. Nevertheless, Cr(III) removal by adsorption is still more efficient at a pH lower than 6, due to the absence of intense Cr precipitation [23]. Mirabedini et al. [24] produced magnetic chitosan hydrogels crosslinked with glyoxal, which showed pH dependent Cr(IV) removal efficiency from water, with optimal values of 80–90% reached at pH 4.

The high affinity of chitosan to heavy metals brings additional interest in the development of chitosan-based membranes for the removal of these metals from wastewaters, taking advantage of both, their adsorptive and water filtration capacity. In a work from Juang et al. [27], chitosan flakes were added to a filtration system to increase the efficiency of regenerated cellulose membranes to remove divalent ions, such as Cu(II), Co(II), Ni(II), and Zn(II). Chitosan flakes were shown to increase the removal degree of these metal ions six to tenfold under acidic conditions. In another work from Li et al. [28], electrospun chitosan membranes were prepared and used as membrane adsorbers for the removal of Cr(IV) from model aqueous solutions. Regarding the Cr(IV) loading capacity of these membranes, their bed saturation and efficiency showed a higher dependence on pH, solution flow rate, fluid flow distribution, and membrane packed patterns, as compared to the Cr(IV) concentration in the feed solution and bed length. The maximum bed loading capacity was found to be 16.5 mg Cr(IV)/g of chitosan under dynamic fluid conditions. Chitosan/polyethylene oxide (PEO) nanofibres (90:10) assembled to spunbonded polypropylene (PP) substrate containing 90% chitosan developed by Desai et al. [29] showed a Cr binding capacity of 35 mg Cr(VI)/g chitosan. The same authors have also shown that the binding capacity of chitosan/PEO nanofibres is dependent on the chitosan content in the chitosan/PEO blend and decreases with the increase of the fibre diameter [29]. In a previous work by Ghaee et al. [30], chitosan/cellulose acetate composite nanofiltration membranes with various amounts of cellulose acetate were prepared and evaluated in terms of their rejection capacity to copper. The membranes obtained exhibited a molecular weight cut-off of 830.74 Da, and a Cu rejection of 81.03%. Cellulose membranes, enriched with chitosan-silver ions, were designed by Căprărescu et al. [31] and used for the removal of iron ions from synthetic wastewaters with an electrodialysis system. The presence of silver ions increased the electric conductivity of the membranes, resulting in iron removal rates > 60% at an applied voltage of 15 V.

In our previous study [12], a polyethersulfone (PES) membrane coated with a thin chitosan layer—a chitosan modified membrane designated for simplicity as cs-PES MF022—

was prepared, and its performance for the treatment of model tannery effluents was compared with that of a SW30 reverse osmosis (RO) membrane in terms of species rejection efficiency and selectivity. The cs-PES MF022 membrane showed rejection coefficients <45% for most of the ionic species present, i.e., Na^+, K^+, Ca^{2+}, Mg^{2+}, Cl^-, NH_4^+, and SO_4^{2-}. These rejection coefficients were significantly lower than those obtained with the SW30 membrane, which provided rejection values above 90% for all these ions. As a result, the cs-PES MF022 membrane was considered unsuitable to produce treated water with the characteristics required to allow the discharge of the effluents into the ecosystem. However, this membrane exhibited an outstanding Cr(III) rejection > 90%, and, thus, a Cr(III) removal selectivity much higher than that obtained with the SW30 membrane. These results evidenced the potential suitability of cs-PES MF022 membranes for the selective recovery of Cr(III) from tannery wastewaters, prompting us to investigate and suggest a two-step process for this purpose. This process consists of the integration of reverse osmosis (RO) (1st step), in which an SW30 membrane is used to obtain a permeate stream with a chemical composition which allows its direct discharge into the environment [12], followed by a diafiltration of the RO concentrate (2nd step) for a selective recovery of Cr(III), using the chitosan modified membrane (cs-PES MF022) mentioned above. Diafiltration involves a solvent (acidic water) consumption higher than that required by membrane processes when operated in conventional mode, and for this reason, it may be regarded as a disadvantageous separation process. However, the possibility to treat complex aqueous streams—such as tannery wastewaters—under conditions of reduced concentration polarization and fouling-related effects [32,33], made us anticipate the suitability of diafiltration for the enhanced removal of contaminants, i.e., other ionic species with a lower affinity to chitosan and low-molecular mass compounds, from these effluents, facilitating the isolation of Cr-containing compounds and, thus, allowing for its subsequent re-use in tannery processes. Therefore, this study aims at evaluating the performance of the cs-PES MF022 membranes for selective removal of Cr(III) from concentrates obtained by RO treatment of synthetic and real tannery effluents by diafiltration, and the possibility to recover Cr(III), by selective desorption, taking advantage of the pH-responsive ability of chitosan [34].

2. Materials and Methods

2.1. Materials

The synthetic tannery effluent was prepared using magnesium chloride ($MgCl_2$; purity 99%) provided by Alfa-Aesar (Kandel, Germany); calcium chloride 2-hydrate ($CaCl_2.2H_2O$; purity > 99%), ammonium sulphate ($(NH_4)_2SO_4$; purity > 99%), and sodium chloride (NaCl; purity 99.8%) supplied by AppliChem PANREAC (Barcelona, Spain); and chromium(III) sulphate basic ($Cr_4(SO_4)_5OH_2$; 26% Cr_2O_3) obtained from Fluka Analytical (Buchs SG, Switzerland). The chitosan (cs) used for the preparation of functional chitosan membranes was provided from Sigma-Aldrich (St. Louis, MO, USA). Acetic acid (CH_3COOH; purity 99.8%), potassium sulphate (K_2SO_4; purity > 99%), and sodium hydroxide (NaOH; purity > 93%) were supplied by Carlo Erba (Val de Reuil, France). Glutaraldehyde (GDA; 25% in H2O) and hydrochloric acid (HCl; purity 35–38%) were purchased from Sigma-Aldrich Chemicals (St. Louis, MO, USA).

2.2. Methods

2.2.1. Concentrates of Real and Synthetic Tannery Effluents Obtained from Reverse Osmosis (RO)

The concentrates of real tannery effluent was obtained by reverse osmosis (RO) treatment of a real tannery effluent supplied by TAMEG-Rouiba-SPA—a Leather Industry located in Rouiba, close to Algiers (Algeria), whereas the concentrate of a synthetic tannery effluent was obtained by RO treatment of a synthetic solution, prepared in the lab, mimicking the composition of the real effluent in terms of the inorganic content (inorganic salts). The treatment of these effluents were performed using an SW30 reverse osmosis (RO) membrane supplied from DOW Chemical Company (Midland, MI, USA). RO showed rejections higher than 95% for all of the inorganic salts (99.2% rejection was obtained for Cr), allowing for the production of a treated water permeate [12] meeting the requirements for a direct discharge of the permeate stream into natural environments, according to the Algerian Legislation [35], and a concentrate enriched in the non-permeable components present in the effluents. The chemical composition of the concentrate of the synthetic and the real tannery effluents obtained by processing with the SW30 RO membrane are summarized in Tables 1 and 2, respectively.

Table 1. Chemical composition of the concentrate obtained by processing of a synthetic effluent with an RO SW30 membrane [12], at pH 3.6 and T = 20 °C.

Parameter [1]	Ca	Cr(III)	K	Mg	Na	S	Cl^-	NH_4^+
Unit	g/L	g/L	mg/L	g/L	g/L	g/L	g/L	g/L
Value	0.64	0.18	0.76	0.33	2.18	0.88	5.01	0.15

[1] The electric charge of the elements determined by ICP-AES was omitted, as ICP-AES quantifies the total amount present in the solution independently of their electric charge.

Table 2. Chemical composition of the concentrate obtained by processing of a real effluent from TAMEG-Rouiba Tannery with an RO SW30 membrane [12], at pH 3.6 and T = 20 °C.

Parameter [1]	Ca	Cr(III)	K	Mg	Na	S	Cl^-	NH_4^+
Unit	g/L	g/L	g/L	g/L	g/L	g/L	g/L	g/L
Value	0.14	0.12	2.77	2.80	1.17	2.09	28.67	0.30

[1] The electric charge of the elements determined by ICP-AES was omitted, as ICP-AES quantifies the total amount present in the solution independently of their electric charge.

2.2.2. Preparation of the Chitosan-Based Membranes (cs-PES MF022)

Functional chitosan membranes were prepared by coating a 2.5% (*w*/*w*) chitosan solution on a polyethersulfone microfiltration membrane (PES MF022, 0.22 μm) from Merck Millipore (Carrigtohill, Ireland) which was used as a support. The 2.5% (*w*/*w*) chitosan solution was prepared by dissolving chitosan in a 5% (*v*/*v*) acetic acid aqueous solution, at room temperature, and then casted on the top surface of the PES MF022 support. The casting was carried out using an Elcometer casting knife film applicator (E.U.), by setting an application air gap of 90 μm, to ensure the formation of a chitosan layer with an uniform and reproducible thickness of ca. 10 μm, as shown in Figure 1 [12]. The resulting membrane was kept and dried at room temperature in a fume hood until complete solvent evaporation.

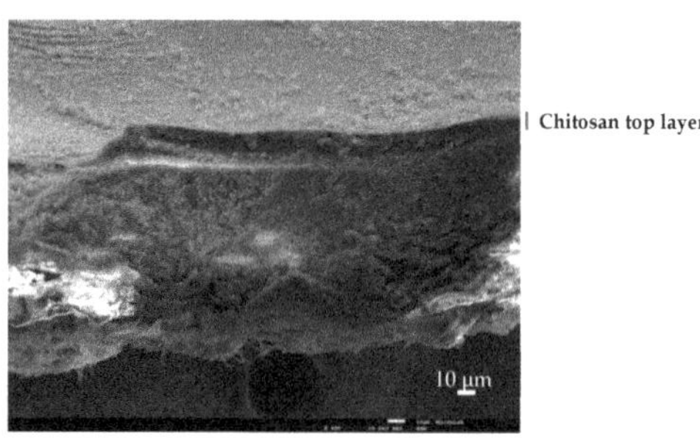

Figure 1. Scanning Electronic Microscopy (SEM) image of the cross-section of a chitosan modified membrane, cs-PES MF022, obtained after filtration of the synthetic tannery effluent (adapted from [12]).

The membrane was then soaked in a 1 M NaOH solution for four hours, immersed in a glutaraldehyde solution (25% in H_2O) overnight, and then washed exhaustively with distilled water. The membrane washing efficiency was accessed by a periodic determination of the UV-Vis spectra (from 200 nm to 400 nm) using a spectrophotometer Thermo Scientific Evolution 201 (Madison, WI, USA), and by determination of the pH and conductivity of the washing solution (supernatant) for inspection of the complete release of the NaOH excess, and the possible release of loosely bound compounds, e.g., chitosan [12]. A pH-meter, type CRISON (Barcelona, Spain), was used to monitor the pH, whereas conductivity was measured by a Schott Lab960 conductivity meter (Mainz, Germany). The washing procedure was considered complete when the pH and conductivity registered for the washing solution were equal to the pH and conductivity values of the distilled water, and the absorbance of the washing was lower than 0.05 within the analysed wavelength range [26]. The membrane was kept in a closed petri dish with some drops of water to ensure the air humidity needed to prevent possible membrane structural alterations.

2.2.3. Diafiltration

Before the permeation experiments, the chitosan modified membrane cs-PES MF022 was mounted in a pressurized stirred filtration cell (MetCell, Greenford, UK), which was then filled with demineralized water acidified with HCl to a pH = 3.6 (acidic water). In the first stage, the cs-PES MF022 membrane was operated in stepwise mode, increasing pressures up to 40 bar for membrane structural compaction, to avoid the possibility of membrane structural changes along the process. In a second stage, the permeation cell was re-filled with acidic water (pH 3.6), and the permeation was conducted by the subsequent increase and decrease of the pressure from 7 bar to 20 bar, in a stepwise way to confirm the structural stability of the membrane. Each pressure step was kept for 10–20 min.

The permeate solution obtained was collected in a reservoir and weighted for a long time using a balance connected to a PC for data acquisition, and then used for determination of the membrane hydraulic permeability.

Each diafiltration experiment was conducted using 35 mL of feed solutions (SW30 concentrate of the synthetic and real tannery effluents). The cell was operated at a constant pressure of 20 bar in a dead-end diafiltration mode, and under a constant stirring speed of 400 rpm. The diafiltration was conducted by the intermittent addition of water (acidified at pH 3.6 with HCl) during the process [36]. As illustrated in Scheme 1, intermittent diafiltration consisted in a series of successive concentration mode filtration steps, interrupted after each 20 mL of permeate collected, for the addition of an identical volume of fresh solvent (20 mL) to the remaining retentate solution in order to restore the initial feed volume in each step. The retentate was sampled (0.5 mL) each time before re-initiation of the diafiltration. This procedure was repeated 7 times during the diafiltration process, leading to a final number of 4 diafiltration volumes (diavolumes).

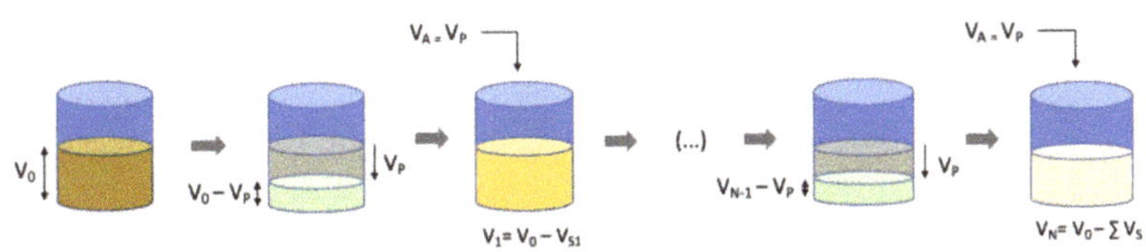

Scheme 1. Schematic representation of the intermittent diafiltration strategy used in the present work. V_0, V_1 and V_N are the feed volumes at the beining of each diafiltration stage, V_P is the permeate volume collected and V_A is the volume of solvent added in each diafiltration stage. V_S corresponds to the volume of the retentate sampled for analysis at each diafiltration stage.

After collecting the final retentate solution, the filtration cell was rinsed with distilled water and pressurized from 20 to 40 bar in order to check the membrane hydraulic permeability after the process.

2.2.4. Data Treatment

Membrane Permeability

The permeate flux, J, L/(m^2·h) was calculated as:

$$J = \frac{Q_P}{A} \quad (1)$$

where Q_P is the volumetric permeate flow (L/h) and A is the surface area of the membrane (m^2). The membrane hydraulic permeability was calculated by registering the permeate flux and applying the Darcy equation:

$$J = L_P * (\Delta P - \Delta\pi) \quad (2)$$

where L_p is the membrane hydraulic permeability, ΔP is the applied pressure, and $\Delta\pi$ is the osmotic pressure difference between the concentrate and the permeate side. The osmotic pressure difference ($\Delta\pi$) was also considered, but given the good permeability of the membrane to most of the species and the reduced amount of species remaining in the retentate along the process, an average $\Delta\pi < 1$ bar was obtained, which was considered negligible compared to the applied operating pressure of 20 bar. Ionic speciation was accessed through the determination of chemical equilibrium diagrams obtained using Make Equilibrium Diagrams Using Sophisticated Algorithms (MEDUSA) software (version: Eq. calcs 32) [37].

Determination of the Compounds Retained in the Membrane and Cr Retention Selectivity

The amounts of each compound, i, retained in the membrane after diafiltration, were calculated by a global mass balance (M_i) to the process, using Equation (3):

$$M_{m,i} = C_{F,i} \times V_F - \sum C_{P,i} \times V_P - \sum C_{S,i} \times V_S - C_{R,i} \times V_R \quad (3)$$

where $M_{m,i}$ is the mass of compound i in the membrane; $C_{F,i}$, $C_{P,i}$, $C_{R,i}$, and $C_{S,i}$ are the concentration of the compound i in the feed, permeate, retentate, and sampling solutions; and V_F, V_P, V_R, and V_S correspond to the volumes of feed, permeate, final retentate, and sampling volume.

The chromium retention selectivity (S_{CR}) was calculated according to the equation:

$$S_{Cr} = R_{Cr}/R_i \quad (4)$$

where R_{Cr} and R_i are the mass of Cr and the mass of each compound i retained in the membrane, relative to that present in the feed, calculated according to Equation (5):

$$R_i = \frac{M_{m,i}}{M_{F,i}} \quad (5)$$

where $M_{m,i}$ and $M_{F,i}$ correspond to the mass of each compound i, retained in the membrane at the end of the process ($N_D = 4$) and that present in the feed solution at the beginning of the process, respectively.

2.2.5. pH Induced Desorption

The cs-PES MF022 membrane used for the diafiltration of the SW30 concentrate from synthetic tannery effluent was cut into identical pieces with an area of 15.9 cm^2. One piece was immerged in a beaker containing 10 mL of demineralized water (pH 5.8), and the other was immerged in an identical volume of demineralized water acidified with HCl to pH 2. The solutions were periodically sampled along 72 h, and the samples were analysed by ICP-AES as described below.

2.2.6. Analytical Methods

Quantification of the Compounds

An inductively coupled plasma atomic emission spectrophotometer, ICP–AES (Ultima model, Horiba Jobin-Yvon, France)—equipped with a radio-frequency (RF) generator of 40.68 MHz—a Czerny–Turner type monochromator with 1.00 m (sequential), an AS500 autosampler, and data acquisition software were employed to determine the concentration of target elements, i.e., Na, K, Ca, Mg, Cr, and S in the feed, permeate, and retentate solutions. Kits (LCK11) from Hach Lange GmbH (Düsseldorf, Germany) were used for chloride determination. The NH_4^+ content was determined colorimetrically using a Skalar SAN++ segmented flow analyser, Skalar Analytical B.V. (AA Breda, The Netherlands).

Characterization of Membrane Structure and Chemical Characterization

The structural characteristics of the cs-PES MF022 membranes were examined using a Field Emission Scanning Electronic Microscopy, FEG-SEM (Jeol JSM–7001F, Tokyo, Japan), upon coating the membrane with an AU/Pd film of 20 nm thickness using a sputter coater from Quorum Technologies (model Q150TES, West Sussex, UK), which were discussed in a previous paper [12]. An energy dispersive X-ray spectrometer (EDS) detector was also used for inspecting the presence of compounds at the membrane surface.

3. Results and Discussion

3.1. Diafiltration of the SW30 Concentrate Solution—Synthetic Tannery Effluent

The concentrate of a tannery synthetic effluent obtained by reverse osmosis (RO) was processed by diafiltration using the chitosan modified membranes (cs-PES MF022), aiming at a selective recovery of chromium.

Prior to filtration of the SW30 concentrate, the structural stability of the membrane was inferred, after its compaction, based on the determination of the membrane hydraulic permeability at increasing and decreasing pressures in a range from 7 bar to 20 bar. As shown in Figure 2, identical permeate fluxes were obtained at the same ΔP values during ascending and descending volumetric flux profiles, resulting in a practically identical hydraulic permeability, *Lp*, of about 0.5 L/(h·m^2·bar). This evidences that the cs-PES MF022 membrane has good structural stability, with an ability to support pressures within the considered ΔP range.

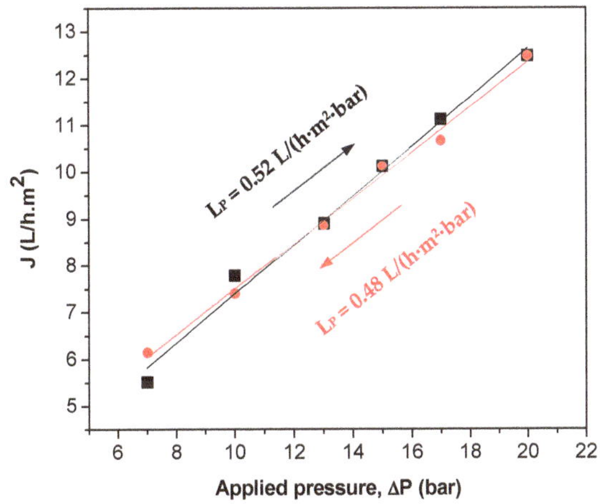

Figure 2. Permeate fluxes, *J*, obtained during filtration of acidic water (pH 3.6) at increasing (squares) and decreasing (circles) pressures using the cs-PES MF022 membrane before diafiltration. *Lp* represents the hydraulic permeability of the membrane determined at increasing and decreasing pressures.

The SW30 concentrate obtained from the RO treatment of a synthetic effluent with the chemical composition indicated in Table 1 was diafiltrated through the cs-PES MF022 membrane, at a ΔP of 20 bar, using a solvent (acidic water) stepwise addition strategy. The permeate flux profile observed during the diafiltration, depicted in Figure 3, shows a constant oscillating profile with an average value of 10.5 $L/(h \cdot m^2)$. A time dependent decline of the permeate flux was not observed, suggesting the absence of significant concentration polarization and fouling phenomena during the process. This may be explained by the attenuation of solutes' accumulation related effects at the membrane surface due to the intermittent addition of solvent and the fluid dynamic conditions established in the filtration cell.

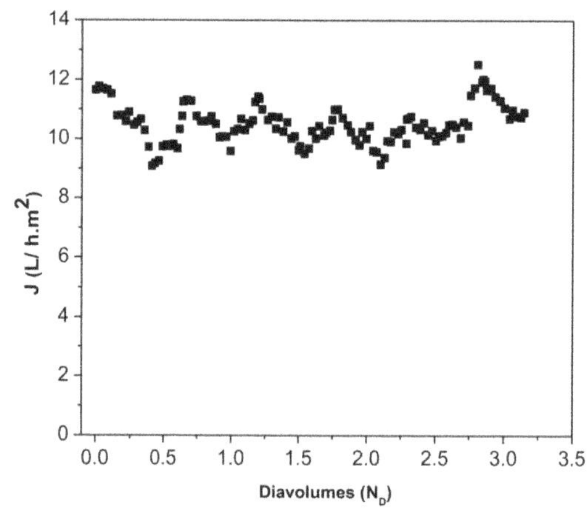

Figure 3. Permeate flux, *J*, obtained during diafiltration of the SW30 concentrate of the synthetic tannery effluent, using the cs-PES MF022 membrane at an applied pressure of 20 bar.

The efficiency of the diafiltration process was evaluated based on the membrane ability to selectively remove Cr(III) (the target compound) in reference to the other ionic species present in the SW30 concentrate, namely Na, K, Ca, Mg, S, Cl^- containing ionic species, and NH_4^+.

As shown in Figure 4, the removal of the different compounds from the SW30 concentrate by diafiltration shows an exponential decay reaching minimum values for most of the species, after permeation of at least 2.5 diavolumes. The mass removal of all the compounds tested fit a negative exponential function, as represented in Equation (6):

$$\frac{M_{Ret,\,i}}{M_{F,i}} = e^{-aN_D} \tag{6}$$

where $M_{Ret,i}$ corresponds to the mass of a given compound, i, in the diafiltration retentate at a specific diavolume number, N_D, whereas $M_{F,i}$ is the mass of compound i in the feed solution ($N_D = 0$). The exponential term a reflects the intensity of the mass decay of each compound i in the retentate solution during diafiltration, which is commonly interpreted as the sieving coefficient of a given molecular species at each time, t, of the process [38]. The $M_{Ret,i}/M_{F,i}$ reflects the apparent mass rejection of the target species, thus, it only accounts for the mass of each species in the retentate solution (excluding that retained in the membrane) at a given N_D. The fitting and exponential parameters obtained for the different compounds are listed in Table 3.

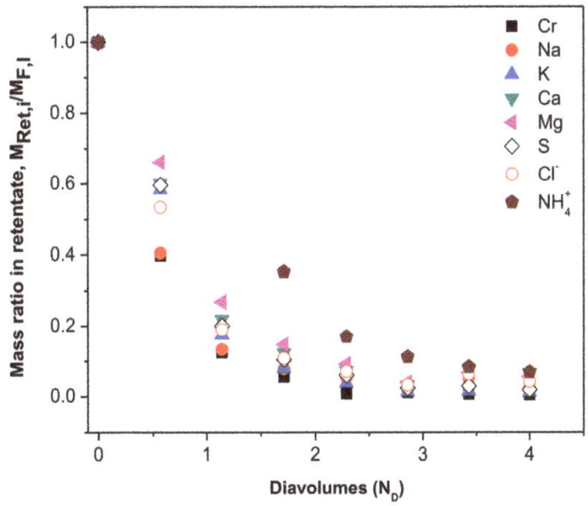

Figure 4. Mass ratio of each compound in retentate during diafiltration with a cs-PES MF022 membrane at 20 bar. Mass ratio was calculated in reference to the total mass of the same compound, $M_{i,x}$, in the SW30 concentrate of a synthetic tannery effluent before diafiltration (diafiltration feed solution). Note that the electric charge of the elements quantified by ICP-AES was omitted, as this technique quantifies the total amount of the element in the solution independently of its electric charge.

Table 3. Fitting parameters to the exponential decay of the mass removal for different compounds in the diafiltration retentate: *a* corresponds to the exponential term, whereas the fitting quality is expressed by the respective R^2.

Compounds [1]	*a*	R^2
Cr	1.69 ± 0.04	0.999
Na	1.60 ± 0.08	0.993
K	1.26 ± 0.10	0.984
Ca	1.11 ± 0.08	0.982
Mg	0.99 ± 0.07	0.982
S	1.18 ± 0.08	0.987
Cl^-	1.24 ± 0.07	0.990
NH_4^+	0.70 ± 0.03	0.993

[1] The electric charge of the elements determined by ICP-AES was omitted, as ICP-AES quantifies the total amount present in the solution independently of their electric charge.

A comparative analysis of the exponential function parameters shows that the removal of all compounds tested fit adequately to Equation (6) with $R^2 \geq 0.982$. The exponential term *a* showed similar values for most ionic species, being slightly higher for Cr and Na, reflecting a faster depletion of these elements (or the respective ion-containing species) from the retentate. On the other hand, a lower value was observed for NH_4^+, indicating a slower removal of this ionic compound from the retentate solution.

The depletion of these compounds from the retentate solution may be explained either by their permeation through cs-PES MF022 membrane or by their adsorption to the membrane, which depends on the chemical affinity of their ionic-containing species to the membrane material.

To elicit the removal mechanism of each compound, a comparative analysis of their permeation and adsorption profiles through/to the membrane was performed.

The permeation profiles (Figure 5A) clearly show that the permeability of the Cr(III)-containing species is significantly lower than that observed for the other species. The mass ratio of Cr(III) in the permeate solution reached only 19.2% after permeation of four diavolumes (N_D = 4), whereas mass ratios higher than 50% were observed for all of the other species tested. The highest mass ratios in the permeate were observed for Mg and K, showing mass ratios of 96.5% and 98.0%, respectively, whereas a mass ratio of 62.6% was observed for Ca, after N_D = 4.

(A)

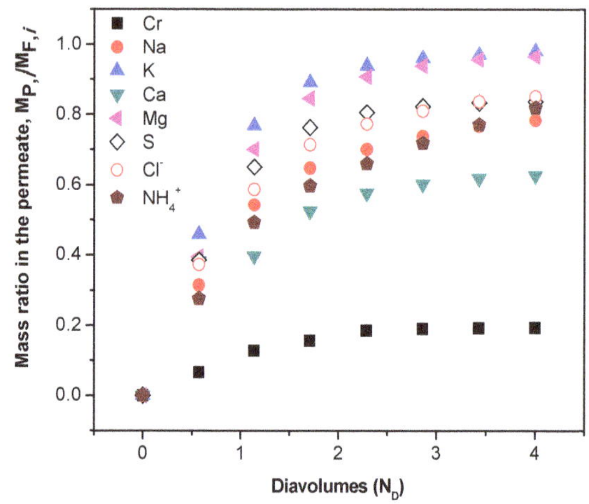

(B)

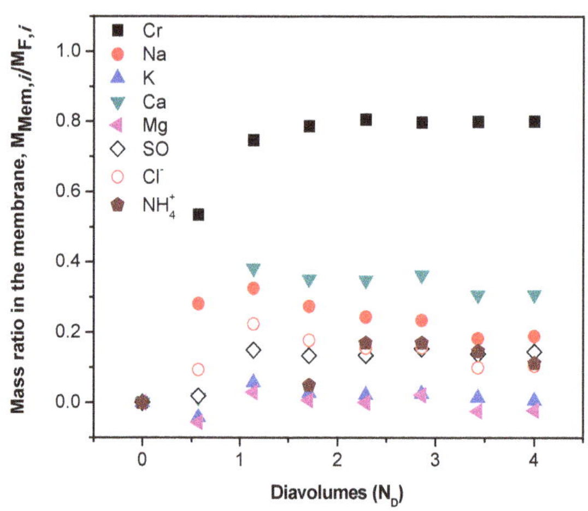

Figure 5. (**A**) Mass ratio of each compound permeated and (**B**) mass ratio of each compound retained in the cs-PES MF022 membrane during diafiltration with the cs-PES MF022 membrane at 20 bar. Mass ratio of each compound was calculated in reference to the total mass of the same compound, $M_{F,I}$, in the SW30 concentrate of a synthetic tannery effluent before diafiltration (diafiltration feed solution). Note that the electric charge of the elements quantified by ICP-AES was omitted, as this technique quantifies the total amount of the element in the solution independently of its electric charge.

The high transmission of these compounds to the permeate resulted in their low accumulation in the membrane, which was found to be inferior to 35% of their total mass, with K and Mg registering residual values close to zero, whereas the mass ratio of Ca in the membrane was equal to 30.7%, at $N_D = 4$. In contrast, the loss of Cr(III) from the retentate solution is mainly explained by the effective retention of Cr-containing compounds in the membrane. As shown in Figure 5 B, a maximum Cr(III) mass ratio of 80% was found in the membrane after permeation of two diavolumes ($N_D = 2$).

In a study developed by Vinodhini et al. [39], UF acetate cellulose and nanochitosan membranes showed Cr rejections varying between 78.5% (initial rejection) and 30.33% (rejection after 90 min of operation) when using feed solutions with 200 ppm of Cr(VI) at pH 5, which were explained by the adsorption of Cr to the amine groups of chitosan. Efficient Cr(VI) and Cr(III) rejections were also reported by Zhang et al. [40] using ultrafiltration membranes decorated with positively charged UiO-66-NH_2 compounds, from feed solutions containing 10.4 mg/L of these ions. The removal of Cr(VI) was explained by the electrostatic attraction to the positive charges in the membrane, whereas concentrations of Cr(III) lower than 1.5 mg/L in the effluent (stable during the operation time) were obtained with MOF@PVDF-0.02 and PEI/MOF@PVDF-0.02 membranes. However, and in contrast to that observed in the present work, the rejection of Cr(III) by these membranes was attributed to the electrostatic repulsion by the positively charged membrane.

The preferential accumulation of Cr-containing compounds at the surface of the chitosan modifying layer upon filtration of the tannery effluent was confirmed in the present study by SEM-EDS analysis. The SEM-EDS results obtained after analysis of chitosan modified membranes (Figure S1, in Supplementary Materials) after permeation of a synthetic tannery effluent (non-processed by reverse osmosis) in concentration mode, showed a significant relative intensity of Cr, in reference to elements such as C, O, and S, which are components of the membrane matrix, and to Au and Pd, which were used to coat the membrane surface, as required for the SEM-EDS analysis.

Despite the high Cr(III) rejection obtained through diafiltration, it was lower than that obtained with the same membrane operated in concentration mode (>99%) or with BW30 and SW30 reverse osmosis (RO) membranes, which led to Cr rejection values of 90% and 99.6%, respectively. However, diafiltration is advantageous if selective removal of Cr is required. Diafiltration produces a higher "cleaning effect", which results in a more efficient removal of compounds non-size excluded by the membrane, in the presence of insignificant fouling effects which prevents the permeate flux decline [12].

Analysis of the Cr(III) retention selectivity in reference to the other compounds tested (Figure 6) shows a retention selectivity of Cr(III)-containing compounds of 2.6 and 4.2 in reference to Na and Ca-containing compounds, respectively; a retention selectivity of 5.5, 7.1, and 7.6 in reference to S-containing compounds, NH_4^+ and Cl^-, respectively; and a selectivity >100 in reference of K-containing compounds, whereas only a vestigial accumulation of Mg-containing compounds in the membrane was detected. The selective retention of Cr(III) in the cs-PES MF022 membrane may be explained by the high chemical affinity of chitosan to heavy metals, via coordination with the amine (NH_2) groups of chitosan [8]. The chemical characterization of cs-PES MF022 membranes by FTIR-ATR was performed and discussed in our previous paper [12]. The presence of chitosan was confirmed based on the detection of the NH_2 stretching band at 1586 cm^{-1}, whereas chitosan crosslinking was confirmed by the decreased intensity of the C=N signal at 1680–1620 cm^{-1} [12]. In particular, the preferential adsorption of Cr(III) to chitosan, in detriment of the other ionic species present in the SW30 concentrate, can be interpreted based on the ability of the NH_2 groups to coordinate heavy metals with f unsaturated orbitals, and its inability to form complexes with alkali and alkali earth metals, such as Na^+, K^+, or Mg^{2+}.

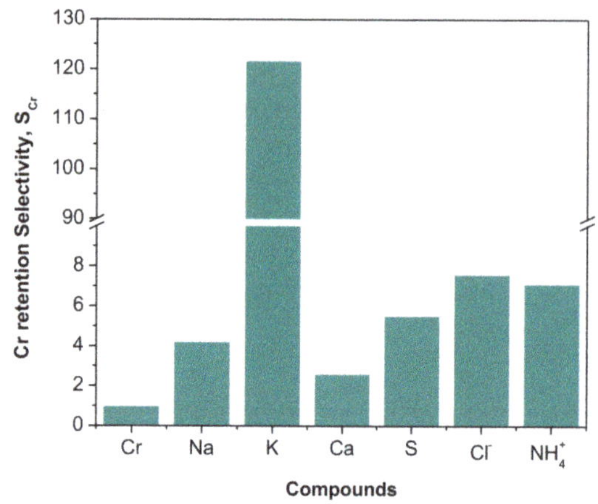

Figure 6. Selective retention of Cr(III), in reference to other species, in the cs-PES MF022 membrane, during diafiltration of the SW30 concentrate of a synthetic tannery effluent at 20 bar.

The selective adsorption of Cr(III) to the cs-PES MF022 membrane (as evidenced in Figure 5B) led to our hypothesis that it is possible to selectively recover the Cr(III) contained in the effluent by membrane desorption after diafiltration, taking advantage of the pH-responsiveness behaviour of chitosan.

3.2. Diafiltration of the SW30 Concentrate Solution—Real Tannery Effluent

The concentrate obtained by RO treatment of a real effluent from the TAMEG-Rouiba tannery with an RO SW30 membrane was also processed through a cs-PES MF022 membrane in a diafiltration mode. Identically to that observed for the diafiltration of the SW30 concentrate from a synthetic effluent, a relatively constant oscillating permeate flux was obtained through the whole process (Figure 7). These results are compatible with the absence of significant concentration polarization and fouling events, revealing that, contrarily to what might be initially expected, the organic content of this effluent (the chemical oxygen demand, COD, of the TAMEG-Rouiba tannery effluent is 92 mg O_2/L after pre-filtration with a microfiltration membrane with an average pore size of 0.45 μm [12]), did not affect the performance of the diafiltration process, at least considering the investigated operation time window.

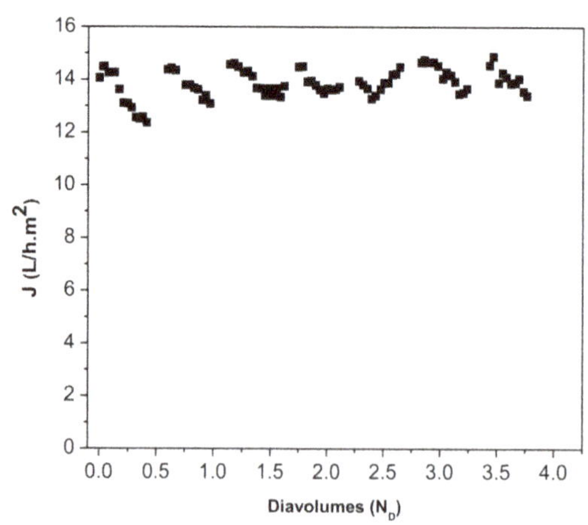

Figure 7. Permeate flux, *J*, obtained during the diafiltration of the SW30 concentrate of the real tannery effluent with the cs-PES MF022 membrane at an applied pressure of 20 bar.

As shown in Figure 8, the diafiltration process led to an exponential loss pattern of all the species studied. However, this exponential decay behaviour was clearly more accentuated for Cr(III), indicating a faster removal of Cr-containing compounds from the diafiltration retentate solutions. Similarly to that observed for the diafiltration of the SW30 concentrate from a synthetic effluent, the removal profiles fit very acceptably to the exponential function represented in Equation (6), as it can be confirmed by the respective R^2 values (Table 4).

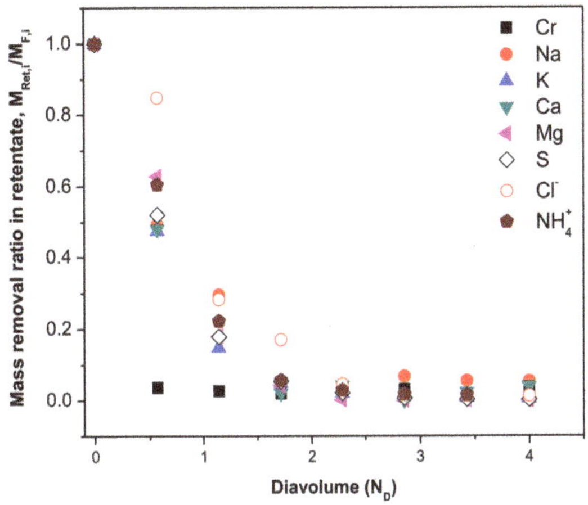

Figure 8. Mass ratio of each compound in the retentate solution during diafiltration with a cs-PES MF022 membrane at 20 bar. Mass ratios were calculated in reference to the total mass of the same compound, $M_{F,i}$, in the SW30 concentrate of the real tannery effluent before diafiltration (diafiltration feed solution).

A comparative analysis of the exponential term, *a*, obtained for the diafiltration of SW30 concentrates from synthetic and real tannery effluents, seems to indicate an identical removal efficiency of most of the compounds studied, with the exception of Cr(III)-containing compounds. The exponential term, *a*, corresponding to the loss of Cr(III)-containing compounds from the SW30 concentrate of the real tannery effluent was equal to 5.85, thus being much higher than that corresponding to the diafiltration of the SW30 concentrate from a synthetic effluent, which was equal to 1.69.

Table 4. Fitting parameters to the exponential decay of the different compounds in diafiltration retentate: *a* is the exponential term, whereas the fitting quality is expressed by the respective R^2.

Compounds [1]	*a*	R^2
Cr	5.85 ± 1.06	0.996
Na	1.13 ± 0.07	0.993
K	1.52 ± 0.07	0.995
Ca	1.45 ± 0.12	0.991
Mg	1.29 ± 0.15	0.968
S	1.41 ± 0.08	0.992
Cl^-	0.93 ± 0.15	0.923
NH_4^+	1.24± 0.11	0.980

[1] The electric charge of the elements determined by ICP-AES was omitted, as ICP-AES quantifies the total amount present in the solution independently of their electric charge.

Such an increase in the exponential term reflects an improved removal of Cr(III) species from the SW30 concentrate from a real tannery effluent, potentially associated to the adsorptive capacity of the organic matter present in real effluent, as discussed in detail below.

In order to understand the removal mechanisms associated with the depletion of each species from the diafiltration retentate, the mass ratio of each compound permeating across the cs-PES MF022 membrane, and that retained in the membrane during diafiltration were determined by respective mass balance calculations.

The profiles depicted in Figure 9A,B show that the cs-PES MF022 membrane exhibits a satisfactory permeability to most of the studied species. Mass ratios higher than 74% were obtained, with Mg and S-containing compounds showing the highest permeability and mass ratios in the permeate of 98% and 99%, respectively, for $N_D = 4$. Oppositely, the mass ratio profiles in the permeate indicate that the Cr-containing compounds are almost completely retained by the cs-PES MF022 membrane, with the Cr(III) mass ratio in the membrane reaching ca. 94% after the permeation of only 0.6 diavolumes. The Cr(III) retention efficiency obtained with the SW30 concentrate of a real tannery effluent was thus much higher than the 80% retention of Cr-containing species obtained by diafiltration of the SW30 concentrate of the synthetic effluent. These results suggest that organic matter present in real tannery effluents contributes positively to the retention of Cr-containing compounds in the membrane. Although the reason for this behaviour is not yet clearly understood, it suggests that the overall adsorption of Cr(III) results from a combined effect of the ability of chitosan to selectively adsorb Cr(III), as previously discussed, and the membrane ability to reject/adsorb organic matter, thus rejecting/adsorbing simultaneously the Cr(III) potentially bound to it.

(A)

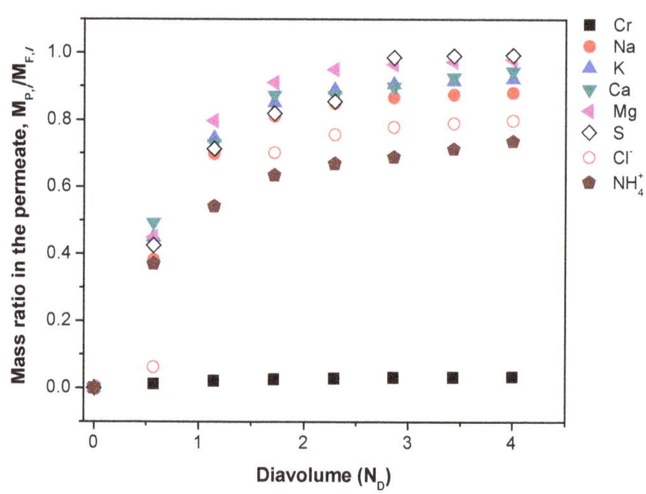

Figure 9. *Cont.*

(**B**)

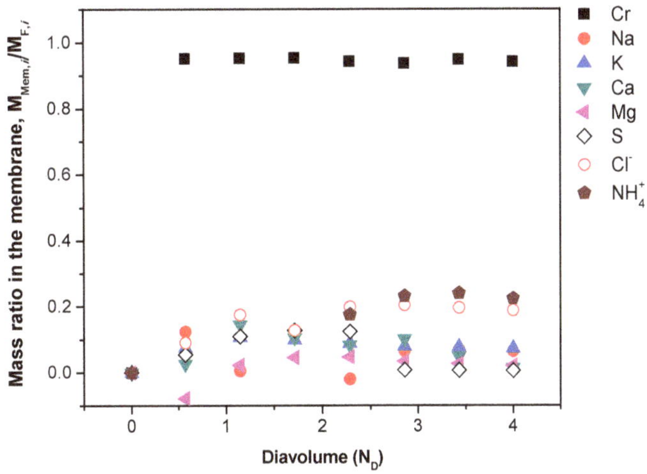

Figure 9. (**A**) Mass ratio of each compound *i* permeated and (**B**) mass ratio of each compound *i* retained in the cs-PES MF022 membrane during diafiltration at 20 bar. Mass ratio for each compound was calculated in reference to the total mass of the same compound, $M_{F,I}$, in the SW30 concentrate of a real tannery effluent before diafiltration (diafiltration feed solution).

The mass ratios of other compounds in the membrane showed maximum values of 18.7% and 22.4% for Cl^- and NH_4^+, respectively, whereas Mg, K, S, and Ca compounds showed mass ratios in the membrane inferior to 7.5%. These differences led to significantly high Cr(III) retention selectivity in the membrane, of 4.2 and 5 in respect to NH_4^+ and Cl^-; 12.9 and 14.6 in reference to K and Na-containing compounds, respectively; and higher than 45 in respect to Mg, Ca, and S -containing compounds, as shown in Figure 10. As shown in Figure S2, the Cr(III) retention selectivity relative to the divalent ions was higher during the diafiltration of the real tannery effluent than that found during the diafiltration of synthetic tannery wastewaters.

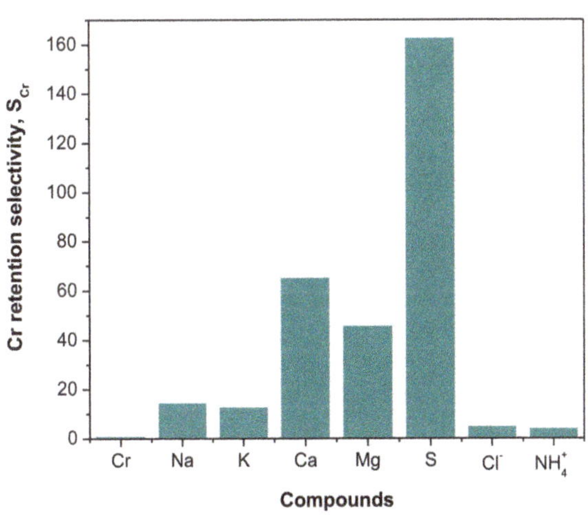

Figure 10. Selective retention of Cr(III) in the cs-PES MF022 membrane during the diafiltration of the SW30 concentrate of a real tannery effluent at 20 bar.

These results suggest the possibility of recovering Cr(III) from tannery effluents by desorption, taking advantage of the pH-responsiveness of chitosan, and thus allowing for Cr(III) reuse in the tannery industry. Moreover, a permeate solution with a residual amount of Cr(III) (only 3% of the Cr(III) mass content of the SW30 concentrate from a real tannery effluent permeate in the cs-PES MF022 membrane) and a retentate solution with a content in ionic species lower than that of the permeate solution are generated by the diafiltration process. Therefore, considering the overall treatment of the tannery effluent through an RO-diafiltration integrated process, illustrated in Scheme 2, the processing costs associated with the disposal of the effluent in the diafiltration could be reduced by recycling the permeate solution to the inlet of the reverse osmosis process.

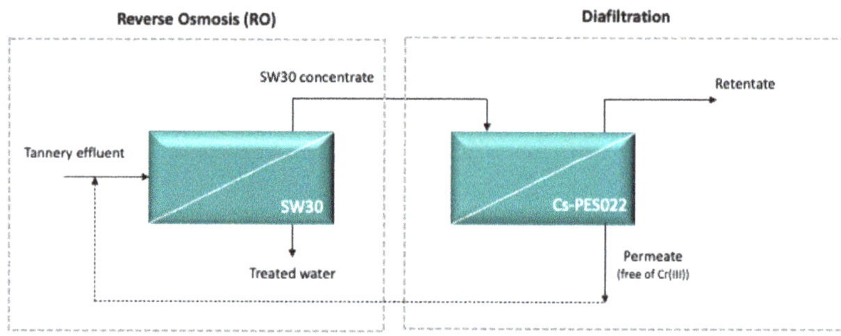

Scheme 2. Schematic representation of the integrated RO-diafiltration process for purification of a real tannery effluent from TAMEG-Rouiba-SPA and selective recovery of Cr(III).

The retentate solution may also be recirculated to the RO inlet for water recovery; however, the low content of ionic species in this processing stream allows for its direct discharge into the ecosystem via Reghaia Lake, respecting the limits for discharge of liquid industrial effluents imposed by the norm No. 06-141 of Rabie El Aouel 1427 (19 April 2006) [35].

3.3. *pH Induced Desorption of Cr from the cs-PES MF022 Membrane*

The membranes used in the diafiltration of the SW30 concentrate from a synthetic effluent were then immersed in water solutions with pH 2 and pH 5.8 in order to demonstrate, as a proof of concept, the possibility to recover Cr(III) by desorption.

The desorption profiles obtained for the different compounds studied are shown in Figure 11.

As shown, a poor desorption of most of the compounds was observed at pH 5.8 (Figure 11A). The desorption of Cr(III) was negligible, while a higher desorption was found for chloride ions (Cl^-). In contrast, as shown in Figure 11B, the desorption of Cr(III) significantly increased at pH 2 in reference to most of the other compounds, except for Cl^- and Na^+ ions (considering that the Na content is all present in the form of Na^+). The mass fraction of Cr(III) equal to 10.7% in the membrane before desorption increased to 20.6% in the solution (leaching solution). Note that the mass fractions in the membrane that were determined only considered the studied compounds (Ca, Cr, Na, K, Mg, S, and Cl^-).

The adsorption of Cr(III) to chitosan is explained by the ability of amine groups (NH_2) to coordinate Cr(III), and also by possible interactions to the hydroxyl groups (OH^-). Amine and hydroxyl groups are mainly neutral at pH 6 (pKa = 6.3 for R-NH_3^+) and protonated at pH 2 (R-NH_3^+ and R-OH_2^+). Therefore, the adsorption ability of chitosan is ruled by pH induced protonation/deprotonation of amine and hydroxyl groups, promoting the desorption of cationic species at lower pH values. It particularly favours those with higher electronic valence, cationic Cr(III) species, such as $Cr(OH)^{2+}$ and $Cr(OH)_2^+$, which are the prevalent Cr(III)-containing species in an aqueous solution at pH 3.6 [6].

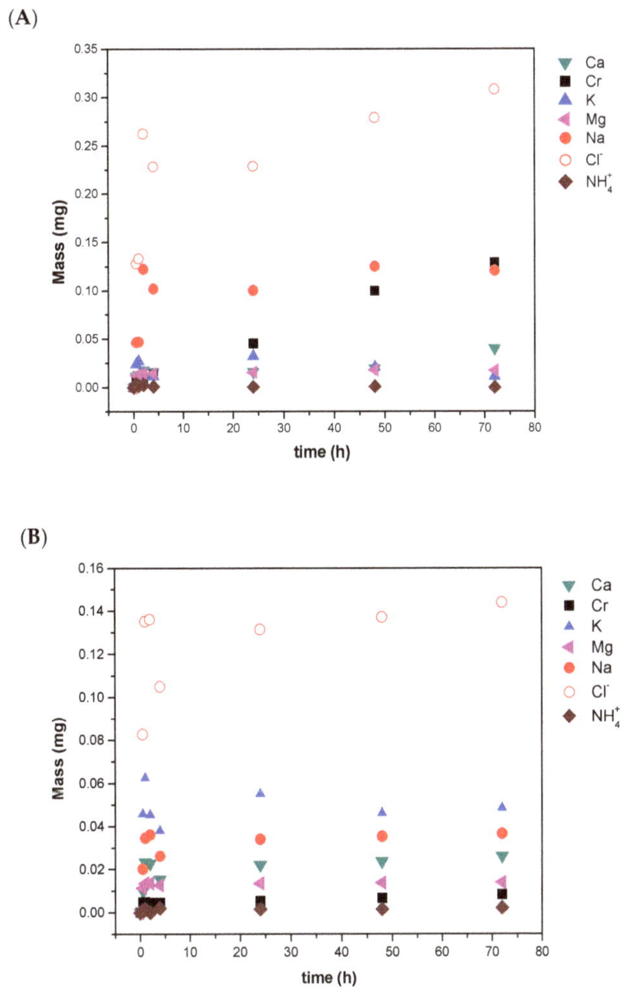

Figure 11. Mass of the compounds desorbed from the cs-PES MF 022 membrane at (**A**) pH 5.8 and (**B**) pH 2, after diafiltration of the SW30 concentrate obtained from filtration of a synthetic tannery effluent.

The high amount of Cl^- and Na^+ ions found in the leaching solution is due to their much higher amount compared to that of Cr(III) in the tannery effluents. Therefore, despite the much lower percentage retention of Na^+ and Cl^- in the membrane, this results in final amounts in the membrane of 6.17 mg and 7.91 mg for Na^+ and Cl^-, respectively, which were higher than the mass of Cr(III) retained in the membrane, equal to 2.91 mg.

As a preliminary proof of concept, the desorption of the target ions was evaluated. A moderate desorption extent (<25%) was obtained for three compounds, Cr(III), Na^+, and Cl^-, in these preliminary desorption assays. Optimization of the desorption process is thus required by the suitable adjustment of desorption parameters, such as temperature, stirring speed, and volume of the leaching solvent for enhanced desorption of these elements. Although desorption optimization is out of the scope of this work, a comparative analysis of the desorption profiles obtained at pH 2 and 5.8 on one hand evidences that Cr(III) desorption selectivity may be improved by conducting the process through successive desorption steps at variable pH values, i.e., a significant Cl^- and Na^+ removal might be attained in a first desorption step at pH 5.8, with good maintenance of the adsorbed Cr(III). This is then followed by a second desorption step at pH 2, for the selective removal of Cr(III) with minimal interference of Na^+ and Cl^-. On the other hand, the desorption profiles obtained for Cr(III), Na^+, and Cl^-, at pH 5.8, show that the Na^+ desorption was faster during the first 5 h of the process, tending to plateau after this period. In contrast, the Cr(III) desorption kinetics showed a more constant behaviour, crossing the Na^+ desorption curve after 72 h, suggesting that process optimization may render a final leaching solution with a more significant Cr(III) content relative to the amount of Na^+.

Nevertheless, if a complete separation of Cr(III) and Na^+ is required. This can be achieved either by an additional nanofiltration operation benefiting from the different size and valence of Cr(III)-containing compounds and Na+, or by using selective cation-exchange resins. The desorption of chloride ions is not actually regarded as a significant problem, as it may lead to the recovery of Cr(III) in the form of chromium chloride.

4. Conclusions

This work evaluates the performance of chitosan modified membranes (cs-PES MF022) for selective recovery of Cr(III) from synthetic and real tannery effluent concentrates obtained by reverse osmosis (RO) with an SW30 membrane, through a diafiltration process. The cs-PES MF022 membrane shows a high ability to retain Cr-containing compounds and a satisfactory permeability to the other compounds studied, which results in a selective removal of Cr-containing compounds of 81% and 97%, in synthetic and real tannery effluents, respectively. The removal of Cr(III)-containing species was mostly explained by Cr(III) retention in the membrane, as only an insignificant amount (<2% of the total amount of Cr in the feed solution) was found in the retentate solution after diafiltration of four diavolumes. The high retention of Cr(III) in the membrane was attributed to the chitosan capacity to adsorb heavy metals with unsaturated *f* orbitals, such as Cr(III), by coordination to the amine groups of chitosan.

It was noteworthy that the permeability of Cr(III)-containing species through the cs-PES MF022 membrane changed significantly during the filtration of real and synthetic tannery effluents. The mass transmission of Cr(III) contained in the real tannery effluent was significantly lower than that obtained by filtration of the synthetic tannery effluents, resulting in a higher and more selective membrane retention of Cr(III)-containing compounds in the former case. This demonstrates the role of the organic matter present in real effluents in the retention of metals.

Moreover, 2.5 diavolumes were found to be the optimal condition for achieving the maximum transmission of most of the compounds studied and the highest retention of Cr(III), without significant variation of permeate flux. This demonstrates the feasibility of a diafiltration strategy for attenuating concentration polarization and membrane fouling effects.

Proof of concept experiments were carried out to demonstrate the possibility to recover Cr(III) by its selective desorption based on the pH-responsive behaviour of chitosan. While a low desorption was obtained for most of the compounds, Cr(III) shows an increased desorption at more acidic pH values (pH = 2), confirming that Cr(III) adsorption to chitosan is mainly due to the coordination to the amine groups. A leaching solution enriched in Cr(III), but still with considerable amounts of Na^+ and Cl^-, was obtained. These results showed that the diafiltration of RO concentrates of tannery effluent with the cs-PES MF022 membrane followed by acidic desorption did not allow for a complete isolation of Cr(III)-containing species. Instead, it rendered a final solution of 92.5% enriched in Cr(III), relative to the tannery effluent RO concentrate (feed solution). Improved Cr(III) purification requires an optimization of the desorption process, or additional treatment of the effluent for separation of Cr(III) from Na^+, e.g., by nanofiltration or by selective Cr(III) adsorption by cation exchange resins. However, this decision depends on the required degree of Cr(III) purity, and should be taken upon a careful analysis of the economic viability of the integrated process.

Supplementary Materials: The following are available online at https://www.mdpi.com/article/10.3390/w13182598/s1, Figure S1: SEM-EDS analysis of the cs-PES MF022 membranes before (top spectrum) and after (bottom spectrum) filtration of the tannery effluent at 20 bar, in concentration mode. Figure S2: Comparative analysis between the selective retention of Cr(III), in reference to other species, by the cs-PES MF022 membrane, during diafiltration of the SW30 concentrate of a synthetic and real tannery effluent at 20 bar.

Author Contributions: Conceptualization, C.A.M.P., J.G.C., and S.V.; methodology, C.A.M.P., J.G.C., and S.V.; data curation, A.Z. and C.A.M.P.; writing—original draft preparation, A.Z. and C.A.M.P.; writing—review and editing, A.Z, F.S., C.A.M.P., J.G.C., and S.V.; supervision, C.A.M.P., J.G.C., and S.V.; funding acquisition, C.A.M.P., J.G.C., and S.V. All authors have read and agreed to the published version of the manuscript.

Funding: This research was supported by the Associate Laboratory for Green Chemistry—LAQV—which is financed by national funds from FCT/MCTES (UIDB/50006/2020 and UIDP/50006/2020).

Institutional Review Board Statement: Not applicable.

Informed Consent Statement: Not applicable.

Data Availability Statement: Data available on request.

Acknowledgments: This work was supported by the Associate Laboratory for Green Chemistry—LAQV—which is financed by national funds from FCT/MCTES (UIDB/50006/2020 and UIDP/50006/2020). Asmaa Zakmout gratefully acknowledges the Erasmus and International Credit Mobility Program for her internship grant at LAQV, Faculty of Science and Technology.

Conflicts of Interest: The authors declare no conflict of interest.

References

1. Saxena, G.; Chandra, R.; Bharagava, R.N. Environmental Pollution, Toxicity Profile and Treatment Approaches for Tannery Wastewater and Its Chemical Pollutants. *Rev. Environ. Contam. Toxicol.* **2016**, *240*, 31–69.
2. Vignati, D.A.L.; Ferrari, B.J.D.; Roulier, J.-L.; Coquery, M.; Szalinska, E.; Bobrowski, A.; Czaplicka, A.; Kownacki, A.; Dominik, J. Chromium bioavailability in aquatic systems impacted by tannery wastewaters. Part 1: Understanding chromium accumulation by indigenous chironomids. *Sci. Total. Environ.* **2019**, *653*, 401–408. [CrossRef] [PubMed]
3. Imran, Q.; Hanif, M.A.; Riaz, M.S.; Noureen, S.; Ansari, T.M.; Bhatti, H.N. Coagulation/Flocculation of Tannery Wastewater Using Immobilized Chemical Coagulants. *J. Appl. Res. Technol.* **2012**, *10*, 79–86.
4. Genawi, N.M.; Ibrahim, M.H.; El-Naas, M.H.; Alshaik, A.E. Chromium Removal from Tannery Wastewater by Electrocoagulation: Optimization and Sludge Characterization. *Water* **2020**, *12*, 1374. [CrossRef]
5. Ma, H.-R.; Li, H.; Wu, W.; Qiao, X.-R. Separation of Fe(III) and Cr(III) from tannery sludge bioleachate using organophosphorus acid extractants. *Res. Chem. Intermed* **2017**, *43*, 2333–2350. [CrossRef]
6. Pietrelli, L.; Francolini, I.; Piozzi, A.; Sighicelli, M.; Silvestro, I.; Vocciante, M. Chromium(III) Removal from Wastewater by Chitosan Flakes. *Appl. Sci.* **2020**, *10*, 1925. [CrossRef]
7. Angelucci, D.M.; Stazi, V.; Andrew, J.; Daugulis, A.J.; Tomei, M.C. Treatment of Synthetic Tannery Wastewater in a Continuous Two-Phase Partitioning Bioreactor: Biodegradation of the Organic Fraction and Chromium Separation. *J. Clean. Prod.* **2017**, *152*, 321. [CrossRef]
8. Kavouras, P.; Pantazopoulou, E.; Varitis, S.; Vourlias, G.; Chrissafis, K.; Dimitrakopulos, G.P.; Mitrakas, M.; Zouboulis, A.I.; Karakostas, T.; Xenidis, A. Incineration of Tannery Sludge under Oxic and Anoxic Conditions: Study of Chromium Speciation. *J. Hazard. Mater.* **2015**, *283*, 672–679. [CrossRef]
9. Kiliç, E.; Font, J.; Puig, R.; Çolak, S.; Çelik, D. Chromium Recovery from Tannery Sludge with Saponin and Oxidative Remediation. *J. Hazard. Mater.* **2011**, *185*, 456–462. [CrossRef] [PubMed]
10. Mendoza-Roca, J.A.; Galiana-Aleixandre, M.V.; Lora-García, J.; Bes-Piá, A. Purification of Tannery Effluents by Ultrafiltration in View of Permeate Reuse. *Sep. Purif. Technol.* **2010**, *70*, 296–301. [CrossRef]
11. Religa, P.; Kowalik, A.; Gierycz, P. A new approach to chromium concentration from salt mixture solution using nanofiltration. *Sep. Purif. Technol.* **2011**, *82*, 114–120. [CrossRef]
12. Zakmout, A.; Sadi, F.; Portugal, C.A.M.; Crespo, J.G.; Velizarov, S. Tannery Effluent Treatment by Nanofiltration, Reverse Osmosis and Chitosan Modified Membranes. *Membranes* **2020**, *10*, 378. [CrossRef]
13. Vinduja, V.; Balasubramanian, N. Electrocoagulation-integrated hybrid membrane processes for the treatment of tannery wastewater. *Environ. Sci. Pollut. Res.* **2013**, *20*, 7441–7449.
14. Deghles, A.; Kurt, U. Treatment of tannery wastewater by a hybrid electrocoagulation/ electrodialysis process. *Chem. Eng. Process.* **2016**, *104*, 43–50. [CrossRef]

15. Dasgupta, J.; Mondal, D.; Chakraborty, S.; Sikder, J.; Curcio, S.; Arafat, H.A. Nanofiltration Based Water Reclamation from Tannery Effluent Following Coagulation Pretreatment. *Ecotoxicol. Environ. Saf.* **2015**, *121*, 22–30. [CrossRef] [PubMed]
16. Ashour, E.A.; Tony, M.A. Eco-friendly removal of hexavalent chromium from aqueous solution using natural clay mineral: Activation and modification effects. *SN Appl. Sci.* **2020**, *2*, 2042S.
17. Mustapha, S.; Ndamitso, M.M.; Abdulkareem, A.S.; Tijani, J.O.; Mohammed, A.K.; Shuaib, D.T. Potential of using kaolin as a natural adsorbent for the removal of pollutants from tannery wastewater. *Heliyon* **2019**, *5*, e029234. [CrossRef]
18. Tammaro, M.; Salluzzo, A.; Perfetto, R.; Lancia, A. A comparative evaluation of biological activated carbon and activated sludge processes for the treatment of tannery wastewater. *J. Environ. Chem. Eng.* **2014**, *2*, 1445–1455. [CrossRef]
19. Ahmad, W.; Qaiser, S.; Ullah, R.; Jan, B.M.; Karakassides, M.A.; Salmas, C.E.; Ikram, G.K.R. Utilization of Tires Waste-Derived Magnetic-Activated Carbon for the Removal of Hexavalent Chromium from Wastewater. *Materials* **2021**, *14*, 34.
20. Ihsanullah; Al-Khaldi, F.A.; Abu-Sharkh, B.; Abulkibash, A.M.; Qureshi, M.I.; Laoui, T.; Atieh, M.A. Effect of acid modification on adsorption of hexavalent chromium (Cr(VI)) from aqueous solution by activated carbon and carbon nanotubes. *Desalination Water Treat.* **2016**, *57*, 7232–7244. [CrossRef]
21. Biswas, S.; Rashid, T.U.; Debnath, T.; Haque, P.; Rahman, M.M. Application of Chitosan-Clay Biocomposite Beads for Removal of Heavy Metal and Dye from Industrial Effluent. *J. Compos. Sci.* **2020**, *4*, 16. [CrossRef]
22. Horzum, N.; Boyacı, E.; Eroglu, A.E.; Shahwan, T.; Demir, M.M. Sorption Efficiency of Chitosan Nanofibers toward Metal Ions at Low Concentrations. *Biomacromolecules* **2010**, *11*, 3301–3308. [CrossRef]
23. Baroni, P.; Vieira, R.S.; Meneghetti, E.; da Silva, M.G.C.; Beppu, M.M. Evaluation of batch adsorption of chromium ions on natural and crosslinked chitosan membranes. *J. Hazard. Mater.* **2008**, *152*, 1155–1163. [CrossRef] [PubMed]
24. Mirabedini, M.; Kassaee, M.Z.; Poorsadeghi, S. Novel Magnetic Chitosan Hydrogel Film, Cross-Linked with Glyoxal as an Efficient Adsorbent for Removal of Toxic Cr (VI) From Water. *Arab. J. Sci. Eng.* **2017**, *42*, 115–124. [CrossRef]
25. Karaer, H.; Kaya, İ. Synthesis, characterization and using at the copper adsorption of chitosan/polyvinyl alcohol magnetic composite. *J. Mol. Liq.* **2017**, *230*, 152–162. [CrossRef]
26. Salehi, E.; Madaeni, S.S.; Rajabi, L.; Derakhshan, A.A.; Daraei, S.; Vatanpour, V. Static and dynamic adsorption of copper ions on chitosan/polyvinyl alcohol thin adsorptive membranes: Combined effect of polyethylene glycol and aminated multi-walled carbon nanotubes. *Chem. Eng. J.* **2013**, *215–216*, 791–801. [CrossRef]
27. Juang, R.-S.; Shiau, R.-C. Metal removal from aqueous solutions using chitosan-enhanced membrane filtration. *J. Membr. Sci.* **2000**, *165*, 159–167. [CrossRef]
28. Li, L.; Li, Y.; Yang, C. Chemical filtration of Cr (VI) with electrospun chitosan nanofiber membranes. *Carbohydr. Polym.* **2016**, *140*, 299–307. [CrossRef]
29. Desai, K.; Kit, K.; Li, J.; Davidson, P.M.; Zivanovic, S.; Meyer, H. Nanofibrous chitosan non-wovens for filtration applications. *Polymer* **2009**, *50*, 3661–3669. [CrossRef]
30. Ghaee, A.; Shariaty-Niassar, M.; Barzin, J.; Matsuura, T.; Ismail, A.F. Preparation of chitosan/cellulose acetate composite nanofiltration membrane for wastewater treatment. *Desalination Water Treat.* **2015**, *57*, 14453–14460. [CrossRef]
31. Căprărescu, S.; Zgârian, R.G.; Tihan, G.T.; Purcar, V.; Totu, E.E.; Modrogan, C.; Chiriac, A.L.; Nicolae, C.A. Biopolymeric Membrane Enriched with Chitosan and Silver for Metallic Ions Removal. *Polymers* **2020**, *12*, 1792. [CrossRef]
32. Wang, L.; Yang, G.; Xing, W.; Xu, N. Mathematic model of the yield for diafiltration processes. *Sep. Purif. Technol.* **2008**, *59*, 206–213. [CrossRef]
33. Veigas, B.; Portugal, C.A.M.; Valério, R.; Fortunato, E.; Crespo, J.G.; Baptista, P.V. Scalable approach for the production of functional DNA based gold nanoprobes. *J. Membr. Sci.* **2015**, *492*, 528–535. [CrossRef]
34. Sabourian, P.; Tavakolian, M.; Yazdani, H.; Frounchi, M.; van de Ven Theo, G.M.; Maysinger, D.; Kakkar, A. Stimuli-responsive chitosan as an advantageous platform for efficient delivery of bioactive agents. *J. Control. Release* **2020**, *317*, 216–231. [CrossRef] [PubMed]
35. Executive Decree, No. 06-141 of 20 Rabie El Aouel 1427 Corresponding to 19 April 2006, Official Journal of the Algerian Republic N 26. Available online: https://and.dz/site/wp-content/uploads/D%C3%A9cret-executif-n%C2%B0-06-141.pdf (accessed on 1 July 2021).
36. Fikar, M.; Kovács, Z.; Czermak, P. Dynamic optimization of batch diafiltration processes. *J. Membr. Sci.* **2010**, *355*, 168–174. [CrossRef]
37. Chemical Equilibrium Diagrams. Available online: https://sites.google.com/site/chemdiagr/ (accessed on 1 July 2021).
38. Shao, J.; Zydney, A.L. Optimization of Ultrafiltration/Diafiltration Processes for Partially Bound Impurities. *Biotechnol. Bioeng.* **2004**, *87*, 286–292. [CrossRef] [PubMed]
39. Vinodhini, P.A.; Sudha, P.N. Removal of heavy metal chromium from tannery effluent using ultrafiltration membrane. *Text. Cloth. Sustain.* **2016**, *2*, 1–5. [CrossRef]
40. Zhang, Y.; Xu, X.; Yue, C.; Song, L.; Lv, Y.; Liu, F.; Li, A. Insight into the efficient co-removal of Cr(VI) and Cr(III) by positively charged UiO-66-NH2 decorated ultrafiltration membrane. *Chem. Eng. J.* **2021**, *404*, 126546. [CrossRef]

water

MDPI

Article

Study on the SBA-15 Silica and ETS-10 Titanosilicate as Efficient Adsorbents for Cu(II) Removal from Aqueous Solution

Doina Humelnicu [1,*], Inga Zinicovscaia [2,3], Ionel Humelnicu [1], Maria Ignat [1], Nikita Yushin [2] and Dmitrii Grozdov [2]

1 Faculty of Chemistry, Alexandru Ioan Cuza University of Iasi, Boulevard Carol I, 11, 700506 Iasi, Romania; ionel@uaic.ro (I.H.); maria.ignat@uaic.ro (M.I.)
2 Joint Institute for Nuclear Research, Joliot-Curie Street, 6, 1419890 Dubna, Russia; zinikovskaia@mail.ru (I.Z.); ynik_62@mail.ru (N.Y.); dsgrozdov@rambler.ru (D.G.)
3 Horia Hulubei National Institute for R&D in Physics and Nuclear Engineering, 30 Reactorului Street MG-6, 077125 Magurele, Romania
* Correspondence: doinah@uaic.ro

Abstract: The efficiency of Cu(II) removal from aqueous solution by two adsorbents, silica SBA-15 and titanosilicate ETS-10, was investigated. Effects of various experimental parameters such as: contact time, pH, initial copper concentration, adsorbent dosage, temperature were investigated in order to determine the maximum adsorption capacity of the adsorbents. The maximum adsorption capacity of silica SBA-15 was achieved at pH 5.0, and of titanosilicate ETS-10 at pH 6.0. The Freundlich, Langmuir, and Temkin isotherm models were applied in order to describe the equilibrium adsorption of Cu(II) by the studied adsorbents. Equilibrium data fitted well to the Langmuir model with a higher adsorption capacity of ETS-10 (172.53 $mg \cdot g^{-1}$) towards Cu(II) than SBA-15 (52.71 $mg \cdot g^{-1}$). Pseudo-first- and pseudo-second-order, Elovich, and Weber–Morris intraparticle diffusion models were used for description of the experimental kinetic data. It was found that the pseudo-first-order and pseudo-second-order kinetic models were the best applicable models to describe the adsorption kinetic data. Thermodynamic parameters that characterize the process indicated that the adsorption of Cu(II) onto the two adsorbents is spontaneous and endothermic.

Keywords: mesoporous silica SBA-15; titanosilicate ETS-10; adsorption; copper

Citation: Humelnicu, D.; Zinicovscaia, I.; Humelnicu, I.; Ignat, M.; Yushin, N.; Grozdov, D. Study on the SBA-15 Silica and ETS-10 Titanosilicate as Efficient Adsorbents for Cu(II) Removal from Aqueous Solution. *Water* **2022**, *14*, 857. https://doi.org/10.3390/w14060857

Academic Editors: Cristina Palet and Julio Bastos-Arrieta

Received: 3 February 2022
Accepted: 6 March 2022
Published: 9 March 2022

1. Introduction

The problem of environment pollution with heavy metal has become one of the serious problems, particularly in the polluted aquatic system.

The release of different pollutants into the environment has increased noticeably as a result of industrialization, and thereby lowered the quality of the environment to an alarming level. Among these pollutants, heavy metals are one of the most dangerous due to their nonbiodegradability, persistence, and toxicity.

There are many situations when low concentration of heavy metals can accumulate to toxic levels through the human food chain and the biosphere from the environment, which can disturb the biochemical processes and human health [1–5]. Heavy metals, such as lead, cadmium, mercury, copper, chromium, zinc, nickel, are used in the different fields of industry such as metal plating, electrolysis, mining, metallurgy, industry fertilizer, pesticide industry, leatherworking, and dyeing industry [2,3,6].

Among the heavy metals, copper is one of the indispensable micronutrients required by organisms at low concentrations. Copper ions play an important role in the enzyme's synthesis, development of tissues and bones for human [7]. At high concentrations, copper toxicity may be observed by a variety of syndromes and effects including renal dysfunction, hypertension, hepatic injury, lung damage and teratogenic effects [8,9].

Due to the mobility and toxicity, the presence of Cu(II) ions in surface water and groundwater represents a real inorganic contamination problem. This ion is one of the most poisonous, whose toxicity is attributed in part to its ability to accumulate in tissues. Human exposure to a high level of copper results in generation of reactive oxygen species (ROS) and free radicals by Fenton-like reaction. These radical species can alter biomolecules like DNA, proteins and lipids.

Therefore, it is important to remove copper from effluents, before discharging them into water bodies. There are various common methods available for the removal of copper ions from wastewaters such as chemical precipitation [10], absorption [11,12] and biosorption [13], membrane separation [14], bioelectrochemical systems [15], ion exchange [16], and electrochemical methods [17]. However, these techniques have limitations such as low efficiency at low metal concentrations or production of secondary sludge, which furthers disposal in a costly process [18].

Among the abovementioned methods, adsorption has been regarded as a cost-effective technology for removal of heavy metals from solutions with low metal concentrations. The main advantages of the technique are low cost of adsorbents, easy desorption, good recycling, highly effective and environmental-friendly nature [1,2,5,6].

The World Health Organization (WHO) recommends a safe amount of Cu(II) of 5 mg/L in drinking water [19]. Therefore, more attention is paid to efficient methods of copper removal from residual waters.

Among the adsorbents used in the literature for the remediation of the wastewaters contaminated with copper are: zeolite [20], chitosan [21], clays [22], graphene nanocomposite [23], carbon nanotubes [24].

One of the adsorbents that is investigated in this study is an ion exchanger belonging to the Engelhard Titanium Silicate (ETS) family. The ETS-10 phase is an extremely interesting titanosilicate microporous material due to its high thermal stability and wide pores (pore size close to 0.8 nm). These materials are useful and can be applied in a variety of fields, such as water purification and heavy metal removal [25,26], gas adsorption [27], and photo-catalysis [28].

The second adsorbent that is used in this research is SBA-type silicas (Santa Barbara Amorphous) that exhibit interesting textural properties, such as large specific surface areas and uniform-sized pores. The advantage of the use of SBA-15 material includes its high surface-to-volume ratio, flexible framework compositions and high thermal stability [29].

An essential condition for an advantageous sorption is an adequately selective sorbent with a high sorption capacity and high level of reusability.

The objectives of the present study were: (i) the investigation of the influence of pH, sorbent dose, copper concentration and temperature on the sorption capacity and removal efficiency in non-competitive conditions; (ii) to model the kinetic and equilibrium of copper adsorption in order to evaluate the kinetic and isotherm parameters; (iii) to establish the level of reusability of the sorbents during consecutive sorption/desorption cycles.

The influence of initial pH of Cu(II) ions solution, contact time, adsorbent dosage and initial concentrations on the Cu(II) ions uptake was studied. The non-linear Langmuir, Freundlich and Temkin isotherm models were used to fit the equilibrium adsorption data. The adsorption rates were determined quantitatively and compared by the pseudo-first-order, pseudo-second-order, Elovich and Weber–Morris intraparticle diffusion models.

2. Materials and Methods

The sorbents have been synthesized by a sol–gel method as described in our previous work [30]. Mesoporous silica SBA-15 was synthesized in acidic conditions using amphiphilic triblock copolymer poly(ethylene glycol)-block-poly(propyleneglycol)-block-poly(ethylene glycol) (Pluronic P123—$EO_{20}PO_{70}EO_{20}$; Sigma-Aldrich, St. Louis, MO, USA) as a surfactant template and TEOS (tetraethyl orthosilicate, Sigma-Aldrich) as silica source. Titanosilicate ETS-10 with the composition $3.4Na_2O:1.5K_2O:TiO_2:5.5SiO_2:150H_2O$

has been prepared from sodium silicate (Sigma-Aldrich) as Si precursor and commercial TiO_2 (Degusa-P25, Sigma-Aldrich) as Ti source.

Sorption Experiments

All chemicals were of analytical reagent grade and no further purification was carried out.

The adsorption experiments were performed in a batch system by stirring at 200 rpm a suspension that contained Cu(II) ions solution and corresponding amounts of the adsorbent. The stock solution containing Cu(II) was prepared from $CuSO_4{\cdot}5H_2O$ (Sigma-Aldrich) and diluted to obtain the appropriate concentrations. The pH varied between 2 and 6, the initial concentration of copper in the solution ranged from 10 to 200 $mg{\cdot}L^{-1}$, at a temperature between 20 °C and 50 °C. The pH of the solution was adjusted with NaOH or HNO_3 0.1 M solution and measured with a HANNA pH/temperature meter HI 991001. About 0.02 g of adsorbent was added into the solution containing Cu(II) ions and was left stirring for a certain period of time. At the end of adsorption experiment, the adsorbents were separated from the solutions using cellulose nitrate membrane filters (0.45 μm pore). The concentrations of Cu(II) ions in the filtrate (before and after adsorption of Cu(II)) were determined using ICP-AES (Analytik Jena, Jena, Germany).

The Cu(II) adsorption q was calculated using the following equation:

$$q = \frac{(C_0 - C_e){\cdot}V}{m} \quad (1)$$

and adsorption removal efficiency, R (%) from the equation:

$$RE = \frac{C_0 - C_e}{C_0}{\cdot}100 \quad (2)$$

where q is the amount of copper ions adsorbed on the adsorbent, mg/g; V is the volume of solution, L; C_0 is the initial concentration of copper in mg/L, C_e is the final copper concentration in the solution, mg/L, and m is the mass of adsorbent, g.

The adsorption capacities of the two adsorbents were analyzed through the use of the Langmuir, Freundlich and Temkin isotherm models. The kinetics of copper adsorption on the ETS-10 and SBA-15 were analyzed using pseudo-first-order, pseudo-second-order, Elovich and Weber–Morris intra-particle diffusion kinetic models.

All batch adsorption experiments were carried out in duplicate and results are presented as arithmetic mean values.

3. Results and Discussion

The adsorbents were characterized by DRX, FT-IR, thermal analysis and SEM-EDX, as we have reported in our previous work [30]. The obtained results indicated that the adsorbents are mesoporous material with a BET surface area of 802.493 s/g for silica SBA-15, and microporous material with a surface area of 31.473 s/g for titanosilicate ETS-10, respectively.

3.1. pH Effect on the Adsorption Process

The adsorption of Cu(II) ions onto the adsorbents varies depending on initial pH, because this parameter causes changes in the charge of adsorbent, the degree of ionization and speciation of the adsorbate.

In this study, the range of initial pH for Cu(II) ions adsorption study was 2.0–6.0. At pH values higher than 7, precipitation of Cu(II) ions as $Cu(OH)_2$ occurs and could lead to the wrong interpretation of adsorption data. On Figure 1 is shown the removal efficiency of the SBA-15 and ETS-10 for Cu(II) ions. As can be seen from Figure 1, the sorption capacity of Cu(II) ions onto the adsorbents increased with an increase in the initial pH value of the solution. The maximum sorption of Cu(II) ions onto adsorbents occurred at pH 6 for ETS-10 and pH 5 for SBA-15, respectively. It was observed that a sharp increase in the copper removal from 8.6% to 99.61% (SBA-15) and from 10% to 99.79% (ETS-10) occurred when the pH values of the solutions changed from 2.0 to 6.0. The low removal efficiency at low pH is apparently due to the presence of a higher concentration of $[H_3O]^+$ in the solution which competes strongly with the Cu(II) ions for the adsorption sites of the SBA-15 and ETS-10 surfaces. With the pH increase, the $[H_3O]^+$ concentration decreases leading to an increase of Cu(II) uptake.

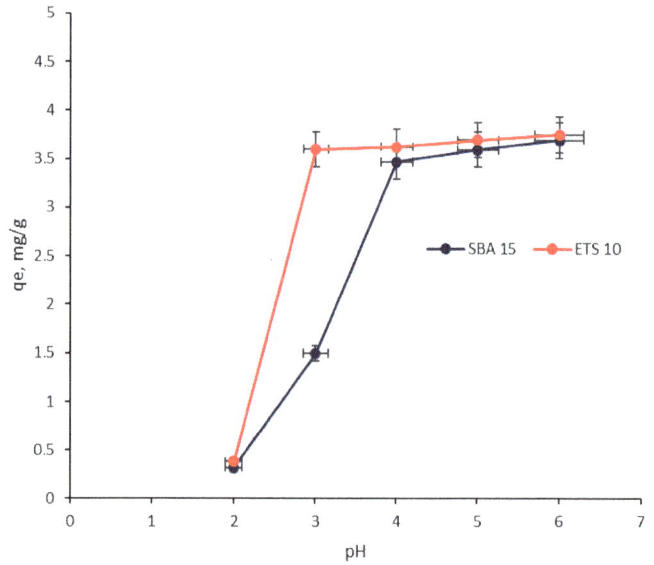

Figure 1. pH influence on the removal efficiency of SBA-15 and ETS-10 adsorbents toward Cu(II). (Ci 10 mg/L, time 60 min, temperature 23 °C, adsorbent dose 20 mg).

3.2. Effect of Adsorbent Dosage

An important parameter that affects the efficiency of adsorption from an economic point of view is the mass of the sorbent. The adsorption process is not effective if it requires a large amount of adsorbent. The influence of the adsorbent mass used on the adsorption of Cu(II) ions was investigated, and the results are shown in Figure 2. The doses of adsorbents varied from 0.010 g to 0.050 g, while the other parameters such as pH, temperature, initial concentration of Cu(II) ions, contact time were kept constant. Based on Figure 2, it is seen that an increase of the adsorbent dose can lead to an increase in the percentage of Cu(II) ions removal from the solution. This is anticipated because, by increasing the adsorbent's dose, the number of adsorption sites available for adsorbent–adsorbate interaction will increase as well. Both adsorbents showed no further increase in the adsorption capacity after a certain amount of adsorbent was added.

3.3. Equilibrium Isotherm, Kinetics and Thermodynamic Studies

3.3.1. Adsorption Equilibrium Isotherm

Adsorption isotherms are fundamental for understanding the mechanism of adsorption and the interaction between sorbent and sorbate. To study the adsorption of Cu(II) onto the sorbent, three of the most commonly used isotherm models were used in this work: Langmuir, Freundlich and Temkin. The Langmuir [31] (Equation (3)) isotherm model characterizes an adsorption monolayer on a surface with a finite number of identical centers that are homogeneously distributed on the surface of sorbent. This model assumes that the binding sites are homogeneously distributed over the adsorbent surface and the binding sites have the same affinity for adsorption of a single molecular layer. The bonding to the adsorption sites can be either chemical or physical in nature, but must be strong enough to avoid displacement of the adsorbed Cu(II) ions.

$$q_e = \frac{q_m \cdot K_L \cdot C_e}{1 + K_L C_e} \quad (3)$$

where q_e is the amount of Cu(II) adsorbed per mass unit of sorbent at equilibrium (mg·g^{-1}), C_e is the equilibrium concentration of remaining Cu(II) ions in the solution (mg·L^{-1}), q_m is a parameter that gives the maximum adsorption capacity of the sorbent (mg·g^{-1}), K_L is a constant that refers to the energy of adsorption/desorption (L·mg^{-1}).

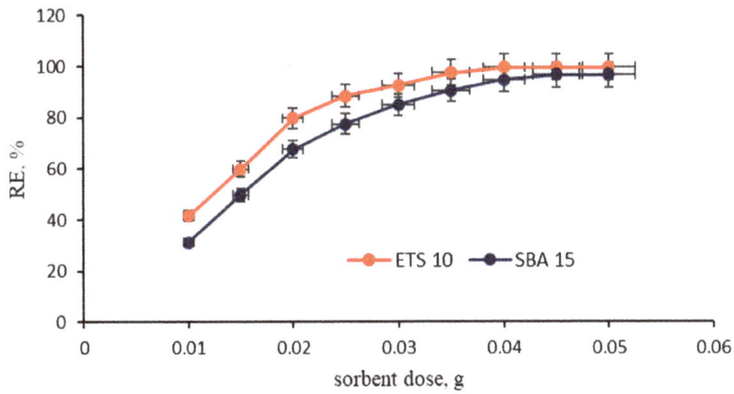

Figure 2. Influence of adsorbent dose on the removal efficiency of SBA-15 and ETS-10 adsorbents toward Cu(II). (C_i 10 mg/L, time 60 min, temperature 23 °C).

The Freundlich [32] (Equation (4)) isotherm is the second mathematical model used to describe the adsorption metal present in solution on solid surface. This model assumes that the adsorbent has an energetically heterogeneous surface and has a different affinity for adsorption.

$$q_e = K_F \cdot C_e^{1/n} \quad (4)$$

where q_e is the amount of Cu(II) adsorbed at equilibrium (mg·g^{-1}); C_e is the concentration of Cu(II) ion in solution at equilibrium (mg·L^{-1}); K_F (L·mg^{-1}) and $1/n$ are the Freundlich constants.

The Temkin isotherm model [33] (Equation (5)) assumes that the adsorption heat of molecules decreases linearly with the increase in coverage of the adsorbent surface, and that adsorption is characterized by a uniform distribution of binding energies.

$$q_e = \frac{RT}{b_T} \cdot \ln(a_T \cdot C_e) \quad (5)$$

where $1/b_T$ represents the sorption potential of the sorbent, a_T is the Temkin constant, R is the universal gas constant (8.314 J K^{-1}·mol^{-1}) and T is the temperature (K).

The interaction of metal ions and adsorbents was further evaluated by the separation factor (R_L). R_L is a dimensionless constant separation factor, an equilibrium parameter derived from the Langmuir isotherm model. The R_L was defined by Hall et al. [34], and is expressed as Equation (6).

$$R_L = \frac{1}{1 + K_L \cdot C_0} \quad (6)$$

K_L is the Langmuir constant and C_0 is the initial concentration of Cu(II) ions. For a favorable adsorption, the R_L value must be between 0 and 1. In this respect, if $R_L > 1$ adsorption is unfavorable, and if $R_L = 0$ adsorption is irreversible. In the present studies, the obtained R_L values were less than one (Table 1), which indicated that the adsorption processes were favorable.

Table 1. Langmuir, Freundlich and Temkin isotherm parameters for the sorption of Cu(II) on silica SBA-15 and titanosilicate ETS-10 adsorbents.

Model	Parameters	Silica SBA-15	Titanosilicate ETS-10
Langmuir	q_m, mg/g	52.71	172.53
	K_L, L/mg	2.04	8.73
	R_L	0.002–0.046	0.0005–0.011
	R^2	0.984	0.999
Freundlich	K_F, mg/g	0.197	0.460
	1/n	0.95	0.86
	R^2	0.977	0.985
Temkin	a_T, L/g	0.066	0.65
	b_T, kJ/mol	0.0118	0.089
	R^2	0.735	0.833

The graphical representation of used models among with experimental data is presented in Figure 3, and the obtained values for Langmuir, Freundlich and Temkin isotherm constants and correlation coefficients are listed in Table 1.

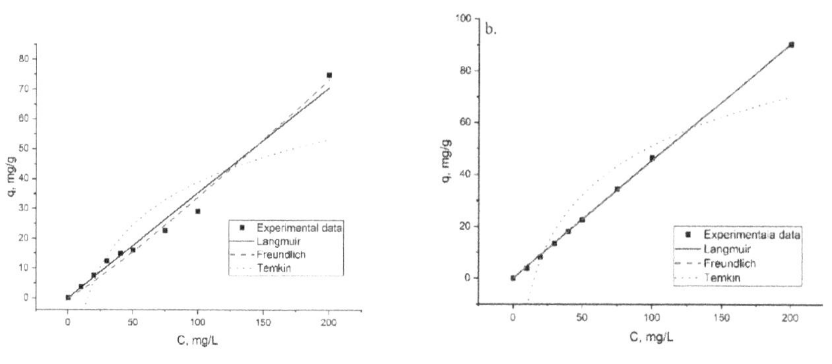

(a) **(b)**

Figure 3. The adsorption isotherms and experimental data for Cu(II) ion sorption on: (**a**) mesoporous silica SBA-15 and (**b**) titanosilicate ETS-10 adsorbents.

The RL values, greater than zero and lower than unit, indicate that the sorption of Cu(II) ions on both adsorbents was favorable and reversible. Sorption of Cu(II) onto SBA-15 and ETS-10 was better described by the Langmuir and Freundlich models according to the high values of R^2 for both adsorbents.

In addition, between two tested adsorbents, in terms of adsorption capacity, the best candidate seems to be ETS-10. The maximum adsorption capacity given by the Langmuir isotherm was 52.71 mg/g for SBA-15 and 172.53 mg/g for ETS-10, respectively.

Monolayer coverage of the surface by the metal ions can be used for the calculation of the specific surface area S according to the following equation [3,35]:

$$S = \frac{q_{max} \cdot N \cdot A}{M} \quad (7)$$

where S is the specific surface area, m^2/g adsorbent; q_{max} the monolayer sorption capacity, g Cu/g adsorbent; N the Avogadro number, $6.023 \cdot 10^{23}$; A the cross-sectional area of metal ion, m^2; M the molecular weight of metal. The molecular weight and the cross-sectional area of Cu(II) are 63.5 and 1.58 $Å^2$ in a close packed monolayer (Cu(II) radius is 0.71 Å), respectively [35,36]. The maximum specific surface area calculated from Equation (7) for Cu(II) adsorption is 7.89 m^2/g for SBA-15, and 25.86 m^2/g for ETS-10, respectively. Comparison of the maximum specific surface area of the adsorbents for Cu(II) adsorption shows that SBA-15 and ETS-10 have a larger specific surface area than other adsorbents [35–37].

The efficiency of the two investigated adsorbents, ETS-10 and SBA-15, for removal of Cu(II) was highlighted by comparison with the values of maximum adsorption capacity presented in the literature for other adsorbents along with testing conditions (Table 2).

Table 2. The comparison of maximum sorption capacity of Cu(II) ions onto different adsorbents.

Adsorbent	Conditions	q, mg/g	Reference
Mesoporous silica SBA-15	pH = 5, t = 23 °C	52.71	Present study
Titanosilicate ETS-10	pH = 5, t = 23 °C	172.53	Present study
Rape straw powders	pH = 4.77, t = 20 °C	34.29	[38]
Sunflower hulls	pH = 5, t = 20 °C	49.74	[39]
Chitosan based ion-imprinted cryo-composites	pH = 4.5	260	[40]
Chemical modified Moringa oleifera leaves powder	pH = 6, t = 50 °C	167.9	[41]
Coconut tree sawdust	pH = 6	3.89	[42]
Eggshell	pH = 6	34.48	[43]
Sugarcane bagasse	pH = 6	3.65	[42]
N-HAP/Chitosan	pH = 7.5, t = 25 °C	113.66	[43]
Chitosan crosslinked with epichlorohydrin-triphosphate	pH = 6, t = 25 °C	130.38	[44]

As can be seen, the highest sorption capacities were reported for some adsorbents, such as: chitosan-based ion-imprinted cryo-composites [40], ETS-10 (present study), chemically modified moringa oleifera [41], natural hydroxyapatite/chitosan composite [43], and chitosan crosslinked with epichlorohydrin-triphosphate [44], respectively.

3.3.2. Adsorption Kinetics

In order to investigate the mechanism of adsorption, the pseudo-first-order (PFO), pseudo-second-order (PSO), Elovich and Weber–Morris models were used to study the experimental data obtained.

The pseudo-first-order model of Lagergren [45] is commonly used for the adsorption of liquid/solid systems and assumes that the rate of variation of surface site concentration is proportional to the amount of surface sites remaining unoccupied.

$$q_t = q_e(1 - e^{-k_1 t}) \tag{8}$$

where q_e and q_t are the amounts of Cu(II) ions adsorbed onto sorbents ($mg \cdot g^{-1}$) at equilibrium and at time t, respectively, and k_1 is the rate constant of first-order adsorption (min^{-1}).

The pseudo-second-order model can be expressed as [46]:

$$q_t = \frac{k_2 \cdot q_e^2 \cdot t}{1 + q_e \cdot k_2 t} \tag{9}$$

where k_2 is the rate constant of second-order adsorption ($g \cdot mg^{-1} \cdot min^{-1}$). This model is more likely to predict the adsorption behavior over the whole range of adsorption. The pseudo-second-order equation assumes that the adsorption behavior was controlled by the rate-controlling step, which can be chemical sorption involving an electronic exchange or distribution between adsorbent and adsorbate. The adsorbate can be transferred from the solution phase to the surface of the adsorbent in several steps. The steps may include film or external diffusion (transfer of adsorbate), pore diffusion, surface diffusion and adsorption on the pore surface. The overall adsorption can occur through one or more steps. The Weber–Morris intraparticle diffusion equation is given by the following equation [47]:

$$q = k_{diff} \cdot t^{0.5} + C_i \tag{10}$$

where k_{diff} is a rate parameter ($mg/g \cdot min^{1/2}$), and C_i is the intercept, which relates to the thickness of the boundary layer

The Elovich kinetic model helps to predict the mass and surface diffusion, activation and deactivation energy of a system [48].

$$q_t = \frac{1}{\beta} \ln(1 + \alpha \cdot \beta \cdot t) \tag{11}$$

where q_t is the sorption capacity at time t (mg/g), α is the initial sorption rate ($mg \cdot g^{-1} \cdot min^{-1}$), β is the desorption constant ($g \cdot mg^{-1}$).

The graphical representation of kinetic models is presented in Figure 4, and the kinetic model constants, along with the correlation coefficient, are given in Table 3.

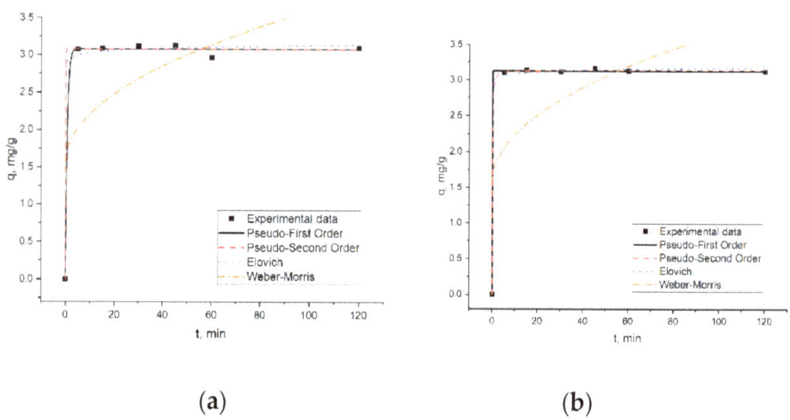

Figure 4. Kinetics of Cu(II) sorption on (**a**) mesoporous silica SBA-15 and (**b**) titanosilicate ETS-10 adsorbents.

Kinetic profiles indicated that the Cu(II) adsorption process was fast for both adsorbents. The adsorbents showed the same behavior regarding the removal of Cu (II), increasing up to 10 min and then were kept almost constant. In other words, the copper adsorption process occurred in two stages: an initial fast stage up to 10 min followed by a second stage in which no significant variation on the adsorption capacity was observed. This observation is due probably to the fact that more adsorption sites are available at the beginning of the experiments, followed by a saturation of the metal on the surface of the adsorbent.

The agreement between experimental data and the model predicted values was expressed by the correlation coefficients (R^2). A relatively high correlation coefficients value indicates that the model successfully describes the kinetics of copper adsorption. The values of R^2, both for PFO and PSO kinetic modes, are comparable, and for both models the theoretically calculated and experimentally obtained values of adsorption capacity were in good agreement. Similar results were found for other adsorbents [5,11,23]. The pseudo-second-order model assumes that the adsorption of adsorbate onto adsorbent supports second-order chemisorptions. The adsorption of copper onto SBA-15 and ETS-10 probably occurred by surface complexation reactions between copper and the sorption sites on the adsorbents.

Table 3. Parameters of the applied kinetic models for the adsorption of Cu(II) on SBA-15 and ETS-10.

	Parameter	Silica SBA-15	Titanosilicate ETS-10
	q_{exp}, mg/g	3.13	3.16
PFO	$q_{e,cal}$, mg/g	3.08	3.13
	k_1, min^{-1}	1.463	194.67
	R^2	0.997	0.999
PSO	$q_{e,\ cal}$, mg/g	3.081	3.135
	k_2, g/mg·min	$2.87 \cdot 10^{44}$	7.321
	R^2	0.997	0.999
Elovich	α, mg/g·min	$6.29 \cdot 10^{26}$	$7.08 \cdot 10^{34}$
	β, g/min	22.133	27.81
	R^2	0.994	0.998
Weber–Morris	k_{diff}	0.204	0.208
	C_i	1.565	1.577
	R^2	0.284	0.295

3.3.3. Thermodynamic Parameters

The thermodynamic parameters could be used to conclude whether the sorption process was spontaneous or not in the behavior of SBA-15 and ETS-10 sorbents for removal of Cu(II).

As reported by Guo et al. [49] and Kumar et al. [50], the values of the distribution coefficient (K_d), calculated using Equation (12) at different temperatures of 20 °C, 30 °C, 40 °C and 50 °C, were used to evaluate the thermodynamic parameters (ΔG°, ΔH°, and ΔS°).

$$K_d = \frac{q_e}{C_e} \cdot M_{adsorbate} \quad (12)$$

where q_e is the amount of Cu(II) retained at equilibrium, mg·g^{-1}, C_e is the concentration of Cu(II) at equilibrium, in the aqueous phase, mg·L^{-1}, and $M_{adsorbate}$ is the mass of Cu(II).

In order to solve the problem of the dimensionless of K_d, the values were multiplied with 55.5 mol·L^{-1} [51], the obtained value being symbolized with K°.

Equation van't Hoff (13) was used for calculation of enthalpy and entropy from the slope and intercept of plot lnK° vs. 1/T (Figure 5).

$$\ln K^0 = \frac{\Delta S^0}{R} - \frac{\Delta H^0}{RT} \quad (13)$$

where R is the universal gas constant (8.314 J·mol^{-1}·K^{-1}), T is the absolute temperature (K).

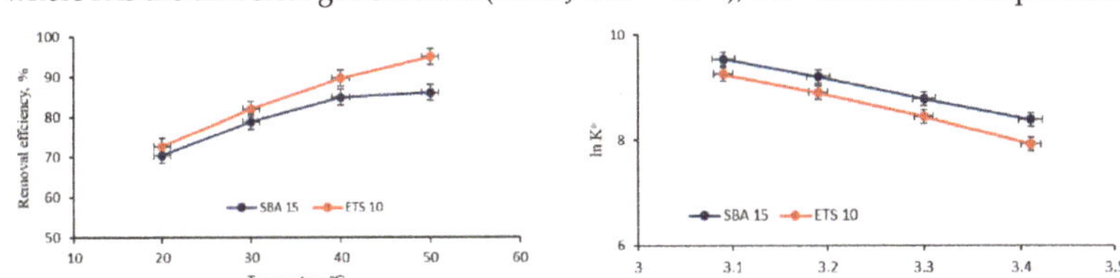

(**a**) (**b**)

Figure 5. Temperature dependence (**a**) of removal efficiency of Cu(II) onto SBA-15 and ETS-10 sorbents and plot for the evaluation of the thermodynamic parameters (**b**).

The standard Gibbs free energy change can be calculated by the following equation:

$$\Delta G^0 = -RT \ln K^0 \quad (14)$$

The obtained results are presented in Table 4.

Table 4. Thermodynamic parameters of the sorption process.

Sorbent	ΔH^0, kJ/mol	ΔS^0, kJ/mol·K	ΔG^0, kJ/mol			
			293	303	313	323
SBA-15	30.09	0.172	−20.31	−22.03	−23.75	−25.47
ETS-10	34.86	0.185	−19.35	−21.19	−23.05	−24.89

The positive values of ΔH^0 (ETS-10 = 34.86 kJ/mol, SBA-15 = 30.09 kJ/mol) denoted that the sorption process was of endothermic nature. The positive values of ΔS^0 (ETS-10 = 0.185 kJ/mol·K, SBA-15 = 0.172 kJ/mol·K) indicate increasing in randomness at the solid–liquid interface or changing the original internal structure of absorbent during the sorption process in Cu(II) aqueous solution. The increase of the negative value of ΔG^0 with the increase of temperature supports the increase of the degree of spontaneity for the sorption of Cu(II) onto both adsorbents.

Thermodynamic parameters revealed that sorption behavior was spontaneous and chemical in nature (almost all values of $\Delta G^0 > -20$ kJ·mol^{-1}) in the process of adsorption of Cu(II) using ETS-10 and SBA-15 as sorbents.

The activation energy of the sorption process (E_a) was obtained from the slope of plotting $\ln(1-\theta)$ vs. $1/T$, where sorbent surface coverage (θ) was calculated using Equation (15) [52]:

$$\theta = \left(1 - \frac{C_e}{C_0}\right) \quad (15)$$

C_e, C_0 are equilibrium and initial concentration of Cu(II) in aqueous solution (mg/L).

According to the modified Arrhenius equation, plotting $\ln(1-\theta)$ vs. $1/T$ (Figure 6) gives a straight line with the slope E_a/R.

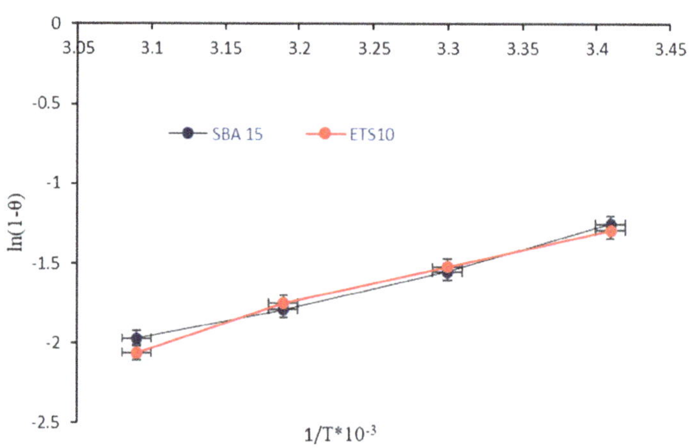

Figure 6. Plot of ln(1-θ) vs. 1/T.

Activation energy values are calculated from the slope of plot and were found to be 43.54 kJ·mol^{-1} and 18.18 kJ·mol^{-1} for ETS-10 and SBA 15, respectively. The positive values of E_a were consistent with the obtained positive values of ΔH° and confirm once more the endothermic nature of the sorption process.

3.4. Modelling of the Interactions

The interactions between the structures of mesoporous silicas SBA-15 and microporous titanosilicate ETS-10 with $CuSO_4 \cdot 5H_2O$ were simulated using the theoretical chemistry methods. The adsorbent macrostructures were represented, in gaseous medium, by a small fragment that respects the atomic arrangement according to the crystalline structures from the Crystallography Open Database (COD). For the sulfate molecule was used a structure with an octahedral configuration for the Cu atom, illustrated in Figure 7a, similar with the one representation in COD.

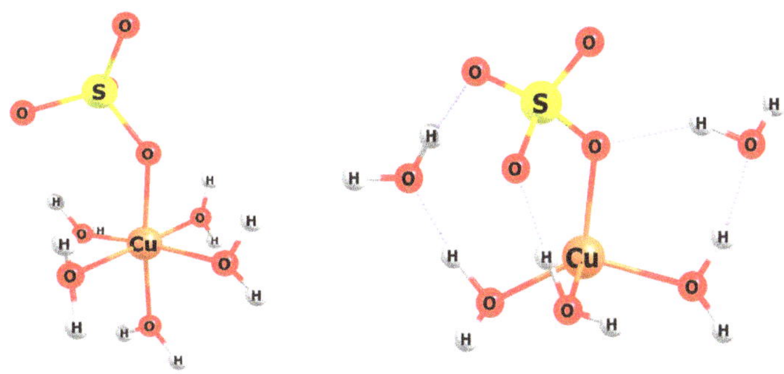

(**a**) (**b**)

Figure 7. Structures of the $CuSO_4 \cdot 5H_2O$ with the metal atom in the configuration: (**a**) octahedral; (**b**) tetrahedral.

The spatial structure of $CuSO_4 \cdot 5H_2O$ in interactions with the porous surface can be found, also, in a structure with the Cu atom in the tetrahedral configuration and with hydrogen-bonding interactions type with two water molecules (Figure 7b). In the octahedral configuration of the crystalline structure, the distance Cu-O is 1.83 Å. This value increases slightly for Cu-O-S to about 1.9 Å and just over 2 Å for Cu-O (H_2O) in the tetrahedral structure. When the cupric structure approaches the surface, an edge or a peak of the porous structure, the water of crystallization will be gradually removed and may leave the area of interest or may still be found around the active center by establishing hydrogen bonds. A direct interaction, a covalent bond, is thus formed between the Cu atom and an oxygen in the adsorbent medium, Si-O-Cu (Figure 8a), or Ti-O-Cu (Figure 8b). The distance between Cu atom and oxygen atom from the adsorbent decreases from 2 Å in crystalline structure to 1.9 Å.

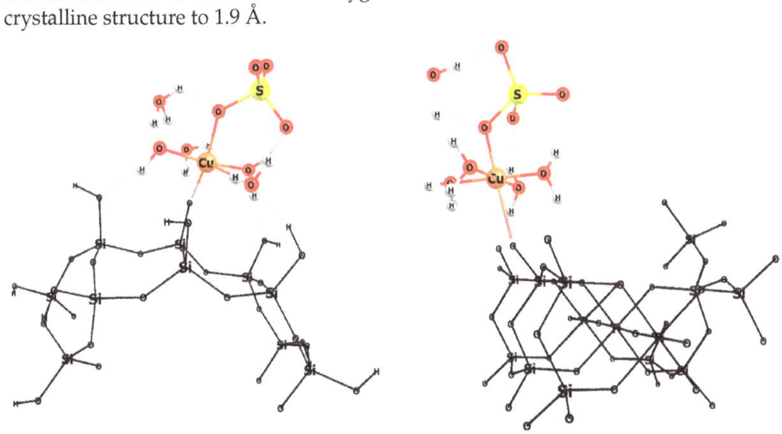

(**a**) (**b**)

Figure 8. The interaction of the structure $CuSO_4 \cdot 5H_2O$ with: (**a**) SBA-15, (**b**) ETS-10.

The removal of some crystallization water molecules and interactions with adsorbent structure determine the square plan hybridization (Figure 9a) or a pyramidal configuration (Figure 9b) for Cu atom, and formation of two covalent bonds with oxygen atom from the porous structures.

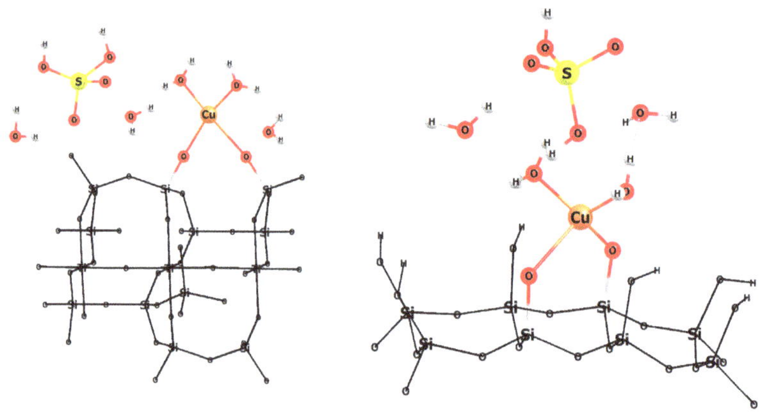

(a) (b)

Figure 9. The configuration of Cu atom in interactions with porous surface: (**a**) square planar; (**b**) tetrahedral.

In an acid environment, the sulfate fragment can form H_2SO_4, which is found in the system under investigation (Figure 8a).

4. Regeneration and Reusability of Sorbents

The applicability of potential sorbents depends on their regeneration under convenient conditions and the possibility of their re-use in successive sorption/desorption cycles. From practical motives, an ideal adsorbent must be reused in successive sorption/desorption cycles with as less as possible loss of the initial adsorption capacity. Therefore, desorption of Cu(II) ions was carried out in batch system by using the adsorbents loaded with copper immediately after the adsorption experiments. As eluent, 0.01 M solution of HCl in five successive sorption/desorption cycles was used and the obtained results are shown in Figure 10.

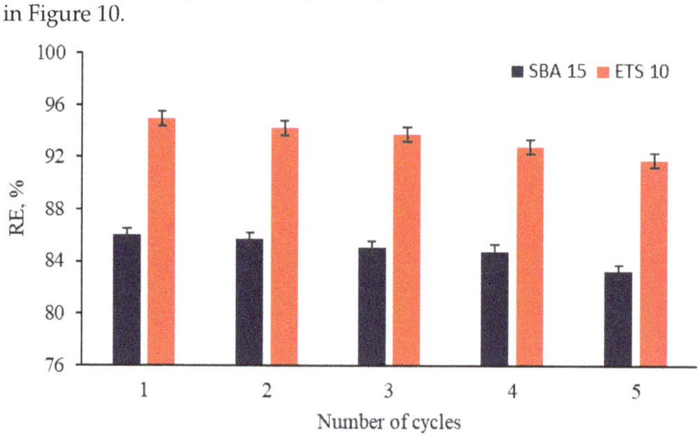

Figure 10. Reusability of the adsorbents under consecutive cycles sorption/desorption.

As can be seen, the removal efficiency slowly decreased during the sorption/desorption cycles, being about 91.75% for ETS-10, and 83.27% for SBA-15, respectively. These values recommend these materials as potential sorbents for efficient removal of Cu(II) from residual waters.

5. Conclusions

The sorption of Cu(II) ions from synthetic wastewaters onto titanosilicate ETS-10 and silica SBA-15 has been studied as a function of contact time, the initial metal ion concentration, adsorbent mass, pH, sorbent dose and temperature. Equilibrium, kinetic and thermodynamic data were applied in order to evaluate the efficiency of the investigated adsorbents for the removal of Cu(II) ions from aqueous solutions. The adsorption of Cu(II) on analyzed adsorbents obeyed the pseudo-second-order kinetics, supporting that the chemisorption would be the rate-determining step. The equilibrium data obtained for the adsorption of copper ions onto investigated adsorbents well fitted the Langmuir model with a maximum theoretical adsorption capacity of 52.71 mg Cu(II)/g for mesoporous silica SBA-15, and 172.53 mg Cu(II)/g for titanosilicate ETS-10, respectively. The adsorption process is endothermic ($\Delta H^\circ > 20$ kJ/mol) and spontaneous (the increase of the negative values of ΔG° with the increase of temperature). The adsorption of Cu(II) on the analyzed sorbents is a reversible process and the adsorbents can be used in five desorption/sorption cycles without significant loss in their adsorption capacities.

Experimental results showed that mesoporous silica SBA-15 and titanosilicate ETS-10 are promising adsorbents for the removal of copper ions from aqueous solutions.

Author Contributions: Conceptualization, I.Z., D.H. and N.Y.; methodology, I.Z., D.H., N.Y., D.G., M.I. and I.H.; software, D.G.; validation, D.G., I.Z. and D.H.; formal analysis, I.Z., D.H. and N.Y.; investigation, I.Z., D.H., N.Y., D.G., M.I. and I.H.; writing—original draft preparation, D.H., I.Z.; writing—review and editing, all authors; project administration, I.Z. and D.H.; funding acquisition, I.Z. and D.H. All authors have read and agreed to the published version of the manuscript.

Funding: This research was funded by IUCN-Dubna project No. 03-4-1128-2020/2022, theme 75. The APC was funded by the Alexandru Ioan Cuza University of Iasi research fund.

Institutional Review Board Statement: Not applicable.

Informed Consent Statement: Not applicable.

Data Availability Statement: The date presented in this study are available on the request from the corresponding author.

Acknowledgments: The authors are thankful to the Romanian Ministry of Research, Innovation and Digitization, within Program 1—Development of the national RD system, Subprogram1.2—Institutional Performance—RDI excellence funding projects, Contract No.11PFE/30.12.2021, for financial support.

Conflicts of Interest: The authors declare no conflict of interest.

References

1. Anastopoulos, I.; Massas, I.; Ehaliotis, C. Use of residues and by-products of the olive-oil production chain for the removal of pollutants from environmental media: A review of batch biosorption approaches. *J. Environ. Sci. Health Part A Toxic/Hazard. Subst. Environ. Eng.* **2015**, *50*, 677–718. [CrossRef] [PubMed]
2. González, A.G.; Pokrovsky, S.O.; Santana-Casiano, J.M.; González-Dávila, M. Bioadsorption of heavy metals. In *Prospects and Challenges in Algal Biotechnology*; Tripathi, B., Kumar, D., Eds.; Springer: Singapore, 2017; pp. 233–255. [CrossRef]
3. Peng, Q.; Liu, Y.; Zeng, G.; Xu, W.; Yang, C.; Zhang, J. Biosorption of copper(II) by immobilizing Saccharomyces cerevisiae on the surface of chitosan-coated magnetic nanoparticles from aqueous solution. *J. Hazard. Mater.* **2010**, *177*, 676–682. [CrossRef] [PubMed]
4. Ofomaja, A.E.; Naidoo, E.B.; Modise, S.J. Biosorption of copper(II) and lead(II) onto potassium hydroxide treated pine cone powder. *J. Environ. Manag.* **2010**, *91*, 1674. [CrossRef] [PubMed]
5. Moyo, M.; Chirinda, A.; Nharingo, T. Removal of copper from aqueous solution using chemically treated potato (*Solanum tuberosum*) leaf powder. *Clean Soil Air Water* **2016**, *44*, 488–495. [CrossRef]
6. Ekere, N.R.; Agwogie, A.B.; Ihedioha, J.N. Studies of biosorption of Pb^{2+}, Cd^{2+} and Cu^{2+} from aqueous solutions using Adansonia digitata root powders. *Int. J. Phytoremediat.* **2016**, *18*, 116–125. [CrossRef]
7. Chen, J.; Jiang, Y.; Shi, H.; Peng, Y.; Fan, X.; Li, C. The molecular mechanisms of copper metabolism and its roles in human diseases. *Pflügers Arch. Eur. J. Physiol.* **2020**, *472*, 1415–1429. [CrossRef]
8. Montazeri, A.; Akhlaghi, M.; Barahimi, A.R.; Jahanbazi Jahan Abad, A.; Jabbari, R. The Role of Metals in Neurodegenerative Diseases of the Central Nervous System. *Shefaye Khatam.* **2020**, *8*, 130–146. [CrossRef]

9. Mathys, Z.; White, A. Copper and Alzheimer's Disease. *Adv. Neurobiol.* **2017**, *18*, 199–216. [CrossRef]
10. Lundström, M.; Liipo, J.; Taskinen, P.; Aromaa, J. Copper precipitation during leaching of various copper sulfide concentrates with cupric chloride in acidic solutions. *Hydrometallurgy* **2016**, *166*, 136–142. [CrossRef]
11. Labidi, A.; Salaberria, A.M.; Fernandes, S.C.M.; Labidi, J.; Abderrabba, M. Adsorption of copper on chitin-based materials: Kinetic and thermodynamic studies. *J. Taiwan Inst. Chem. Eng.* **2016**, *65*, 140–148. [CrossRef]
12. Mushtaq, M.; Bhatti, H.N.; Iqbal, M.; Noreen, S. Eriobotrya japonica seed biocomposite efficiency for copper adsorption: Isotherms, kinetics, thermodynamic and desorption studies. *J. Environ. Manag.* **2016**, *176*, 21–33. [CrossRef]
13. Abdelfattah, I.; Ismail, A.A.; Sayed, F.A.; Almedolab, A.; Aboelghait, K.M. Biosorption of heavy metals ions in real industrial wastewater using peanut husk as efficient and cost effective adsorbent. *Environ. Nanotechnol. Monit. Manag.* **2016**, *6*, 176–183. [CrossRef]
14. Duan, H.; Wang, S.; Yang, X.; Yuan, X.; Zhang, Q.; Huang, Z.; Guo, H. Simultaneous separation of copper from nickel in ammoniacal solutions using supported liquid membrane containing synergistic mixture of M5640 and TRPO. *Chem. Eng. Res. Des.* **2017**, *117*, 460–471. [CrossRef]
15. Ntagia, E.; Rodenas, P.; Ter Heijne, A.; Buisman, C.J.N.; Sleutels, T.H.J.A. Hydrogen as electron donor for copper removal in bioelectrochemical systems. *Int. J. Hydrog. Energy* **2016**, *41*, 5758–5764. [CrossRef]
16. Ntimbani, R.N.; Simate, G.S.; Ndlovu, S. Removal of copper ions from dilute synthetic solution using staple ion exchange fibres: Dynamic studies. *J. Environ. Chem. Eng.* **2016**, *4*, 3143–3150. [CrossRef]
17. Lambert, A.; Drogui, P.; Daghrir, R.; Zaviska, F.; Benzaazoua, M. Removal of copper in leachate from mining residues using electrochemical technology. *J. Environ. Manag.* **2014**, *133*, 78–85. [CrossRef]
18. Ahluwalia, S.S.; Goyal, D. Removal of heavy metals by waste tea-leaves from aqueous solution. *Eng. Life Sci.* **2005**, *5*, 158–162. [CrossRef]
19. Awual, M.R.; Hasan, M.M.; Khaleque, M.A.; Sheikh, M.C. Treatment of copper(II) containing wastewater by a newly developed ligand based facial conjugate materials. *Chem. Eng. J.* **2016**, *288*, 368–376. [CrossRef]
20. Holub, M.; Balintova, M.; Kovacova, Z. Evaluation of Zeolite Adsorption Properties for Cu(II) Removal from Acidic Aqueous Solutions in Fixed-Bed Column System. *Proceedings* **2018**, *2*, 1293. [CrossRef]
21. Xiao, F.; Cheng, J.; Cao, W.; Yang, C.; Chen, J.; Luo, Z. Removal of heavy metals from aqueous solution using chitosan—Combined magnetic biochars. *J. Colloid. Interface. Sci.* **2019**, *540*, 579–584. [CrossRef]
22. Hussain, S.T.; Khaleefa Ali, S.A. Removal of heavy metal by ion exchange using bentonite clay. *J. Ecol. Eng.* **2021**, *22*, 104–111. [CrossRef]
23. Ali, I.; Burakov, A.E.; Melezhik, A.V.; Babkin, A.V.; Burakova, I.V.; Neskomornaya, M.E.A.; Galunin, E.V.; Tkachev, A.G.; Kuznetsov, D.V. Removal of Copper(II) and Zinc(II) Ions in Water on a Newly Synthesized Polyhydroquinone/Graphene Nanocomposite Material: Kinetics, Thermodynamics and Mechanism. *ChemistrySelect* **2019**, *4*, 12708–12718. [CrossRef]
24. Wang, Z.; Xu, W.; Jie, F.; Zhao, Z.; Zhou, K.; Liu, H. The selective adsorption performance and mechanism of multiwall magnetic carbon nanotubes for heavy metals in wastewater. *Sci. Rep.* **2021**, *11*, 16878. [CrossRef] [PubMed]
25. De Luca, P.; Bernaudo, I.; Elliani, R.; Tagarelli, A.; Nagy, J.; Macario, A. Industrial Waste Treatment by ETS-10 Ion Exchanger Material. *Materials* **2018**, *11*, 2316. [CrossRef]
26. Lv, L.; Su, F.; Zhao, X.S. Microporous titanosilicate ETS-10 for the removal of divalent heavy metals. *Stud. Surf. Sci. Catal.* **2005**, *156*, 933–940. [CrossRef]
27. De Luca, P.; Poulsen, T.G.; Salituro, A.; Tedeschi, A.; Vuono, D.; Kònya, Z.; Madaràsz, D.; Nagy, J.B. Evaluation and comparison of the ammonia adsorption capacity of titanosilicates ETS-4 and ETS-10 and aluminotitanosilicates ETAS-4 and ETAS-10. *J. Therm. Anal. Calorim.* **2015**, *122*, 1257–1267. [CrossRef]
28. De Luca, P.; Chiodo, A.; Nagy, J.B. Activated ceramic materials with deposition of photocatalytic titano-silicate micro-crystals. *Sustain. Chem.* **2011**, *154*, 155–165. [CrossRef]
29. Awual, M.R.; Rahman, I.M.M.; Yaita, T.; Khaleque, M.A.; Ferdows, M. pH dependent Cu(II) and Pd(II) ions detection and removal from aqueous media by an efficient mesoporous adsorbent. *Chem. Eng. J.* **2014**, *236*, 100–109. [CrossRef]
30. Zinicovscaia, I.; Yushin, N.; Humelnicu, D.; Grozdov, D.; Ignat, M.; Demcak, S.; Humelnicu, I. Sorption of Ce(III) by silica SBA-15 and titanosilicate ETS-10 from aqueous solution. *Water* **2021**, *13*, 3263. [CrossRef]
31. Langmuir, I. The adsorption of gases on plane surfaces of glass, mica and platinum. *J. Am. Chem. Soc.* **1918**, *143*, 1361–1403. [CrossRef]
32. Freundlich, H.M.F. Over the adsorption in solution. *J. Phys. Chem.* **1906**, *57*, 385–471.
33. Temkin, M. Adsorption equilibrium and the kinetics of processes on non-homogeneous surfaces and in the interaction between adsorbed molecules. *Zh. Fiz. Chim.* **1941**, *15*, 296–332.
34. Hall, K.R.; Eagleton, L.C.; Acrivos, A.; Vermeulen, T. Pore- and Solid-Diffusion Kinetics in Fixed-Bed Adsorption under Constant-Pattern Conditions. *Ind. Eng. Chem.* **1966**, *5*, 212–223. [CrossRef]
35. Ho, Y.S.; Huang, C.T.; Huang, H.W. Equilibrium sorption isotherm for metal ions on tree fern. *Process Biochem.* **2002**, *37*, 1421–1430. [CrossRef]
36. Keskinkan, O.; Göksu, M.Z.L.; Yüceer, A.; Basıbüyük, M.; Forster, C.F. Heavy metal adsorption characteristics of a submerged aquatic plant (*Myriophylum spicatum*). *Process Biochem.* **2003**, *39*, 179–183. [CrossRef]

37. Özer, A.; Özer, D.; Özer, A. The adsorption of copper(II) ions on to dehydrated wheat bran (DWB):determination of the equilibrium and thermodynamic parameters. *Process Biochem.* **2004**, *39*, 2183–2191. [CrossRef]
38. Liu, X.; Chen, Z.-Q.; Han, B.; Su, C.-L.; Han, Q.; Chen, W.-Z. Biosorption of copper ions from aqueous solution using rape straw powders: Optimization, equilibrium and kinetic studies. *Ecotoxicol. Environ. Saf.* **2018**, *150*, 251–259. [CrossRef]
39. Witek-Krowiak, A. Analysis of temperature-dependent biosorption of Cu^{2+} ions on sunflower hulls: Kinetics, equilibrium and mechanism of the process. *Chem. Eng. J.* **2012**, *192*, 13–20. [CrossRef]
40. Dinu, M.V.; Dinu, I.A.; Lazar, M.M.; Dragan, E.S. Chitosan-based ion-imprinted cryo-composites with excellent selectivity for copper ions. *Carbohydr. Polym.* **2018**, *186*, 140–149. [CrossRef]
41. Reddy, K.H.D.; Seshaiaha, K.; Reddy, A.V.R.; Lee, S.M. Optimization of Cd(II), Cu(II) and Ni(II) biosorption by chemically modified Moringa oleifera leaves powder. *Carbohydr. Polym.* **2012**, *88*, 1077–1086. [CrossRef]
42. Putra, W.P.; Kamari, A.; Yusoff, S.N.M.; Fauziah Ishak, C.; Mohamed, A.; Hashim, N.; Md Isa, I. Biosorption of Cu(II), Pb(II) and Zn(II) Ions from Aqueous Solutions Using Selected Waste Materials: Adsorption and Characterisation Studies. *J. Encapsulation Adsorpt. Sci.* **2014**, *4*, 25–35. [CrossRef]
43. Bazargan-Lari, R.; Zafarani, H.R.; Bahrololoom, M.E.; Nemati, A. Removal of Cu(II) ions from aqueous solutions by low-cost natural hydroxyapatite/chitosan composite: Equilibrium, kinetic and thermodynamic studies. *J. Taiwan Inst. Chem. Eng.* **2014**, *45*, 1642–1648. [CrossRef]
44. Laus, R.; Costa, T.G.; Szpoganicz, B.; Favere, V.T. Adsorption and desorption of Cu(II), Cd(II) and Pb(II) ions using chitosan crosslinked with epichlorohydrin-triphosphate as the adsorbent. *J. Hazard. Mater.* **2010**, *183*, 233–241. [CrossRef]
45. Langergren, S. About the theory of so-called adsorption of soluble substance. *Kung. Sven. Veten Hand.* **1898**, *24*, 1–39.
46. Ho, Y.S.; Mckay, G. Pseudo-second order model for sorption processes. *Process Biochem.* **1999**, *34*, 451–465. [CrossRef]
47. Weber, W.J.; Morris, J.C. Kinetics of adsorption on carbon from solution. *ASCE Sanit. Eng. Div. J.* **1963**, *89*, 31–39. [CrossRef]
48. Juang, R.S.; Chen, M.L. Application of the Elovich equation to the kinetics of metal sorption with solvent-impregnated resins. *Ind. Eng. Chem. Res.* **1997**, *36*, 813–820. [CrossRef]
49. Guo, Z.; Liu, X.; Huang, H. Kinetics and thermodynamics of reserpine adsorption onto strong acidic cationic exchange fiber. *PLoS ONE* **2015**, *10*, e0138619. [CrossRef]
50. Kumar, A.; Rout, S.; Ghosh, M.; Singhal, R.K.; Ravi, P.M. Thermodynamic parameters of U (VI) sorption onto soils in aquatic systems. *SpringerPlus* **2013**, *2*, 530. [CrossRef]
51. Lütke, S.F.; Igansi, A.V.; Pegoraro, L.; Dotto, G.L.; Pinto, L.L.A.; Cadaval, T.R.S. Preparation of activated carbon from black wattle bark waste and its application for phenol adsorption. *J. Environ. Chem. Eng.* **2019**, *7*, 103396. [CrossRef]
52. Sundaram, C.S.; Viswanathan, N.; Meenakshi, S. Defluoridation chemistry of synthetic hydroxyapatite at nano scale: Equilibrium and kinetic studies. *J. Hazard. Mater.* **2008**, *155*, 206–215. [CrossRef] [PubMed]

water

MDPI

Article

Time Evolution Study of the Electric Field Distribution and Charge Density Due to Ion Movement in Salty Water

Vasileios Bartzis [1] and Ioannis E. Sarris [2,*]

1 Department of Food Science and Technology, Faculty of Food Science, Campus Alsos Egaleo, University of West Attica, Ag. Spyridonos 28, 12243 Athens, Greece; vbartzis@uniwa.gr
2 Department of Mechanical Engineering, University of West Attica, 12244 Athens, Greece
* Correspondence: sarris@uniwa.gr

Abstract: Desalination and water purification through the ion drift of salted water flow due to an electric field in a duct is perhaps a feasible membrane-free technology. Here, the unsteady modulation of ion drift is treated by employing the Poison–Nernst–Plank (PNP) equations in the linear regime. Based on the solution of the PNP equations, the closed-form relationships of the charge density, the ion concentration, the electric field distribution and its potential are obtained as a function of position and time. It is found that the duration of the ion drift is of the order of one second or less. Moreover, the credibility of various electrical circuit models is examined and successfully compared with our solution. Then, the closed form of the surface charge density and the potential that are calculated without the linear approximation showed that the compact layer is crucial for the ion confinement near the duct walls. To test this, nonlinear solutions of the PNP equations are obtained, and the limits of accuracy of the linear theory is discussed. Our results indicate that the linear approximation gives accurate results only at the fluid's bulk but not inside the double layer. Finally, the important issue of electric field diminishing at the fluid's bulk is discussed, and a potential method to overcome this is proposed.

Keywords: electric field; salt ion drift; water duct flow; diffuse layer thickness

Citation: Bartzis, V.; Sarris, I.E. Time Evolution Study of the Electric Field Distribution and Charge Density Due to Ion Movement in Salty Water. *Water* **2021**, *13*, 2185. https://doi.org/10.3390/w13162185

Academic Editors: Cristina Palet and Julio Bastos-Arrieta

Received: 15 July 2021
Accepted: 8 August 2021
Published: 10 August 2021

1. Introduction

Research on desalination and ion removal methods from a water solution, in general, is of crucial importance for freshwater sustainability. Except for the classical methods, such as the multi-stage flash distillation (MSF), multiple effect distillation and vapor compression distillation [1–11], other methods based on special membranes or electrodes, such as the electrodialysis and the reverse osmosis (RO) electro-deionization, electrophoresis, electroosmosis and capacitive deionization [12–36], have recently been developed. The bottleneck of all the above methods, which are based on the ion drift due to electric fields, is the attempt to resolve the formation of a double layer, i.e., a boundary layer adjacent to the electrodes. Upon its formation, due to ion confinement at the higher potential region of the electrodes, no electric force remains to attract other ions from the fluid's bulk.

Recently, we proposed the electric ion drift (EID) method, which works similarly to the capacitive deionization method but without the use of membranes or special electrodes [37]. Our method is based on the ion drift due to the application of an electric field from a capacitor externally placed from the solution. A similar configuration to the EID method is studied by molecular dynamics simulations and proved that the proposed method can successively drift salt ions in nanochannels [27]. This method works similarly to the capacitive deionization method [38–45], but in its original concept, it is more elegant since no porous electrodes are required. The EID method is of the same energy consumption as the capacitive deionization method of about 0.2 kWh/m^3 and can only be efficient in low salt concentrations as it is presented in our following work [46] for solution equilibrium but without the use of membranes or special electrodes. There, the spatial distributions of the electric field and potential, the ion concentration and also the electric charge confined adjacent to the walls are presented for various applied external fields and duct widths.

An in-depth investigation is performed here for the salty water solution flow in a duct that is submerged into an external electric field using the Poisson–Nernst–Planck (PNP) equations and considering the double-layer formation bottleneck. Both the temporal and the spatial distributions are studied. The model equations and details are described in Section 2. By considering the Stern model, the closed-form solutions of charge density, ion concentration, electric field intensity and potential are presented in the linear regime in Section 3. The spatial and temporal distributions found are compared to existing electric circuit models and the equilibrium distributions of Reference [46].

The linear approximation is incapable of providing reasonable results neither for the surface charge density nor for the electric field or potential in the electrodes nearby area (adjacent to the duct walls), i.e., inside the double layer. This analysis is made in Section 4, and a novel method to calculate the surface charge density and the other field quantities is proposed. Lastly, the non-linear PNP equation is deployed in Section 5, and the conditions under which the linear model is valid are investigated. Finally, in Sections 6 and 7, we proceed to compare our results with any experimental data available and to draw the most important conclusions.

2. Model Equations and Mathematical Evaluation

It is assumed that positive and negative ions of equal concentrations are diluted in a water solution. The salted water flows in a duct, while the duct is subjected to a steady-state and homogeneous electric field created by a pair of plate electrodes, which are under voltage V (Figure 1). The external electric field intensity E has a direction normal to the plate electrodes from positive to negative along the y-axis. The fluid is flowing in the duct perpendicular to the external electric field with velocity $\vec{v}$. Thus, ions are moving due to the combined motion of the water flow along the streamwise direction and along the y-axis due to the electric force with velocity $\vec{v}_y$. The y-axis ionic flux of the positive ions is given by the relation $J_+ = C_+ v_y$, where C_+ is the concentration of the positive ions. From the analysis that is presented in Appendix A and Equation (A5), we have

$$J_+ = -C_+ ze\mu_+ \frac{\partial \varphi}{\partial y} - D_+ \frac{\partial C_+}{\partial y} \tag{1}$$

where z is the number of overflow protons; $e = 1.6 \times 10^{-19}$ Cb; $\mu_+ \equiv \frac{1}{6\pi v r}$ (see Appendix A) is the mobility of the positive ions, with v being the dynamic viscosity of water and r being the effective radius of the ions; φ is the electric potential in the fluid's bulk; and $D_+ = \mu_+ kT$ is the diffusion coefficient of the positive ion. $k \equiv \frac{R}{N_A}$ is the Boltzmann constant; R = 8.314 J/(mol K); N_A is the Avogadro constant; and T is the absolute temperature, which is considered T = 300 K throughout this work.

In the above Equation (1), the electric field is substituted by $E = -\frac{\partial \varphi}{\partial y}$. Following the same procedure, the ionic flux of the negative ions is read as $J_- = +C_- ze\mu_- \frac{\partial \varphi}{\partial y} - D_- \frac{\partial C_-}{\partial y}$, where z is the number of overflow electrons. Considering that both positive and negative ions have the same z, the same mobility $\mu_+ = \mu_- = \mu$ and, thus, $D_+ = D_- = D$, the ionic fluxes can be read more simply as

$$J_\pm = \mp C_\pm ze\mu \frac{\partial \varphi}{\partial y} - D \frac{\partial C_\pm}{\partial y} \tag{2}$$

where the sign of the ionic fluxes indicates the direction of motion and z is considered positive.

The concentration conservation equation may read as

$$\frac{\partial C_\pm}{\partial t} = -\frac{\partial}{\partial y}(J_\pm) = -\frac{\partial}{\partial y}(\mp C_\pm ze\mu \frac{\partial \varphi}{\partial y} - D\frac{\partial C_\pm}{\partial y}) \tag{3}$$

where the concentration $C_{\pm}$ is in (mol/m^3) and the ionic fluxes J in $\text{mol}/(m^2{\cdot}s)$.

The charge density due to the above ion concentration reads as

$$\rho = C_+ zF - C_- zF = zF(C_+ - C_-) \tag{4}$$

and F = 96,485.34 C/mol is the Faraday constant.

By applying the 1D Poisson equation in the y-direction of the duct, it is found that

$$\frac{\partial^2\varphi}{\partial y^2} = -\frac{\rho}{\varepsilon} \rightarrow \frac{\partial^2\varphi}{\partial y^2} = -\frac{zF(C_+ - C_-)}{\varepsilon} \tag{5}$$

The above Equations (3) and (5) constitute the so-called Poisson–Nernst–Planck (PNP) equations system, which is solved in the next Section.

It should be noticed that the present study is in accordance with the Debye–Huckel theory [47–50], also used in References [37,46], where water is considered as a continuous dielectric medium. Thus, the external electric field effect in water is to initiate change in its electric permittivity in the solution to $\varepsilon = \varepsilon_r\varepsilon_0$, where $\varepsilon_r \approx 80$ is the relative permittivity of the water and $\varepsilon_0 = 8.85 \times 10^{-12}\ F/m$.

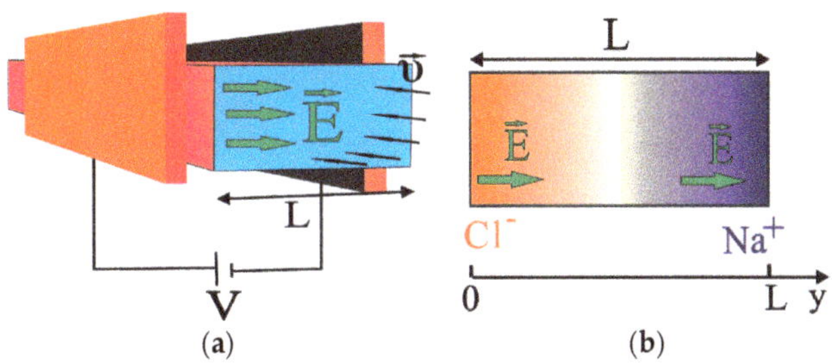

Figure 1. Flow configuration (**a**) and indicative salt ion distribution (**b**) of the present model.

3. PNP Equations Solution for the Stern Model

3.1. Boundary Conditions

For the study of the model, the general Stern model can be used where the ions have finite dimensions and thus can only approach to a small distance from the wall of the duct. The contribution of the Stern model is analytically discussed in Appendix B.

Before discussing the solution of the PNP equations, we must define the appropriate boundary conditions. In the center of the duct, at $y = \frac{L}{2}$, the electric potential is kept equal to zero, $\varphi\left(y = \frac{L}{2}\right) = 0$, while $\varphi(0)$ and $-\varphi(0)$ are the potential at $y = 0$ and $y = L$, respectively. Thus, by considering a linear relation of distance for the potential at the compact part of the double layer adjacent to the two electrodes, we have

$$\varphi = \pm\varphi(0) \pm \lambda_s \frac{\partial\varphi}{\partial y} \quad \text{for } y = 0, L \tag{6}$$

where λ_s is the effective thickness of the compact part of the double layer [51,52] (see Appendix B).

Moreover, since the process is a non-Faradaic one [52,53], because no charge is transferred through the electrodes, the ion fluxes and the current density should be zero at the boundaries, i.e.,

$$J_{\pm} = \mp C_{\pm} ze\mu \frac{\partial\varphi}{\partial y} - D\frac{\partial C_{\pm}}{\partial y} = 0 \text{ for } y = 0, L \tag{7}$$

$$i = zeN_A(J_+ - J_-) = 0 \text{ for } y = 0, L \tag{8}$$

Finally, the initial conditions of the model are such that a uniform ion distribution is applied at $t < 0$, together with a zero potential.

3.2. Linearized Solution of the PNP Equation

We considered that the ion concentration in the fluid's bulk is supposed to slightly change linearly along the y-direction from 0 to L. As the ion concentrations are linearly increased or decreased to opposite walls, the total concentration is constant; i.e., the total charge density is lower than the density of each ion (either positive or negative) [52,54–56]. Thus, most of the solution bulk is quasi-neutral, except in the double layer, i.e., a boundary layer of width of the order of 1 μm or less. Thus, from Equation (3), it is found that

$$\frac{\partial C_+}{\partial t} = +D\frac{\partial^2 C_+}{\partial y^2} + ze\mu C_+\frac{\partial^2 \varphi}{\partial y^2} \quad (9)$$

$$\frac{\partial C_-}{\partial t} = +D\frac{\partial^2 C_-}{\partial y^2} - ze\mu C_-\frac{\partial^2 \varphi}{\partial y^2} \quad (10)$$

Considering that $C_+ + C_- = 2C_M$, where C_M is the concentration of each ion in the center of the duct (equal for the two ions due to symmetry), subtracting the above equations and by recalling that the charge density is given by the relationship:

$$\rho = zF(C_+ - C_-) \quad (11)$$

it is found that

$$\frac{1}{D}\frac{\partial \rho}{\partial t} = \frac{\partial^2 \rho}{\partial y^2} - \kappa^2\rho \quad (12)$$

where

$$\kappa = \sqrt{\frac{2z^2 C_M F^2}{\varepsilon RT}} \quad (13)$$

Thus, the width κ^{-1} corresponds to the so-called Debye length due to the capacitor that may simulate the diffuse layer according to the Gouy–Chapman theory at the edges of the duct. The analytical solution of Equation (12) is given in Appendix C and results in the following relations:

$$\rho = \alpha \sinh\left(\kappa\left(y - \frac{L}{2}\right)\right)\left(1 - e^{-\frac{t}{\tau}}\right) \quad (14)$$

$$\frac{\partial \varphi}{\partial y} = -\frac{\alpha}{\varepsilon\kappa}\cosh\left(\kappa\left(y - \frac{L}{2}\right)\right)\left(1 - e^{-\frac{t}{\tau}}\right) - \alpha\frac{1}{\tau}e^{-\frac{t}{\tau}}\frac{\cosh\left(\kappa\frac{L}{2}\right)}{\varepsilon D\kappa^3} \quad (15)$$

$$\varphi = -\alpha\frac{\sinh\left[\kappa\left(y - \frac{L}{2}\right)\right]}{\varepsilon\kappa^2}\left(1 - e^{-\frac{t}{\tau}}\right) - \alpha\frac{1}{\tau}e^{-\frac{t}{\tau}}\frac{\left(y - \frac{L}{2}\right)}{\varepsilon D\kappa^3}\cosh\left(\kappa\frac{L}{2}\right) \quad (16)$$

where

$$\alpha = \frac{\varphi(0)\varepsilon\kappa^2}{\cosh\left(\kappa\frac{L}{2}\right)\left(\lambda_s\kappa + \tanh\left(\kappa\frac{L}{2}\right)\right)} \quad (17)$$

$$\tau = \frac{L\left(1 + \frac{2\lambda_s}{L}\right)}{2D\kappa\left(\lambda_s\kappa + \tanh\left(\kappa\frac{L}{2}\right)\right)} \quad (18)$$

and

$$\kappa^2 = \frac{2z^2 C_M F^2}{\varepsilon RT} \quad (19)$$

Thus, from the charge density, it is found that the concentrations of the positive and negative ions are varied as

$$C_{+} = C_M + \frac{\alpha \sinh\left(\kappa\left(y - \frac{L}{2}\right)\right)}{2zF}\left(1 - e^{-\frac{t}{\tau}}\right) \tag{20}$$

$$C_{-} = C_M - \frac{\alpha \sinh\left(\kappa\left(y - \frac{L}{2}\right)\right)}{2zF}\left(1 - e^{-\frac{t}{\tau}}\right) \tag{21}$$

Validation of the Electric Circuit Models for Long Ducts

In case of a duct of width of mm order or wider, $L \geq 10^{-3}$ m, and since $\lambda_s \sim 1 - 10\ \dot{A}$, it is found for the rate $\frac{\lambda_s}{L} \leq 10^{-6}$ and, thus, may be considered almost equal to zero. Moreover, for such length of the duct and since $\kappa \sim 10^8\ m^{-1}$, it can also be considered that $\tanh\left(\kappa\frac{L}{2}\right) \approx 1$. Thus, after these simplifications:

$$\tau = \frac{L}{2D\kappa(\lambda_S \kappa + 1)} \tag{22}$$

$$\alpha = \frac{\varphi(0)\varepsilon\kappa^2}{\cosh\left(\kappa\frac{L}{2}\right)(\lambda_S \kappa + 1)} \tag{23}$$

As it may be observed, the quantities in Equations (14)–(21) are exponentially varied, and τ is a characteristic time, the so-called time constant. Here, the above theory fits to the Gouy–Chapman and Stern theories that consider the solution as an RC circuit (see Appendix B), because the characteristic time of the above analysis is exactly the same as the time of the RC circuit (compare Equations (22) and (A10)). Since the fluid bulk is in quasi-neutral condition, its resistance per unit surface is equal to $R = \frac{L(\kappa^{-1})^2}{D\varepsilon}$ (Equation (A7)), and for the time constant effective capacitor $\tau = RC^{tot}$ (due to the charge confinement at the edges), we have

$$C^{tot} = \frac{\varepsilon}{2(\lambda_S + \kappa^{-1})} \tag{24}$$

As it may be observed, Equation (24) is exactly the same as the Stern's model (see Appendix B) as would be expected. Thus, we have shown that the electric model in long ducts is equal to the linear PNP equation solution, and it is valid under the same assumptions for the solution. The time duration to the end of the phenomenon is plotted in Figure 2 for the case of Na ions, for the salted water solution of NaCl ($D_{Na} = 1.3 \times 10^{-9}\ \frac{m^2}{s}$), as a function of the initial concentration C_M and various duct widths. As always for the exponentially decay variations, equilibrium is expected after 5τ. Thus, here it will also be achieved in less than a second at the most as it is observed in Figure 2.

3.3. Long Time Behavior

As the time increases, then $(1 - e^{-\frac{t}{\tau}}) \to 1$ and $e^{-\frac{t}{\tau}} \to 0$, and Equations (14)–(21) have their equilibrium forms as

$$\rho = \alpha \sinh\left(\kappa\left(y - \frac{L}{2}\right)\right) \tag{25}$$

$$\frac{\partial \varphi}{\partial y} = -\frac{\alpha}{\varepsilon\kappa}\cosh\left(\kappa\left(y - \frac{L}{2}\right)\right) \tag{26}$$

$$\varphi = -\alpha \frac{\sinh\left[\kappa\left(y - \frac{L}{2}\right)\right]}{\varepsilon\kappa^2} \tag{27}$$

$$C_{+} = C_M + \frac{\alpha \sinh\left(\kappa\left(y - \frac{L}{2}\right)\right)}{2zF} \tag{28}$$

$$C_- = C_M - \frac{\alpha \sinh\left(\kappa\left(y - \frac{L}{2}\right)\right)}{2zF} \tag{29}$$

The equilibrium solutions of Equations (25)–(29) can be easily verified by the recent results of Reference [46] for the present water solution with $\kappa = 1.027 \times 10^8 z C_M^{\frac{1}{2}}$ (in S.I. (System International)) (from Equation (19)) and $z_{Na} = z_{Cl} = 1$. For $L = 0.0015$, 0.015, 0.15 m, where $\tanh\left(\kappa\frac{L}{2}\right) \approx 1$ and thus $\left(\sinh\left(\kappa\frac{L}{2}\right) \approx \cosh\left(\kappa\frac{L}{2}\right)\right)$ and for $C_M = 8.6$, 100, 400 mol/m^3, and by considering the effect as negligible due to the compact layer, i.e., $\lambda_S\kappa \to 0$, Equations (25)–(27) are exactly the same as in Reference [46]. However, the concentration Equations (28) and (29) are different than the ones of Reference [46] that read as

$$C_+ = C_M e^{-\frac{F\varphi(0)}{RT}\frac{\sinh[\kappa(\frac{L}{2}-y)]}{\sinh(\kappa\frac{L}{2})}} \tag{30}$$

$$C_- = C_M e^{+\frac{F\varphi(0)}{RT}\frac{\sinh[\kappa(\frac{L}{2}-y)]}{\sinh(\kappa\frac{L}{2})}} \tag{31}$$

The first observation is that Equations (28) and (29) are expected to differ from Equations (30) and (31) of Reference [46] due to the linearization of the exponential term ($e^x = 1 + x + \frac{x^2}{2!} + \frac{x^3}{3!} + \cdots$). This happens when $C_+ + C_- = 2C_M$, which is the same as the linearization of the exponential function when only the first two terms of the Taylor expansion are kept. Thus, Equations (30) and (31) are more accurate and are used for comparison in the next sections where nonlinear terms are considered for the PNP equation. Moreover, the linear approximation is valid along the duct width, Reference [46], for

$$\left|\frac{F}{RT}\varphi\right| < 1 \text{ or } \varphi(0) \le 0.026 \text{ V} \tag{32}$$

which is also applied for the linearized PNP solution. By using Equations (25)–(29) and after some calculus, it can easily be determined that the electric potential and electric field become zero at the fluid bulk, as the width of the duct, the initial concentration and the potential $\varphi(0)$ are increased as is also discussed in Reference [46]. As it is illustrated below, the linear approximation gives reasonable results at the fluid's bulk.

Furthermore, the surface charge density (in C/m^2) can be found by integrating Equation (25) of the linearized PNP solution as

$$\sigma_l = \int_0^y \rho dy = \frac{\varphi(0)\varepsilon\kappa}{(\lambda_S\kappa + 1)}\left[\frac{\cosh\left(\kappa\left(y - \frac{L}{2}\right)\right)}{\cosh\left(\kappa\frac{L}{2}\right)} - 1\right],$$

and its absolute value adjacent to each electrode is given for $y = \frac{L}{2}$ as

$$|\sigma_l| = \frac{\varphi(0)\varepsilon\kappa}{(\lambda_s\kappa + 1)}\left(\frac{1}{\cosh\left(\kappa\frac{L}{2}\right)} - 1\right) \tag{33}$$

Based on the above approximations, $\frac{1}{\cosh\left(\kappa\frac{L}{2}\right)} \to 0$, the surface charge density is found to be exactly the same as the electric circuit RC model as

$$|\sigma_l| = \varphi(0)\frac{\varepsilon}{(\lambda_S + \kappa^{-1})} = \varphi(0)C \tag{34}$$

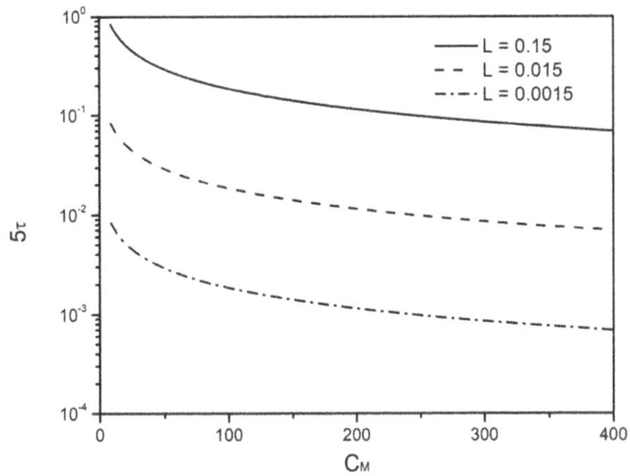

Figure 2. Time constant variation 5τ as a function of C_M for various duct widths.

4. Exact Evaluation of the Electric Field and the Surface Charge Density

The calculation of the electric field and the surface charge density under steady-state conditions is presented below without the linear approximation. Starting from the electric field definition:

$$E = \frac{-\partial\varphi}{\partial y} \rightarrow \frac{\partial E}{\partial y} = -\frac{\partial^2\varphi}{\partial y^2} \rightarrow \frac{\partial E}{\partial \varphi}\frac{\partial \varphi}{\partial y} = \frac{\rho}{\varepsilon} \rightarrow -2E\frac{\partial E}{\partial \varphi} = \frac{2\rho}{\varepsilon} \rightarrow d\left(E^2\right) = -\frac{2\rho}{\varepsilon}d\varphi,$$

and by considering $\varphi\left(y = \frac{L}{2}\right) = 0$ and E_M the electric field intensity is in the center of the duct, and by integrating it, it is found that

$$\int_{E_M}^{E} d(E^2) = -\int_0^{\varphi} \tfrac{2\rho}{\varepsilon} d\varphi \rightarrow E^2 - E_M^2 = \tfrac{4}{\varepsilon}C_M F \int_0^{\varphi} \sinh(\tfrac{zF}{RT}\varphi)d\varphi \rightarrow$$
$$E = \sqrt{E_M^2 + \tfrac{4C_M RT}{\varepsilon}\left[\cosh(\tfrac{zF}{RT}\varphi) - 1\right]}$$

where

$$\rho = C_+ zF - C_- zF \rightarrow \rho = -2C_M F \sinh(\frac{zF}{RT}\varphi)$$

which is valid only inside the diffuse layer and not in the compact layer, thus, in regions where the Boltzmann distribution is valid. In this region, E_M is almost diminished relative to the values of the second part of the undersquare quantity; thus,

$$E \cong \sqrt{\frac{4C_M RT}{\varepsilon}[\cosh(\frac{zF}{RT}\varphi) - 1]} = \sqrt{\frac{8C_M RT}{\varepsilon}}\left|\sinh\left(\frac{zF}{2RT}\varphi\right)\right|$$

and close to the positive electrode:

$$E = \sqrt{\frac{8C_M RT}{\varepsilon}}\sinh(\frac{zF}{2RT}\varphi) = -\frac{\partial\varphi}{\partial y} \text{ for } \lambda_s \le y \le \frac{L}{2} \tag{35}$$

The integration of the 1D Poisson equation from $y = \lambda_S$ to $y = \frac{L}{2}$ can give the surface charge density near the walls as $\int_{\lambda_S}^{\frac{L}{2}} \frac{d^2\varphi}{dy^2}dy = -\int_{\lambda_S}^{\frac{L}{2}} \frac{\rho}{\varepsilon}dy \rightarrow \left.\frac{d\varphi}{dy}\right|_{\frac{L}{2}} - \left.\frac{d\varphi}{dy}\right|_{\lambda_S} = -\frac{\int_{\lambda_S}^{L/2}\rho dy}{\varepsilon} =$

$-\frac{\sigma}{\varepsilon}$, where σ is the surface charge density (in C/m^2) near the positive electrode. Applying that $E_M = \left.\frac{d\varphi}{dy}\right|_{\frac{L}{2}} = 0$, it is $-\left.\frac{d\varphi}{dy}\right|_{\lambda_S} = -\frac{\sigma}{\varepsilon} \rightarrow \sigma = -\varepsilon E_{\lambda_S}$, which due to Equation (35) gives

$$\sigma = -\sqrt{8\varepsilon C_M RT}\sinh(\frac{zF}{2RT}\varphi_s) \quad (36)$$

where φ_s is the potential at the outer Helmholtz plane (OHP) at distance λ_S.

It should be noticed that due to continuity, the electric field magnitude is equal at both sides of the OHP, i.e., of the diffuse layer but also of the compact layer side. Moreover, electric field magnitude is almost constant along the compact layer. Thus, for the positive electrode, it is $\varphi_S = \varphi(0) + \lambda_s \left.\frac{d\varphi}{dy}\right|_{\lambda_S} = \frac{\sigma\lambda_s}{\varepsilon} + \varphi(0)$, and by replacing φ_S in Equation (36), we have

$$\sigma = -\sqrt{8\varepsilon C_M RT}\sinh[\frac{zF}{2RT}(\frac{\sigma\lambda_s}{\varepsilon} + \varphi(0))] \quad (37)$$

or after substituting

$$\sigma = -3.76 \cdot 10^{-3} C_M^{\frac{1}{2}} \sinh[z(19.34\varphi(0) + 13.65\sigma)] \text{ (S.I.)} \quad (38)$$

The distribution of the surface charge density as a function of $\varphi(0)$ is shown in Figure 3 for various concentrations considering the case of salty water, $z_{Na} = z_{Cl} = 1$. As may be seen by comparing these results with the those from Reference [46], where no compact layer was considered, the deployment of the compact layer results in a reduced charge. Moreover, the charge is practically independent from the solution salt concentration.

As it may be observed, the relation (38) is different than Equation (34) due to linearized PNP equations or the electric models. However, it collapses to this upon linearization, i.e., for $\sinh\left[\frac{zF}{2RT}\left(\frac{\sigma\lambda_s}{\varepsilon} + \varphi(0)\right)\right] \approx \frac{zF}{2RT}\left(\frac{\sigma\lambda_s}{\varepsilon} + \varphi(0)\right)$, something that can not be applied in general at the double layer. Now, the potential φ_s at the outer Helmholtz plane is

$$\varphi_S = \varphi(0) - \lambda_s\sqrt{\frac{8C_M RT}{\varepsilon}}\sinh(\frac{zF}{2RT}\varphi_s) \quad (39)$$

or

$$\varphi_S = \varphi(0) - 2.65 \times 10^{-3} C_M^{\frac{1}{2}} \sinh(19.34 \cdot z\varphi_s) \text{ (S.I)} \quad (40)$$

The potential at the outer Helmholtz plane is shown in Figure 4 as a function of $\varphi(0)$ for various concentrations considering the case of salty water, $z_{Na} = z_{Cl} = 1$. As it can be observed, φ_S is scaled almost linear with $\varphi(0)$ for values up to 0.1 V. However, as $\varphi(0)$ is increased further, φ_S is saturated to values lower than 1 V independently of $\varphi(0)$.

Now, the potential distribution along y in the diffuse layer for $\lambda_s \leq y \leq \frac{L}{2}$ can be found from Equation (35) as

$$\int_{\varphi_s}^{\varphi} \frac{d\varphi}{\sinh(\frac{F}{2RT}\varphi)} = -\sqrt{\frac{8C_M RT}{\varepsilon}} \int_{\lambda_s}^{y} dy \rightarrow \varphi = \frac{4RT}{zF}\tanh^{-1}\{\tanh(\frac{zF\varphi_s}{4RT}) \cdot e^{-\kappa(y-\lambda_s)}\} \quad (41)$$

Thus, the electric field intensity inside the compact layer is given by

$$E_{\lambda_S} = -\frac{\sigma}{\varepsilon} \rightarrow E_{\lambda_S} = \sqrt{\frac{8C_M RT}{\varepsilon}}\sinh[\frac{zF}{2RT}(-\lambda_s E_{\lambda_S} + \varphi(0))] \quad (42)$$

or

$$E_{\lambda_S} = 5.31 \times 10^6 C_M^{\frac{1}{2}} \sinh[z\left(19.34\varphi(0) - 9.67 \times 10^{-9}\ E_{\lambda_S}\right)] \text{ (S.I)} \quad (43)$$

As it is observed from Figure 5 for the case of salty water, $z_{Na} = z_{Cl} = 1$, the electric field intensity at the compact layer takes huge values independently of the ion concentration, especially as the potential increases.

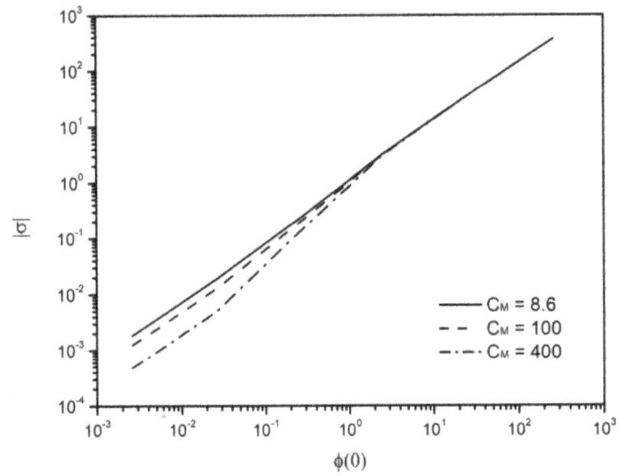

Figure 3. Surface charge density dependance on $\varphi(0)$ at the outer Helmholtz plane for various C_M.

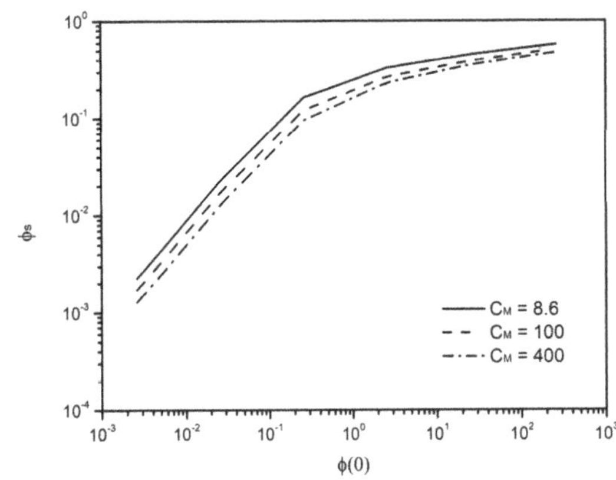

Figure 4. Potential φ_S dependance on $\varphi(0)$ at the outer Helmholtz plane for various C_M.

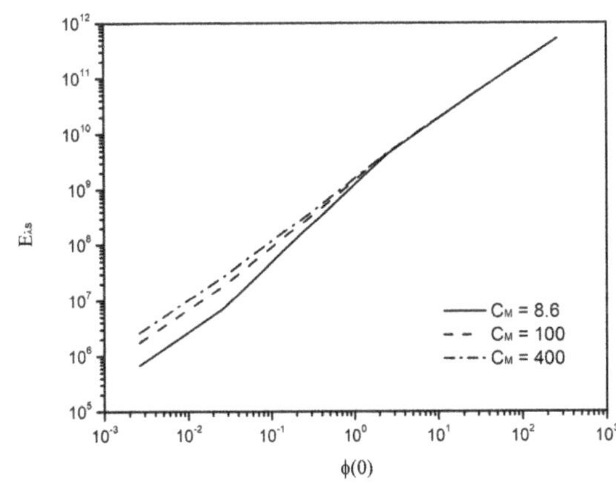

Figure 5. Electric field intensity inside the compact layer dependance on $\varphi(0)$ for various C_M.

5. Nonlinear PNP Solution

Starting from $\rho = zF(C_+ - C_-)$ and setting Equation (3) in the form

$$\frac{\partial C_+}{\partial t} = +D\frac{\partial^2 C_+}{\partial y^2} + ze\mu C_+\frac{\partial^2 \varphi}{\partial y^2} + ze\mu\frac{\partial \varphi}{\partial y}\frac{\partial C_+}{\partial y} \tag{44}$$

$$\frac{\partial C_-}{\partial t} = +D\frac{\partial^2 C_-}{\partial y^2} - ze\mu C_-\frac{\partial^2 \varphi}{\partial y^2} - ze\mu\frac{\partial \varphi}{\partial y}\frac{\partial C_-}{\partial y} \tag{45}$$

and by subtracting, we have

$$\frac{1}{D}\frac{\partial \rho}{\partial t} = \frac{\partial^2 \rho}{\partial y^2} - \frac{z^2F^2(C_+ + C_-)}{\varepsilon RT}\rho - \frac{z^2F^2}{RT}E\frac{\partial(C_+ + C_-)}{\partial y} \tag{46}$$

The above Equation (46) can be transformed into the linear Equation (12) subject to $C_+ + C_- = constant$. Since along the y-direction the ion concentration is changing, the decrease in the positive ion needs to correspond to the increase in the negative ion for their summary to be constant and equal to $2C_M$, i.e., $C_+ + C_- = 2C_M$.

This is valid throughout the duct width (for the case of salty water, $z_{Na} = z_{Cl} = 1$) when $\left|\frac{F}{RT}\varphi\right| < 1$ or $\varphi \leq 0.026$ V, i.e., in the case of very weak external electric fields as already proved in Reference [46]. However, this is true only in the solution's bulk as the applied potential increases and faults inside the double layer. Since we proved that the time scale of the problem is of the order of a second, we study the final equilibrium state of the ion drift. Similar to Section 3.3, we can assume that the final concentration profiles upon equilibrium that came after the solution of Equations (44)–(46) are of the form

$$C_+ = C_M e^{+\frac{Fz\varphi(0)}{RT(\lambda_S\kappa + \tanh(\kappa\frac{L}{2}))}\frac{\sinh[\kappa(y-\frac{L}{2})]}{\cosh(\kappa\frac{L}{2})}} \tag{47}$$

and

$$C_- = C_M e^{-\frac{Fz\varphi(0)}{RT(\lambda_S\kappa + \tanh(\kappa\frac{L}{2}))}\frac{\sinh[\kappa(y-\frac{L}{2})]}{\cosh(\kappa\frac{L}{2})}} \tag{48}$$

The effect of the nonlinearity can be investigated when a third term in the Taylor series of the exponential function, $e^x = 1 + x + \frac{x^2}{2!} + \frac{x^3}{3!} + \cdots$ is considered. Then, for the additional term, it is

$$C_+ + C_- = C_M(2 + W),\text{ and } \frac{\partial(C_+ + C_-)}{\partial y} = C_M Q\kappa\sinh[2\kappa\left(y - \frac{L}{2}\right)], \tag{49}$$

where $W = [\frac{Fz\varphi(0)\sinh\left[\kappa\left(y-\frac{L}{2}\right)\right]}{RT\left(\lambda_s\kappa+\tanh\left(\kappa\frac{L}{2}\right)\right)\cosh\left(\kappa\frac{L}{2}\right)}]^2$, and $Q = [\frac{Fz\varphi(0)}{RT\left(\lambda_s\kappa+\tanh\left(\kappa\frac{L}{2}\right)\right)\cosh\left(\kappa\frac{L}{2}\right)}]^2$.

Thus, from Equation (46) and by setting $\frac{\partial \rho}{\partial t} = 0$ for the steady-state, it is

$$0 = \frac{\partial^2 \rho}{\partial y^2} - \frac{z^2F^2}{\varepsilon RT}C_M(2 + W)\rho - YEC_M,\text{ where } Y = \frac{z^2F^2}{RT}Q\kappa\sinh[2\kappa\left(y - \frac{L}{2}\right)] \tag{50}$$

The shift of the linearity is due to W and Y. It will be useful to study their profiles along y for various potentials $\varphi(0)$ (for the case of salty water, $z_{Na} = z_{Cl} = 1$). In the range of 0.026, 0.2 and 2 V, concentrations are in the range of 8.6, 100 and 400 mol/m^3 with a duct width of L = 0.015 m, while the effect of L is insignificant.

As it is observed in Figures 6 and 7, only inside the compact layer can a deviation from linearity be encountered, while the difference is very negligible outside this layer as one would expect. Moreover, the width of the deviation is found to be inversely proportional to C_M as it is also discussed in Reference [46].

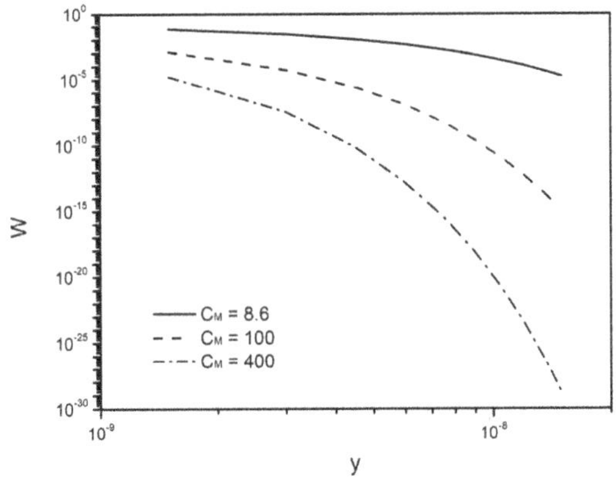

(**a**)

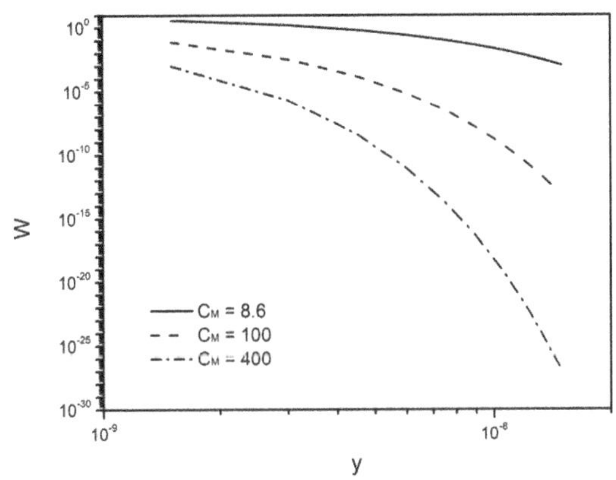

(**b**)

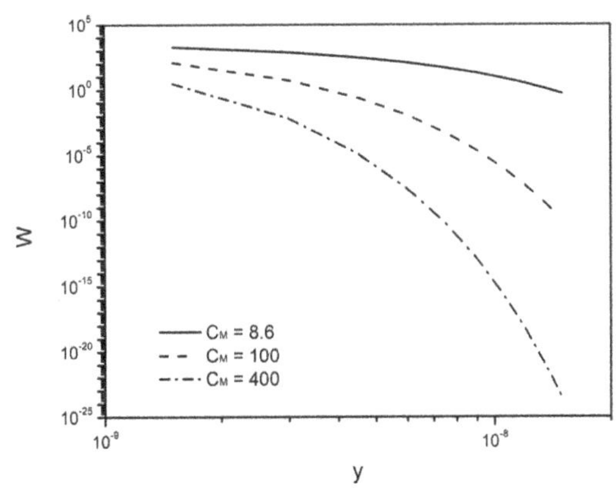

(**c**)

Figure 6. Effect of the nonlinearity on W at $\varphi(0) = 0.026$ (**a**), 0.2 (**b**) and 2 (**c**) along y for various C_M.

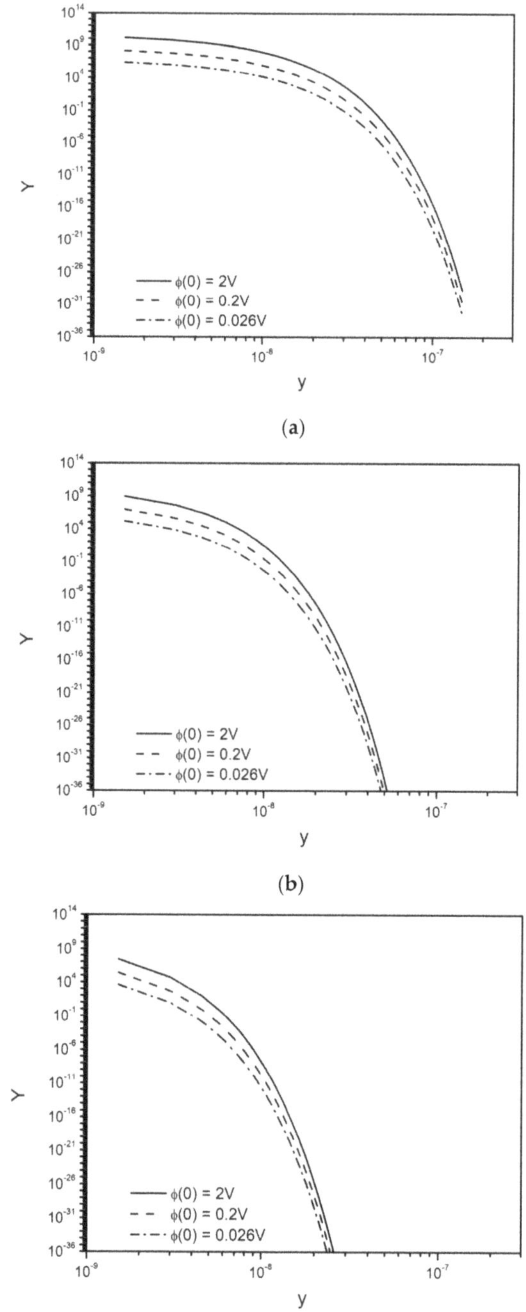

(c)

Figure 7. Effect of the nonlinearity on Y along y for various $\varphi(0)$ at $C_M = 8.6$ mol/m^3 (**a**), $C_M = 100$ mol/m^3 (**b**) and $C_M = 400$ mol/m^3 (**c**).

6. Comparison with Experimental Results

The only available and relevant experimental data are from the capacitive deionization method and are obtained for the water desalination case. A typical duct width of the order L = 1.5 mm is proposed and constructors indicate a performance of about 65% for low salt concentrations. In a similar approach, the authors of Reference [40] succeeded in reducing an initial ion concentration of $C_{bef} = 1.686$ gr/L $= 29$ mole/m^3 to the final concentration of $C_{after} = 0.497$ gr/L $= 8.6$ mole/m^3 with V = 0.4 V; thus, the electric charge that it is removed per surface area is about $|\sigma| = 2967$ Cb/m^2. This large ion drift can only be achieved by the existence of an electric field in the fluid bulk. This is due to the micropores in the capacitive deionization method. However, this is not possible here due to the non-Faradaic conditions to which our analysis refers. Clearly, the phenomenon of ion drift stops long before we have a noticeable reduction in ion in the bulk of the solution. This is due to the zeroing of the electric field in the bulk due to the creation of the double layer on the side walls of the duct. Thus, we need to continuously absorb the compact layer in order to permit additional ions to move close to the duct sidewalls. This can be conducted in an almost continuous way because as we showed, the time to reach the final equilibrium state is of the order of one second or less. This is equivalent to maintaining a non-zero electric field at the fluid's bulk. This mechanism of ion removal from the salted water is discussed in Reference [37] since in this case, the solution had a constant electric field, which is impossible upon creation of the double layer, and in order for the electric field to exist, we have to continuously eliminate this layer. Thus, electric field existence in the duct is the crucial parameter to ensure that an efficient desalination can be achieved by this method. The absorption of the double layer replaces the porous electrodes and if achieved experimentally may be the solution for continuous desalination without the inevitable interruptions that must occur in capacitive deionization to discharge the electrodes.

7. Conclusions

This study examined the ion concentration, electric field, electric potential profiles and surface charge density inside a salt solution when it is under the effect of an external electric field as a function of time and position. The compact layer is discussed as it forms an additional mechanism to diminish the electric field in the bulk of the fluid. The conditions and the region where the linear approximation gives satisfactory results are examined.

The energy consumption of conventional methods using membranes is in the order of 0.30–2.11 kWh/m^3, quite costly but provides desalination of a large scale. With a capacitive deionization method, we have a smaller energy consumption of about 0.2 kWh/m^3. At present with this method, we can manage smaller amounts of water, and it can be used as a secondary desalination method. Research continues intensively on this method mainly with regard to electrodes in order to achieve greater absorbency. With the EID method, we have the same energy consumption, but we are exempt from the cost of electrodes.

The importance of maintaining a non-zero electric field in the fluid's bulk is shown, which can only be succeeded by the continuous absorption of the double layer.

Author Contributions: Conceptualization, V.B.; methodology, V.B.; software, I.E.S.; investigation, V.B.; writing—original draft preparation, V.B.; writing—review and editing, I.E.S.; visualization, I.E.S. All authors have read and agreed to the published version of the manuscript.

Funding: This research received no external funding.

Institutional Review Board Statement: Not applicable.

Informed Consent Statement: Not applicable.

Data Availability Statement: Not applicable.

Conflicts of Interest: The authors declare no conflict of interest.

Nomenclature

$C_{\pm}$	Concentration, mole/m^3
μ	Mobility, s/Kg
D	Diffusion coefficient, m^2/s
J	Ionic flux, mol/(m^2·s)
i	Current density, A/m^2
T	Absolute temperature, K
z	Number of overflow protons or electrons
E	Electric field intensity V/m
y	y-axis coordinate, m
L	Width of the duct, m
σ	Electric charge surface density C/m^2
C	Capacity, F
Greek symbols	
ε	electric permittivity, F/m
φ	Electric potential, V
ρ	Charge density, C/m^3
v	Velocity, m/s
ν	Dynamic viscosity, Pa s
σ	Surface charge density, C/m^2
Subscripts	
y	Along y-axis
bef	Before
after	After

Constants

$N_A = 6.023 \times 10^{23}$ mol^{-1}
$F = 96,485.34$ C/mol
$R = 8.314 \frac{\text{J}}{\text{mol·K}}$
$\varepsilon_0 = 8.85 \times 10^{-12}$ F/m
$\varepsilon_r \approx 80$
$e = 1.6 \times 10^{-19}$ Cb

Appendix A. General Theory of Ion Movement

It is supposed that a positive ion of concentration C_+ is dissolved in a water solution that is flowing in the streamwise direction of a duct (Figure 1). Moreover, a uniform electric field of magnitude E is applied normal to the flow, and the corresponding force acting on the ions is

$$f^{E}_{(1\ ion)} = zeE \tag{A1}$$

Then, the result is that ions start to drift. Thus, a spatial concentration distribution along the y-direction is formed, which causes the concentration gradient $\frac{\partial C_+}{\partial y}$. The pseudoforce per ion due to this is given by the relation:

$$f^{\nabla C}_{(1\ ion)} = -\frac{RT}{N_A C_+}\frac{\partial C_+}{\partial y} \tag{A2}$$

The negative sign indicates that the force direction is opposite to the concentration gradient direction, along the y-direction. Moreover, during the ion movement, a friction force also exists between the moving ion and the solvent. This force is opposite to the ion velocity $\vec{v}_y$, which is given approximately by the relation (we consider as usual the ion shape as a small sphere with radius r that moves with a small velocity inside a fluid of dynamic viscosity ν):

$$f^{fr}_{(1ion)} = -6\pi\nu r \cdot v_y \tag{A3}$$

Quickly, equilibrium is established; thus,

$$\Sigma f^{y}_{(1ion)} = 0 \quad \text{(A4)}$$

By using Equations (A1)–(A3), the velocity v_y can be found as

$$v_y = \frac{zeE}{6\pi v r} - \frac{RT}{C_+ N_A 6\pi v r}\frac{\partial C_+}{\partial y} \quad \text{(A5)}$$

Appendix B. The RC Models—Time Scale

In Electrochemistry, there is an old theoretical approximation that when an electrolyte is under the effect of an external electric field, it is modeled as an effective electric circuit constituted by one capacitor and one resistance connected inline [52]. In the following, the most important models are reviewed.

Appendix B.1. Gouy-Chapman Model

The very first model used was attributed to Gouy–Chapman [52] in which only the diffuse layer that is formed from the ions' confinement nearby the electrodes' surface is considered. This thin layer can be simulated by a capacitor with a plate separation width: $\lambda_D = \kappa^{-1} = \sqrt{\frac{\varepsilon RT}{2F^2 z^2 C_M}}$. Thus, the capacity per unit area is equal to $C_D = \frac{\varepsilon}{\kappa^{-1}}$. Since an additional capacitor is formed at the other wall of the duct, the total capacity per unit area is then

$$C_D^{tot} = \frac{\varepsilon}{2(\kappa^{-1})} \quad \text{(A6)}$$

Assuming that the bulk of the solution is in an almost uniform initial ion concentration, the electric force due to the external electric field on the ions is fast balanced by the friction force, and their asymptotic velocity can be found from (A4) when $\frac{\partial C_+}{\partial y} = 0$ for the positive ions as in $v_y = -\frac{ze}{6\pi v r}\frac{\partial \varphi}{\partial y}$. Their ionic flux is also given by $J_+ = C_M v_y = -C_M \frac{ze}{6\pi v r}\frac{\partial \varphi}{\partial y}$, while $J_- = C_M \frac{ze}{6\pi v r}\frac{\partial \varphi}{\partial y}$ is similarly given for the negative one. The electric current per unit surface is given by $i = zeN_A(J_+ - J_-)$ or substituting $i = -\frac{\varepsilon D}{(\kappa^{-1})^2}\frac{\partial \varphi}{\partial y}$. Comparing this relationship with the Ohm's law $i = -\sigma\frac{\partial \varphi}{\partial y}$, the special conductivity is given by the relation $\sigma = \frac{\varepsilon D}{(k^{-1})^2}$. The resistance per unit surface is given by $R = \frac{L}{\sigma}$ and, thus,

$$R = \frac{L(\kappa^{-1})^2}{D\varepsilon} \quad \text{(A7)}$$

The time constant of the *RC* circuit determines that characteristic time of the phenomenon as

$$\tau = RC_D^{tot} = \frac{L\kappa^{-1}}{2D} \quad \text{(A8)}$$

while 5τ is its practical duration according to the RC circuit theory. In the above Gouy–Chapman theory, only the diffuse layer and no compact layer is considered.

Appendix B.2. Stern's Model

Assuming that ions have finite dimensions and, thus, that they can approach in an ionic radius distance to the walls of the duct that is of the order $1\dot{A}$ (further increased by the solvent molecules that surround the ion), the Stern model needs to be used. According to this model, the double layer is formed by two areas:

a. The compact part that is simulated by a Helmholtz capacitor of effective width of the order of molecule that is almost constant. The term effective is due to the possibility that the solvent (i.e., the water here) may not be considered in this region as a continuous media, and the permittivity ε is considered to be similar to the fluid's bulk. Thus, the width of the compact part may be considered the size of the ionic radius, i.e., $\lambda_s \sim 1 - 10\ \dot{A}$ [51], while in the present calculations, it is assumed to be equal to $\lambda_s \approx 5\dot{A}$. Thus, the capacity of the compact part per unit area is given by $C_H = \frac{\varepsilon}{\lambda_S}$.

b. The diffuse layer that is simulated by a Gouy–Chapman-type capacitor with effective width of the diffuse part $\lambda_D = \kappa^{-1}$, and the known capacity per area unit is given by $C_D = \frac{\varepsilon}{\kappa^{-1}}$.

The above two capacitors are both inline connected and connected with the two capacitors existed at the other wall side (near the second electrode). The two electrodes constitute the Stern model [52]. Thus, the total capacity per surface unit is then

$$C^{tot} = \frac{\varepsilon}{2(\lambda_S + \kappa^{-1})} \tag{A9}$$

The time constant of the RC circuit of the Stern model is then

$$\tau = \frac{L}{2D\kappa(\lambda_s\kappa + 1)} \tag{A10}$$

Appendix C. Solution of the Linear PNP Equation

Using the Laplace transformation in Equation (12) and assuming $\rho(t = 0) = 0$ for the initial current density, we have

$$\frac{\partial^2}{\partial y^2}\overline{\rho} - M^2\overline{\rho} = 0 \tag{A11}$$

where

$$M^2 = \kappa^2 + \frac{S}{D} \tag{A12}$$

and $\overline{\rho} = \int_0^\infty dte^{-St}\rho dt$ is the Laplace transformation of ρ. The solution of Equation (A11) is of the form

$$\overline{\rho} = B\cosh(My') + A\sinh(My') \tag{A13}$$

where $y' = y - \frac{L}{2}$, and A, B are constants. Due to antisymmetric conditions around the duct center at $y = \frac{L}{2}$, it is $\overline{\rho}(y') = -\overline{\rho}(-y')$, and, thus, $B = 0$, i.e.,

$$\overline{\rho} = A\sinh\left(M\left(y - \frac{L}{2}\right)\right) \tag{A14}$$

where

$$M^2 = \kappa^2 + \frac{S}{D} \tag{A15}$$

In order to determine the A constant, the Laplace transformed Poisson Equation (5) is used, $\frac{\partial^2\overline{\varphi}}{\partial y^2} = -\frac{\overline{\rho}}{\varepsilon}$, and integrated once:

$$\frac{\partial\overline{\varphi}}{\partial y} = -\frac{A}{\varepsilon M}\cosh\left(M\left(y - \frac{L}{2}\right)\right) + \Gamma \tag{A16}$$

where Γ is a constant to be determined by considering the current density, $i = zeN_A(J_+ - J_-)$, and it vanishes at the duct walls. Using Equation (12), the current density can be found as

$$i = -\kappa^2\varepsilon D\frac{\partial\varphi}{\partial y} - D\frac{\partial\rho}{\partial y} \tag{A17}$$

While the Laplace transformed Equation (A17) reads as

$$\bar{i} = -\kappa^2 \varepsilon D \frac{\partial \overline{\varphi}}{\partial y} - D \frac{\partial \overline{\rho}}{\partial y} \quad \text{(A18)}$$

and $\bar{i}(y=0) = 0$. Substituting $\overline{\rho}$ and $\frac{\partial \overline{\varphi}}{\partial y}$ from Equations (A14) and (A16), respectively, the constant Γ can be found as

$$\Gamma = \frac{AM}{\varepsilon}\left(M^{-2} - \kappa^{-2}\right)\cosh\left(M\frac{L}{2}\right) \quad \text{(A19)}$$

Moreover, integrating Equation (A16) and using the antisymmetric condition, $\overline{\varphi}(y) = \overline{\varphi}(L-y)$, we find $\overline{\varphi} = -\frac{A}{\varepsilon M^2}\sinh\left[M\left(y-\frac{L}{2}\right)\right] + \Gamma(y-L)$, and from Equation (A19):

$$\overline{\varphi} = -\frac{A\cosh\left(M\frac{L}{2}\right)}{\varepsilon M^2}\left\{\frac{\sinh\left[M\left(y-\frac{L}{2}\right)\right]}{\cosh\left[M\frac{L}{2}\right]} + \frac{MS\left(y-\frac{L}{2}\right)}{D\kappa^2}\right\} \quad \text{(A20)}$$

The constant A can be obtained by the boundary condition at $y = 0$ from the transformed Equation (6), $\overline{\varphi} = +\varphi(0)s^{-1} + \lambda_s\frac{\partial \overline{\varphi}}{\partial y}$, when $\overline{\varphi}$ and $\frac{\partial \overline{\varphi}}{\partial y}$ are substituted by Equations (A20) and (A16) for $y = 0$. Thus, A is found as

$$A = \frac{\varphi(0)s^{-1}\varepsilon M^2}{\cosh\left(M\frac{L}{2}\right)}\left[\frac{1}{\frac{MSL}{2D\kappa^2}\left(1+\frac{2\lambda_s}{L}\right) + \lambda_s M + \tanh\left(M\frac{L}{2}\right)}\right] \quad \text{(A21)}$$

The Debye time is defined as $\tau_D = \frac{\lambda_D^2}{D} = \left(\kappa^2 D\right)^{-1}$ and is of the order $\tau_D \sim 10^{-7}$s, while we are interested in system responses at higher time scales. From the Laplace transformation, $\overline{f}(s) = \int_0^\infty d\tau\, e^{-s\tau} f(\tau)$, and it may be observed that for the function not to be zero as time increases, S should be less than a threshold value. Given that $\tau_D^{-1} = \kappa^2 D$, it is found that $S \ll \kappa^2 D$, and from Equation (A12), it appears that $M \approx \kappa$. Then,

$$A = \alpha\frac{S^{-1}}{1+\tau S} \quad \text{(A22)}$$

$$\text{where } \alpha = \frac{\varphi(0)\varepsilon\kappa^2}{\cosh\left(\kappa\frac{L}{2}\right)\left(\lambda_s\kappa + \tanh\left(\kappa\frac{L}{2}\right)\right)} \quad \text{(A23)}$$

$$\text{and } \tau = \frac{L\left(1+\frac{2\lambda_s}{L}\right)}{2D\kappa\left(\lambda_s\kappa + \tanh\left(\kappa\frac{L}{2}\right)\right)} \quad \text{(A24)}$$

Thus,

$$\overline{\rho} = \alpha\sinh\left(\kappa\left(y-\frac{L}{2}\right)\right)\frac{S^{-1}}{1+\tau S} \quad \text{(A25)}$$

$$\frac{\partial \overline{\varphi}}{\partial y} = -\frac{\alpha}{\varepsilon\kappa}\cosh\left(\kappa\left(y-\frac{L}{2}\right)\right)\frac{S^{-1}}{1+\tau S} - \alpha\frac{1}{1+\tau S}\frac{\cosh\left(\kappa\frac{L}{2}\right)}{\varepsilon D\kappa^3} \quad \text{(A26)}$$

$$\overline{\varphi} = -\alpha\frac{\cosh\left(\kappa\frac{L}{2}\right)}{\varepsilon\kappa^2}\left\{\frac{\sinh\left[\kappa\left(y-\frac{L}{2}\right)\right]}{\cosh\left[\kappa\frac{L}{2}\right]}\frac{S^{-1}}{1+\tau S} + \frac{1}{1+\tau S}\frac{\left(y-\frac{L}{2}\right)}{D\kappa}\right\} \quad \text{(A27)}$$

The inverse Laplace transformation of functions is $\frac{S^{-1}}{1+\tau S}$ and $\frac{1}{1+\tau S}$ is $[L]^{-1}\left[\frac{S^{-1}}{1+\tau S}\right] = 1 - e^{-\frac{t}{\tau}}$ and $[L]^{-1}\left[\frac{1}{1+\tau S}\right] = \frac{1}{\tau}e^{-\frac{t}{\tau}}$, respectively, resulting then in Equations (14)–(21).

References

1. Setiawan, A.A.; Zhao, Y.; Nayar, C.V. Design, Economic analysis and environmental considerations of mini-grid hybrid power system with reverse osmosis desalination plant for remote areas. *Renew. Energy* **2009**, *34*, 374–383. [CrossRef]
2. Cipollina, A.; Micale, G. Coupling sustainable energy with membrane distillation processes for seawater desalination. In Proceedings of the 2010 1st International Nuclear & Renewable Energy Conference (INREC), Amman, Jordan, 21–24 March 2010.
3. García-Rodríguez, L. Renewable energy applications in desalination: State of the art. *Sol. Energy* **2003**, *75*, 381–393. [CrossRef]
4. Wilf, M.; Awerbuch, L.; Bartels, C.; Mickley, M.; Pearce, G.; Voutchkov, N. *The Guidebook to Membrane Desalination Technology. Reverse Osmosis, Nanofiltration and Hybrid Systems. Processes, Design, Applications*; Balaban Desalination Publications: L'Aquila, Italy, 2011.
5. Guoling, R.; Houjun, F. Technical Progress in Seawater Desalination Technology at Home and Abroad. *China Water Wastewater* **2008**, *20*, 86–90.
6. Mathioulakis, E.; Belessiotis, V.; Delyannis, E. Desalination by using alternative energy: Review and state-of-the-art. *Desalination* **2007**, *203*, 346–365. [CrossRef]
7. Mohamed, E.; Papadakis, G.; Mathioulakis, E.; Belessiotis, V. The effect of hydraulic energy recovery in a small sea water reverse osmosis desalination system; experimental and economical evaluation. *Desalination* **2005**, *184*, 241–246. [CrossRef]
8. Mohamed, E.; Papadakis, G.; Mathioulakis, E.; Belessiotis, V. An experimental comparative study of the technical and economic performance of a small reverse osmosis desalination system equipped with an hydraulic energy recovery unit. *Desalination* **2006**, *194*, 239–250. [CrossRef]
9. Mohamed, E.S.; Papadakis, G.; Mathioulakis, E.; Belessiotis, V. A direct coupled photovoltaic seawater reverse osmosis desalination system toward battery based systems—A technical and economical experimental comparative study. *Desalination* **2008**, *221*, 17–22. [CrossRef]
10. Li, N.; Fane, A.; Winston, W.H.; Matsuura, T. *Advanced Membrane Technology and Applications*; Wiley: Hoboken, NJ, USA, 2008.
11. Mezher, T.; Fath, H.; Abbas, Z.; Khalid, A. Techno-economic assessment and environmental impacts of desalination technologies. *Desalination* **2011**, *266*, 263–273. [CrossRef]
12. Maerzke, K.A.; Siepmann, J.I. Effects of an Applied Electric Field on the Vapor−Liquid Equilibria of Water, Methanol, and Dimethyl Ether. *J. Phys. Chem. B* **2010**, *114*, 4261–4270. [CrossRef] [PubMed]
13. Bakker, H.J. Structural Dynamics of Aqueous Salt Solutions. *Chem. Rev.* **2008**, *108*, 1456–1473. [CrossRef] [PubMed]
14. Morro, A.; Drouot, R.; Maugin, G.A. Thermodynamics of Polyelectrolyte Solutions in an Electric Field. *J. Non-Equilib. Thermodyn.* **1985**, *10*, 131–144. [CrossRef]
15. Wright, M. *An Introduction to Aqueous Electrolyte Solutions*; Wiley: Hoboken, NJ, USA, 2007.
16. Gordon, M.J.; Huang, X.; Pentoney, S.L.; Zare, R.N. Capillary Electrophoresis. *Science* **1988**, *242*, 224–228. [CrossRef] [PubMed]
17. Strathmann, H. *Ion-Exchange Membrane Separation Processes*; Elsevier: Amsterdam, The Netherlands, 2004; Volume 9.
18. Ganzi, C.; Egozy, Y.; Giuffrida, A.; Jha, A. High purity water by electro-deionization. *Ultrapure Water J.* **1987**, *4*, 43–50.
19. Tironi, I.; Sperb, R.; Smith, P.; van Gunsteren, W. A generalized reaction field method for molecular dynamics simulations. *J. Chem. Phys.* **1995**, *102*, 5451–5459. [CrossRef]
20. Berendsen, H.; Postma, J.; van Gunsteren, W. *Intermolecular Forces*; Pullman, B., Ed.; Reidel Publishing Co.: Dordrecht, The Netherlands, 1981; pp. 331–342.
21. Wulfsberg, G. *Principles of Descriptive Inorganic Chemistry*; University Science Books: South Orange, NJ, USA, 1991; pp. 23–31.
22. Rasaiah, J.; Lynden-Bell, R. Computer simulation studies of the structure and dynamics of ions and non–polar solutes in water. *Philos. Trans. R. Soc. Lond. Ser. A Math. Phys. Eng. Sci.* **2001**, *359*, 1545–1574. [CrossRef]
23. Belch, A.C.; Berkowitz, M.; McCammon, J.A. Solvation structure of a sodium chloride ion pair in water. *J. Am. Chem. Soc.* **1986**, *108*, 1755–1761. [CrossRef]
24. Das, A.; Tembe, B. The pervasive solvent-separated sodium chloride ion pair in water-DMSO mixtures. *Proc. Indian Acad. Sci. (Chem. Sci.)* **1999**, *111*, 353–360.
25. Murad, S. The role of external electric fields in enhancing ion mobility, drift velocity, and drift–diffusion rates in aqueous electrolyte solutions. *J. Chem. Phys.* **2011**, *134*, 114504. [CrossRef] [PubMed]
26. Sheikholeslami, M.; Sheremet, M.A.; Shafee, A.; Li, Z. CVFEM approach for EHD flow of nanofluid through porous medium within a wavy chamber under the impacts of radiation and moving walls. *J. Therm. Anal. Calorim.* **2019**, *138*, 573–581. [CrossRef]
27. Sofos, F.; Karakasidis, T.E.; Spetsiotis, D. Molecular dynamics simulations of ion separation in nano-channel water flows using an electric field. *Mol. Simul.* **2019**, *45*, 1395–1402. [CrossRef]
28. Suss, M.E.; Porada, S.; Sun, X.; Biesheuvel, P.; Yoon, J.; Presser, V. Water desalination via capacitive deionization: What is it and what can we expect from it? *Energy Environ. Sci.* **2015**, *8*, 2296–2319. [CrossRef]
29. Azamat, J. Functionalized Graphene Nanosheet as a Membrane for Water Desalination Using Applied Electric Fields: Insights from Molecular Dynamics Simulations. *J. Phys. Chem. C* **2016**, *120*, 23883–23891. [CrossRef]
30. Drahushuk, L.W.; Strano, M.S. Mechanisms of Gas Permeation through Single Layer Graphene Membranes. *Langmuir* **2012**, *28*, 16671–16678. [CrossRef]
31. Sun, C.; Boutilier, M.S.H.; Au, H.; Poesio, P.; Bai, B.; Karnik, R.; Hadjiconstantinou, N.G. Mechanisms of Molecular Permeation through Nanoporous Graphene Membranes. *Langmuir* **2014**, *30*, 675–682. [CrossRef] [PubMed]
32. Cohen-Tanugi, D.; Grossman, J.C. Water Desalination across Nanoporous Graphene. *Nano Lett.* **2012**, *12*, 3602–3608. [CrossRef]

33. Humplik, T.; Lee, J.; O'Hern, S.C.; A Fellman, B.; Baig, M.A.; Hassan, S.F.; Atieh, M.; Rahman, F.; Laoui, T.; Karnik, R.; et al. Nanostructured materials for water desalination. *Nanotechnology* **2011**, *22*, 292001. [CrossRef] [PubMed]
34. Suk, M.E.; Aluru, N. Water Transport through Ultrathin Graphene. *J. Phys. Chem. Lett.* **2010**, *1*, 1590–1594. [CrossRef]
35. Wang, E.; Karnik, R. Water Desalination: Graphene Cleans Up Water. *Nat. Nanotechnol.* **2012**, *7*, 552–554. [CrossRef] [PubMed]
36. Ninos, G.; Bartzis, V.; Merlemis, N.; Sarris, I. Uncertainty quantification imlementations in human hemodynamic. *Comput. Methods Programs Biomed.* **2021**, *203*, 106021. [CrossRef] [PubMed]
37. Bartzis, V.; Sarris, I.E. A theoretical model for salt ion drift due to electric field suitable to sea water desalination. *Desalination* **2020**, *473*, 114163. [CrossRef]
38. Oren, Y. Capacitive deionization (CDI) for desalination and water treatment past, present and future (a review). *Desalination* **2008**, *228*, 10–29. [CrossRef]
39. Choia, J.; Dorjib, P.; Shonb, H.; Honga, S. Applications of capacitive deionization: Desalination, softening, selective removal, and energy efficiency. *Desalination* **2019**, *449*, 118–130. [CrossRef]
40. Minhas, M.; Jande, Y.; Kim, W. Hybrid Reverse Osmosis-Capacitive Deionization versus Two-Stage Reverse Osmosis: A Comparative Analysis. *Chem. Eng. Technol.* **2014**, *37*, 1137–1145. [CrossRef]
41. Peters, P.B.; van Roij, R.; Bazant, M.; Biesheuvel, P. Analysis of electrolyte transport through charged nanopores. *Phys. Rev. E* **2016**, *93*, 053108. [CrossRef] [PubMed]
42. Biesheuvel, P.M.; Bazant, M. Analysis of ionic conductance of carbon nanotubes. *Phys. Rev. E* **2016**, *94*, 050601. [CrossRef] [PubMed]
43. Tsouris, C.; Mayes, R.; Kiggans, J.; Sharma, K.; Yiacoumi, S.; DePaoli, D.; Dai, S. Mesoporous carbon for capacitive deionization of saline water. *Envirom. Sci. Tech.* **2011**, *45*, 10243–10249. [CrossRef] [PubMed]
44. He, F.; Biesheuvel, P.M.; Bazant, M.; Hatton, T.A. Theory of water treatment by capacitive deionization with redox active porous electrodes. *Water Res.* **2018**, *132*, 282. [CrossRef] [PubMed]
45. Guyes, E.N.; Shocron, A.N.; Simakovski, A.; Biesheuvel, P.M.; Suss, M. A one dimensional model for water desalination by flow-through electrode capacitive deionization. *Desalination* **2017**, *415*, 8. [CrossRef]
46. Bartzis, V.; Sarris, I.E. Electric field distribution and diffuse layer thickness study due to salt ion movement in water desalination. *Desalination* **2020**, *490*, 114549. [CrossRef]
47. Atkins, P.; Paola, J. *Atkin's Physical Chemistry*, 8th ed.; Oxford University Press: New York, NY, USA, 2006; Volume 5, pp. 136–169.
48. Mortimer, R.G. *Physical Chemistry*, 3rd ed.; Elsevier: Amsterdam, The Netherlands, 2008; Volume 8, pp. 351–378.
49. Brett, C.A.; Arett, A. *Electrochemistry Principles, Methods, and Applications*; Oxford University Press: New York, NY, USA, 1994.
50. Debye, P.; Hückel, E. The theory of electrolytes. I. Lowering of freezing point and related phenomena. *Phys. Z.* **1923**, *24*, 185–206.
51. Bonnefont, A.; Argoul, F.; Bazant, M. Analysis of diffuse-layer effects on time-dependent interfacial kinetics. *J. Electroanal. Chem.* **2001**, *500*, 52–61. [CrossRef]
52. Bazant, M.; Thornton, K.; Ajdari, A. Diffuse-charge dynamics in electrochemical systems. *Phys. Rev. E* **2004**, *70*, 021506. [CrossRef] [PubMed]
53. Biesheuvel, P.M.; Dykstra, J.E. The difference between Faradaic and Nonfaradaic processes in Electrochemistry. *arXiv* **2021**, arXiv:1809.02930.
54. Golovnev, A.; Trimper, S. Exact solution of the Poisson–Nernst–Planck equations in the linear regime. *J. Chem. Phys.* **2009**, *131*, 114903. [CrossRef] [PubMed]
55. Nekrasov, S.A. Calculating the electrostatic field in the bulk of an aqueous solution. *Russ. J. Phys. Chem. A* **2012**, *86*, 1732–1735. [CrossRef]
56. Han, J.-H.; Muralidhar, R.; Waser, R.; Bazant, M.Z. Resistive Switching in Aqueous Nanopores by Shock Electrodeposition. *Electrochim. Acta* **2016**, *222*, 370–375. [CrossRef]

water MDPI

Article

Evaluation of Adsorption Mechanism of Chromium(VI) Ion Using Ni-Al Type and Ni-Al-Zr Type Hydroxides

Fumihiko Ogata [1], Noriaki Nagai [1], Ayako Tabuchi [1], Megumu Toda [2], Masashi Otani [2], Chalermpong Saenjum [3,4], Takehiro Nakamura [1] and Naohito Kawasaki [1,5,*]

1 Faculty of Pharmacy, Kindai University, 3-4-1 Kowakae, Higashi-Osaka, Osaka 577-8502, Japan; ogata@phar.kindai.ac.jp (F.O.); nagai_n@phar.kindai.ac.jp (N.N.); 1711710039w@kindai.ac.jp (A.T.); nakamura@phar.kindai.ac.jp (T.N.)
2 Kansai Catalyst Co. Ltd., 1-3-13, Kashiwagi-cho, Sakai-ku, Sakai, Osaka 590-0837, Japan; megumu.toda@kansyoku.co.jp (M.T.); masashi.ootani@kansyoku.co.jp (M.O.)
3 Faculty of Pharmacy, Chiang Mai University, Suthep Road, Muang District, Chiang Mai 50200, Thailand; chalermpong.saenjum@gmail.com
4 Cluster of Excellence on Biodiversity-Based Economics and Society (B.BES-CMU), Chiang Mai University, Suthep Road, Muang District, Chiang Mai 50200, Thailand
5 Antiaging Center, Kindai University, 3-4-1 Kowakae, Higashi-Osaka, Osaka 577-8502, Japan
* Correspondence: kawasaki@phar.kindai.ac.jp; Tel.: +81-6-4307-4012

Abstract: To evaluate the feasibility of nickel–aluminum (the Ni^{2+}:Al^{3+} molar ratios of 1.0:1.0 and 1.0:2.0 are denoted as NA11 and NA12, respectively) and nickel–aluminum–zirconium type (the Ni^{2+}:Al^{3+}:Zr^{4+} molar ratios of 0.9:1.0:0.09 and 0.9:2.0:0.09 are denoted as NAZ1 and NAZ2, respectively) hydroxides for Cr(VI) removal from aqueous media, the adsorption capability and adsorption mechanism of Cr(VI) using the above-mentioned adsorbents were investigated in this study. The quantity of Cr(VI) adsorbed onto NA11, NA12, NAZ1, and NAZ2 was 25.5, 25.6, 24.1, and 24.6 mg g^{-1}, respectively. However, the quantity of aluminum (base metal) released from NA11 (approximately 0.14 mg g^{-1}) was higher than that from NAZ1 (approximately 1.0 μg g^{-1}), indicating that NAZ1 was more suitable for Cr(VI) removal than NA11. In addition, the effects of pH, contact time, and temperature on the adsorption of Cr(VI) were evaluated. Moreover, to elucidate the adsorption mechanism of Cr(VI) using NA11 and NAZ1, the elemental distribution, X-ray photoelectron spectrometry spectra, and ion exchange capability were also determined. Cr(VI) adsorbed onto the NAZ1 surface was easily desorbed using a sodium hydroxide solution under our experimental conditions. The information regarding this study can be useful for removing Cr(VI) from aqueous media.

Keywords: nickel–aluminum complex hydroxide; nickel–aluminum–zirconium complex hydroxide; chromium(VI); adsorption

Citation: Ogata, F.; Nagai, N.; Tabuchi, A.; Toda, M.; Otani, M.; Saenjum, C.; Nakamura, T.; Kawasaki, N. Evaluation of Adsorption Mechanism of Chromium(VI) Ion Using Ni-Al Type and Ni-Al-Zr Type Hydroxides. *Water* **2021**, *13*, 551. https://doi.org/10.3390/w13040551

Academic Editors: Cristina Palet and Julio Bastos-Arrieta

Received: 22 January 2021
Accepted: 18 February 2021
Published: 21 February 2021

1. Introduction

In 2015, the United Nations announced the 2030 agenda for sustainable development. Among the 17 Sustainable Development Goals (SGDs), Goal No. 6 (clean water and sanitation) and Goal No. 14 (life below water), which are directly and strongly related to the water environment, focused on resolving global aquatic environmental problems. Exposure to high level of water contamination by heavy metals is considered as a serious variety of adverse health outcomes [1]. As one of the most important contaminants, two main oxidation states including Cr(VI) and Cr(III) is present in the environment. Importantly, Cr(VI) compounds are widely used in various industrial processes. On the other hand, Cr(III) is used as a nutritional supplement [2]. Cr(VI) is usually considered more toxic and carcinogenic than Cr(III) [3,4]. Therefore, Cr(VI) compounds were classified as a group 1 carcinogenic to humans by the International Agency for Research on Cancer (IARC). In addition, the World Health Organization (WHO) and U.S. Environmental Protection Agency (USEPA) has established the maximum permissible Cr(VI) content for industrial wastewater, surface water, and drinking water in the values of 0.25, 0.10, and 0.05 mg L^{-1}, respectively [5,6]. Conversely, chromium is stockpiled in Japan because it is a relatively rare metal with several applications. Japan is one of the main consumers of rare metals worldwide. However, the recycling technology of rare metals, including chromium, has not yet been developed. Therefore, the development of removal and/or recycling techniques for Cr(VI) from aqueous media is very important for establishing a sustainable society [7,8].

Numerous techniques have been studied and applied to evaluate the Cr(VI) removal including membrane filtration, chemical precipitation, solvent extraction, adsorption, electrochemical reduction and oxidation, and photoreduction [9–12]. Adsorption is one of the best choice and most cost-effective technique of heavy metals removal such as Cr(VI), and the adsorption process indicates that adsorbents bind heavy metals by chemical binding, ion exchange, and physical attractive forces [1].

Currently, the physicochemical properties and feasibility of metal complex hydroxides have received more attention, and their application has been evaluated for the capability on heavy metals removal from aqueous media [13]. Various studies have reported previously for the capability of Mg-Al-CO_3-hydrotalcite [14,15], hydrotalcite-hydroxyapatite material doped with carbon nanotubes [16], Mg-Al hydrotalcite kaolin clay [17], hydrotalcite/carbon [18], sulfide assembled hydrotalcite compounds [19], calcined nano-Mg/Al hydrotalcite [20], and metal oxide as dual-functional adsorbents [1] on Cr(VI) removal from aqueous media. In addition, our previous studies reported that Fe-Mg-type hydrotalcite and nickel–aluminum complex hydroxide showed excellent adsorption capability for Cr(VI) from aqueous media [8,21]. However, there are few reports on the adsorption/removal of heavy metals using ternary metal complex hydroxides from aqueous media. In addition, multi-complex metal hydroxides are useful adsorbents for removing heavy metals from aqueous media [8,21]. Multi-complex metal hydroxides exhibit quite different physicochemical properties compared to their single or double metal hydroxides.

Additionally, our previous studies reported that nickel–aluminum complex hydroxide and/or nickel–aluminum–zirconium complex hydroxide are useful for the removal of oxyanions, such as arsenic and phosphate ions, from aqueous media [22,23]. The incorporation of zirconium (Zr^{4+}) into a metal complex hydroxide induces an increase in the charge in the metal layer and strongly affects its physicochemical properties. These changes affect to the capability on adsorption and removal of heavy metals from aqueous media. We have already evaluated the Cr(VI) adsorption capability using nickel–aluminum complex hydroxide [21]. Additionally, Cr(VI) (oxyanion form) is ubiquitously present in the water environment. Therefore, the incorporation of zirconium in the nickel–aluminum complex hydroxide might be useful adsorbent for removal of Cr(VI) from aqueous media or might show the superior adsorption capability of Cr(VI). However, no information is available regarding Cr(VI) adsorption using the incorporation of zirconium in the nickel–aluminum complex hydroxide.

Therefore, this study mainly focused on Cr(VI) adsorption using the incorporation of zirconium in the nickel–aluminum complex hydroxide. Additionally, various parameters including the effects of initial concentration, temperature, pH, and contact time on adsorption were also evaluated here.

2. Materials and Methods

2.1. Materials and Chemicals

Standard solutions of Cr(VI) ($K_2Cr_2O_7$ in 0.1 mol L^{-1} HNO_3) was purchased from FUJIFILM Wako Pure Chemical Co., Osaka, Japan. Different molar ratios of nickel–aluminum complex hydroxides and nickel–aluminum–zirconium complex hydroxide were prepared in the molar ratios of Ni^{2+} to Al^{3+} of 1.0, and 0.5, referred as NA11 and NA12, respectively. Additionally, the molar ratios of 0.9:1.0:0.09 and 0.9:2.0:0.09 of Ni^{2+}:Al^{3+}:Zr^{4+} were denoted as NAZ1 and NAZ2, respectively. The synthesis method and physicochemical characteristics were reported in our previous studies [22]. Briefly, the materials were synthesized by the following method. $NiSO_4{\bullet}6H_2O$, $Al_2(SO_4)_3{\bullet}18H_2O$, and $Zr(SO_4)_2{\bullet}4H_2O$ were mixed and then the reaction solution was heated. After that, the reaction solution was added to the distilled water at pH 9.0 for 800 rpm at 25 °C. After mixing for 2 h, the suspension was filtered, washed, and dried at 110 °C for 12 h. The adsorbent morphology and crystal structure were measured using SU1510 (Hitachi Ltd., Tokyo, Japan) and Mini Flex II (Rigaku, Tokyo, Japan), respectively. NOVA4200*e* instrument (Yuasa Ionics, Osaka, Japan) was used to measure the specific surface area. Amount of surface hydroxyl groups and pH_{pzc} of the adsorbents were measured by the fluoride ion adsorption method [24] and the method reported by Faria et al. [25], respectively.

To elucidate the adsorption mechanism of Cr(VI), elemental analysis and electron spectroscopy were analyzed using JXA-8530F (JEOL, Tokyo, Japan) and AXIS-NOVA instruments (Shimadzu Co., Ltd., Kyoto, Japan), respectively. Finally, the amount of sulfate ions released from the adsorbent in the adsorption experiment was also measured using a DIONEX ICS-900 (Thermo Fisher Scientific Inc., Tokyo, Japan). The measurement conditions were already reported in our previous study [23]. Finally, the amount of base metal released from NA11 and NAZ1 was measured. NA11 or NAZ1 (0.05 g) was added to the distilled water (50 mL, pH 7.0). The suspension was shaken at 100 rpm and 25 °C for 24 h. The suspension was filtered through a 0.45 μm membrane filter. The concentration of aluminum released from NA11 and NAZ1 were measured using an iCAP-7600 Duo instrument (Thermo Fisher Scientific Inc., Tokyo, Japan).

2.2. *Quantity of Cr(VI) Adsorbed*

Each adsorbent (0.05 g) and the Cr(VI) solution at 50 mg L^{-1} (50 mL, pH 7.0) were mixed and then shaken at 100 rpm and 25 °C for 24 h. Cr(VI) concentrations of the obtained solutions after filtrated through a 0.45 μm membrane filter were measured using an iCAP-7600 Duo instrument (Thermo Fisher Scientific Inc., Tokyo, Japan). The adsorption capability on Cr(VI) of each tested adsorbent was calculated by the difference between Cr(VI) concentration before and after adsorption. All data are presented as mean ± standard error from triplicate experiments.

2.3. *Effect of pH, Contact Time, and Temperature on the Removal of Cr(VI)*

Initially, to investigate the effect of pH, 0.05 g of each tested adsorbent (NA11 and NAZ1), and different pH conditions of 50 mL Cr(VI) solution in the concentration of 50 mg L^{-1} were mixed. The solution pH was adjusted between 3, 5, 7, 9, and 11, using either nitric acid or sodium hydroxide solutions (FUJIFILM Wako Pure Chemical Co., Osaka, Japan). The suspension was shaken at 100 rpm and 25 °C for 24 h. The suspension was filtered through a 0.45 μm membrane filter. Second, to elucidate the contact time effect, 0.05 g of same adsorbents and the 50 mL Cr(VI) solution at 50 mg L^{-1} were mixed, and then shaken at 100 rpm and 25 °C for 30 s, 1, 5, 10, and 30 min, and 1, 3, 6, 9, 15, 18, 21, 24, 30, 42, and 48 h. Third, to elucidate the effect of temperature, 0.05 g of same adsorbents and the Cr(VI) solution at different concentrations (10, 20, 30, 40, and 50 mg L^{-1}) were mixed, and then shaken at 100 rpm and 5, 25, and 45 °C for 24 h. The quantity of Cr(VI) adsorbed was also calculated using the above-mentioned method. The student's *t*-test was used for a comparative analysis of two groups. A minimum *p*-value of 0.05 ($p < 0.05$) was chosen as the significant level. All data are presented as mean ± standard error from triplicate experiments.

2.4. *Recovery of Cr(VI) From NAZ1 using Desorption Solutions*

To investigate the recovery of Cr(VI) adsorbed onto the NAZ1 surface, the adsorption/desorption performance of NAZ1 was evaluated in this section. First, NAZ1 (0.1 g) and the Cr(VI) solution at 100 mg g^{-1} (50 mL) were mixed, and then the suspension was shaken at 100 rpm and 25 °C for 24 h. Quantity of Cr(VI) adsorbed was also calculated as described the above. Second, NAZ1 was collected after adsorption, and then dried at 25 °C for 24 h. Collected NAZ1 was added to the sodium hydroxide solution at 1, 10, and 100 mmol/L (50 mL). The suspension was shaken at 100 rpm and 25 °C for 24 h. Cr(VI) concentrations of the obtained solutions after filtrated through a 0.45 μm membrane filter were measured using an iCAP-7600 Duo instrument. The quantity of Cr(VI) desorbed was calculated from the difference between the concentration of Cr(VI) before and after adsorption. All data are presented as mean ± standard error from triplicate experiments.

3. Results and Discussion

3.1. Physicochemical Properties

First, NA11 and NA12 were selected for Cr(VI) removal from aqueous media. Because our previous study elucidated the optimal molar ratios condition of nickel and aluminum in the metal complex hydroxide for removal of Cr(VI) [21]. Additionally, previous studies showed that the incorporation of Zr^{4+} into a Ni-Al type hydroxide significantly improved the adsorption capability of oxyanions such as arsenic ions [22] and phosphate ions [23]. Therefore, the compositions of NA11, NA12, NAZ1, and NAZ2 were selected in this study. The physicochemical characteristics of NA11, NA12, NAZ1, and NAZ2 used in this study have already been reported in our previous studies [22]. Therefore, we briefly describe these properties. The prepared adsorbents were not perfectly spherical in shape under our experimental conditions. Additionally, no significant differences between each adsorbent were observed in this study. NA11 and NA12 have an amorphous nature [23]. The XRD patterns of NAZ1 and NAZ2 showed similar trends to those of NA11 and NA12. In addition, the XRD patterns of NAZ1 and NAZ2 showed that the incorporation of zirconium in the NA series could be successful, and then the distances between metals and between the octahedral layers changed in this study [22,26,27]. The specific surface areas of NA11, NA12, NAZ1, and NAZ2 were 22.8, 26.4, 51.9, and 27.8 $m^2\ g^{-1}$, respectively. Additionally, the amount of surface hydroxyl groups was 1.92, 1.62, 1.08, and 1.51 mmol g^{-1}, respectively. Therefore, the incorporation of Zr^{4+} into a Ni-Al type hydroxide affects the increasing specifics surface area and the decreasing amount of surface hydroxyl groups under our experiment conditions. The pH_{pzc} value of each adsorbent was between 6.2 and 6.4 under our experimental conditions.

3.2. Quantity of Cr(VI) Adsorbed

The quantity of Cr(VI) adsorbed onto NA11, NA12, NAZ1, and NAZ2 was 25.5 ± 0.25 mg g^{-1} (1.1 ± 0.011 mg m^{-2}), 25.6 ± 0.12 mg g^{-1} (0.97 ± 0.005 mg m^{-2}), 24.1 ± 0.45 mg g^{-1} (0.46 ± 0.001 mg m^{-2}), and 24.6 ± 0.13 mg g^{-1}(0.88 ± 0.005 mg m^{-2}), respectively. No significant difference was observed in Cr(VI) removal using the NA and NAZ series under our experimental conditions. The adsorption capability of Cr(VI) from aqueous media was not improved by incorporating zirconium in the nickel–aluminum complex hydroxide. In addition, the relationship between the quantity of Cr(VI) adsorbed and the physicochemical characteristics was evaluated. The correlation coefficient between the quantity of Cr(VI) adsorbed, and the surface hydroxyl groups was 0.890, which indicates that this parameter was related to the adsorption of Cr(VI) using the NA and NAZ series.

The base metal released from the adsorbent is one of the most important issues in adsorption treatment. This phenomenon directly and strongly affects the adsorption capability of Cr(VI) using an adsorbent. Therefore, we evaluated the quantity of base metal released from NA11 and NAZ1 in a preliminary experiment. As a result, the quantity of aluminum released from NA11 (approximately 0.14 mg g^{-1}) was greater than that from NAZ1 (approximately 1.0 μg g^{-1}), which suggests that NAZ1 was more suitable for Cr(VI) removal than NA11. The incorporation of Zr^{4+} into NA11 increases the positive charge, and then more occupies the octahedral holes in a close-packed configuration of hydroxide ions. This phenomenon indicates that NAZ1 is more suitable compared to NA11 in this study. This result is very important for the application of NAZ1 in the field. Therefore, NAZ1 was used in the following adsorption experiment, and NA11 was also selected for comparison in this study.

Table 1 shows the comparison of the Cr(VI) adsorption capacities of NA11 and NAZ1 with those of other reported adsorbents. The adsorption capability of Cr(VI) using NA11 and NAZ1 exhibited a similar trend to that of other reported adsorbents (except for calcined nano-Mg/Al hydrotalcite, and NiAl-LDH). Therefore, NAZ1 can be applied for the removal of Cr(VI) from aqueous media.

Table 1. Comparison of Cr(VI) adsorption capacity of NA11 and NAZ1 with other reported adsorbents.

Adsorbents	Adsorption Capability (mg/g)	pH	Temp. (°C)	Initial Concentration (mg/L)	Contact Time (h)	Adsorbent (g/L)	Ref.
Calcined Mg-Al-CO_3 hydrotalcite	Approximately 24	6.0	30	10	4	0.2	[15]
Calcined nano-Mg/Al hydrotalcite	52.4	3.0	22	110	2	1	[20]
MgAl-LDH	30.28	6–7	25	30–55	24	0.2	[28]
NiAl-LDH	57.50	6–7	25	30–55	24	0.2	[28]
Ni-Fe-LDH	26.78	-	-	4–20	4–5	0.2	[29]
Mg-Al-Cl LDH	20.10	4.0	15	40	2	2	[30]
Mg-Al-CO_3 hydrotalcite	17.00	6	30	10	24	1	[31]
NA11	25.5	7	25	100	24	1	This study
NAZ1	24.1	7	25	100	24	1	This study

3.3. *Effect of pH, Contact Time, and Temperature on the Removal of Cr(VI)*

Solution pH is one of the most important parameters that directly and strongly affects the adsorption capability of Cr(VI) from aqueous media. Figure 1 exhibited the effect of pH on Cr(VI) adsorption using NA11 and NAZ1. In this experiment, the optimal pH condition for Cr(VI) removal using NA11 and NAZ1 was approximately between 7.0 and 9.0. Similar trends were reported by previous studies using Mg-Al-Cl LDH and Mg-Al LDH [30,32]. Cr(VI) oxyanion species in an aqueous solution depend on the solution pH and chromium concentration. The Cr(VI) oxyanion species in solution are chromate ($CrO_4{}^{2-}$), dichromate ($Cr_2O_7{}^{2-}$), and hydrogen chromate ($HCrO_4{}^{-}$) [33]. The surfaces of NA11 and NAZ1 are protonated in acidic condition and therefore acquire positive charges. However, under strongly acidic condition (pH 3.0), low Cr(VI) adsorption was observed. These phenomena can be explained by the dissolution of the adsorbent at low pH solution. Similar trends were reported by previous studies [17,32]. On the other hand, only chromate is stable in solution, and there are many hydroxide ions above a pH of 6.8. Therefore, competitive adsorption between hydroxide ions and chromate ions onto NA11 and NAZ1 occurred easily [20,33,34]. The removal efficiency of Cr(VI) using NA11 and NAZ1 decreased pH from 9.0 to 11. Considering the adsorption capability of Cr(VI) using NA11 and NAZ1, an initial pH of 7.0 was selected in the following experiments in this study.

Figure 2 demonstrated the contact time effect on Cr(VI) adsorption using NA11 and NAZ1 with adsorption kinetics. The data showed that the rapidly adsorption occurred during the initial 10 min, and then it gradually became slower over time, reaching equilibrium in approximately 20 h. Previous studies reported the adsorption capability of metal complex hydroxide on heavy metals was a very rapid process [29,32]. It takes within 2 h for Cr(VI) solution with an initial concentration of from 5 to 200 mg g^{-1} to reach equilibrium. However, the quantity of Cr(VI) adsorbed using NA11 and NAZ1 was slightly higher compared to the reported adsorbent. Additionally, amount adsorbent (adsorbents reported in previous studies) was more needed for Cr(VI) removal from aqueous media. Therefore, NA11 and NAZ1 were useful for removal of Cr(VI) from aqueous media. The process of Cr(VI) adsorption from aqueous media using an adsorbent can be elucidated by kinetic models, such as the pseudo-first-order and pseudo-second-order models. These kinetic models can explain the rate-controlling mechanism of the Cr(VI) adsorption process, such as chemical reaction, diffusion control, and mass transfer [32]. Therefore, the adsorption kinetic data were analyzed in terms of the pseudo-first-order and pseudo-second-order models (Table 2).

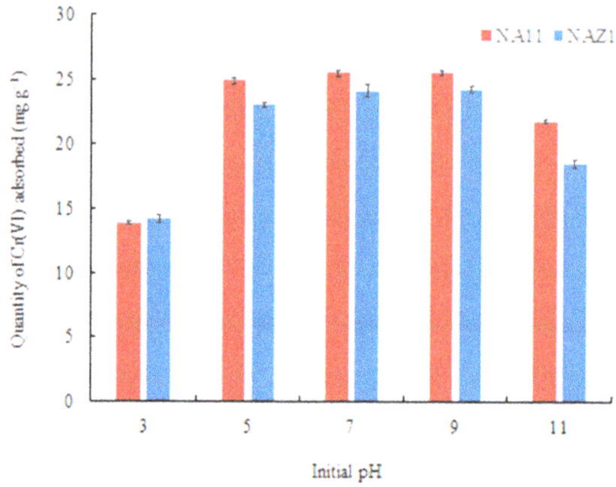

Figure 1. Effect of pH on the adsorption of Cr(VI) using NA11 and NAZ.

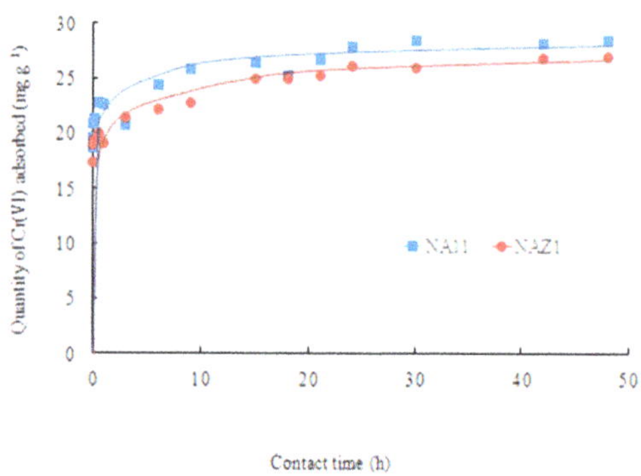

Figure 2. Effect of contact time on the adsorption of Cr(VI) using NA11 and NAZ1.

Table 2. Kinetic parameters for the adsorption of Cr(VI).

Adsorbents	q_e (mg g^{-1})	Pseudo-First-Order Model			Pseudo-Second-Order Model		
		k_1 (h^{-1})	q_e (mg g^{-1})	r	k_2 (g mg^{-1} h^{-1})	q_e (mg g^{-1})	r
NA11	28.5	0.22	8.0	0.870	0.04	28.4	0.999
NAZ1	27.0	0.09	8.8	0.972	0.04	26.8	0.999

The pseudo-first-order and pseudo-second-order models were based on physical sorption and chemical sorption, respectively [35,36]. These equations are expressed as follows:

$$\ln(q_e - q_t) = \ln q_e - k_1 t, \quad (1)$$

$$\frac{t}{q_t} = \frac{t}{q_e} + \frac{1}{k_2 \times {q_e}^2}, \quad (2)$$

where q_e and q_t are the quantity of Cr(VI) adsorbed (mg g^{-1}) at equilibrium and given time t, and k_1 and k_2 are the pseudo-first-order (h^{-1}) and pseudo-second-order (g mg^{-1} h^{-1}) rate constants, respectively. The correlation coefficient (r) in the pseudo-second-order model

(0.999) was higher than that in the pseudo-first-order model (0.870–0.972). In addition, closer values of q_e in the calculation (26.8 ± 0.01–28.4 ± 0.02 mg g^{-1}) and experiment (27.0 ± 0.01–28.5 ± 0.02 mg g^{-1}) indicate that the Cr(VI) adsorption data fit well with the pseudo-second-order model compared to the pseudo-first-order model. These results suggest that Cr(VI) oxyanions diffuse through the solution to the adsorbent's external surface and boundary layer [29].

Previous studies reported that metal hydroxide complexes can remove heavy metals from aqueous media by various mechanisms [32]. First, the adsorption onto the adsorbent external surface. Second, intercalation by ion exchange mechanism with anions in the interlayer adsorbent. Usually, a combination of the above-mentioned mechanisms easily occurs to remove heavy metals from aqueous media. We have already described the relationship between the amount of Cr(VI) adsorbed and the surface functional group in the above section, which indicates that the adsorption mechanism of Cr(VI) using NA11 and NAZ1 was related to the adsorbent surface. Additionally, the conditions of NA11 and NAZ1 surfaces before and after adsorption were observed (Figure 3). Warm color and cold color shows high concentration and low concentration, respectively. The value of chromium (Cr) intensity was from 2 to over 15 and from 2 to 15 for NA11 and NAZ1 before and after adsorption. Therefore, the chromium quantity clearly increased on the NA11 and NAZ1 surfaces after Cr(VI) adsorption.

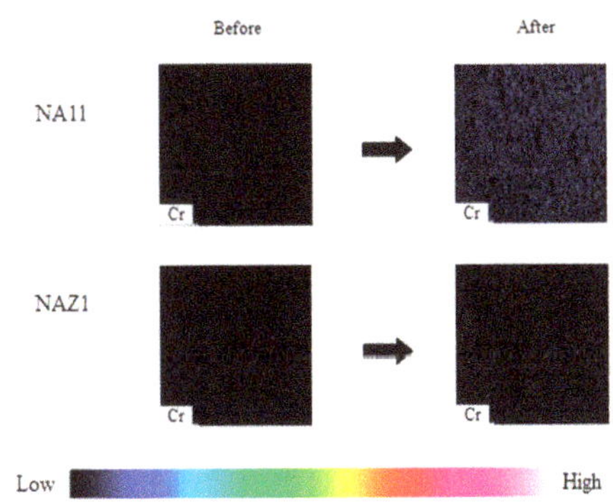

Figure 3. Elemental distribution of Cr(VI) onto NA11 and NAZ1 surface before and after adsorption.

The adsorption mechanism of Cr(VI) from aqueous media using NA11 and NAZ1 was also investigated in detail, anion exchange in the adsorption treatment (from Figure 2) was evaluated in this study. We measured the X-ray photoelectron spectra of the NA11 and NAZ1 surfaces before and after adsorption (Figure 4). Cr(2p) peaks at 575–577 eV were detected after adsorption, which was not detected before adsorption. On the other hand, the intensity of sulfur (165 eV for S(2p)) peak slightly or significantly decreased after adsorption of Cr(VI) using NA11 or NAZ1 in this study. In addition, Figure 5 shows the relationship between the quantity of Cr(VI) adsorbed and the quantity of sulfate ions released from the interlayer of NA11 and NAZ1. As shown in Figure 5, a positive linear relationship (correlation coefficient was 0.918 and 0.945 for NA11 and NAZ1, respectively) was observed under our experimental conditions. This result was proposed here that the sulfate ions in the interlayer of NA1 and NAZ1 were exchanged with Cr(VI).

Figure 6 exhibited the effect of temperature and the adsorption isotherms on Cr(VI) adsorption using NA11 and NAZ1. The quantity of Cr(VI) adsorbed using NA11 and NAZ1 was not different and slightly increased, respectively, at adsorption temperatures between 5 and 45 °C. Therefore, the adsorption temperature did not strongly affect the adsorption capability of Cr(VI) using NA11 and NAZ1 under our experimental conditions.

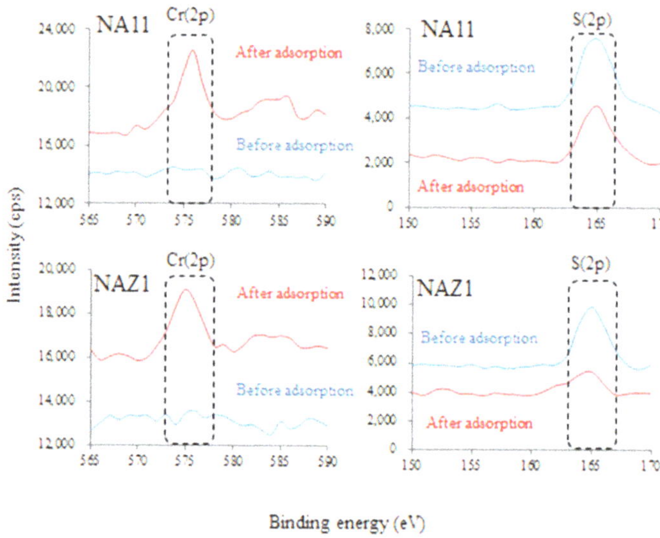

Figure 4. X-ray photoelectron spectrometry spectra of NA11 and NAZ1 before and after adsorption.

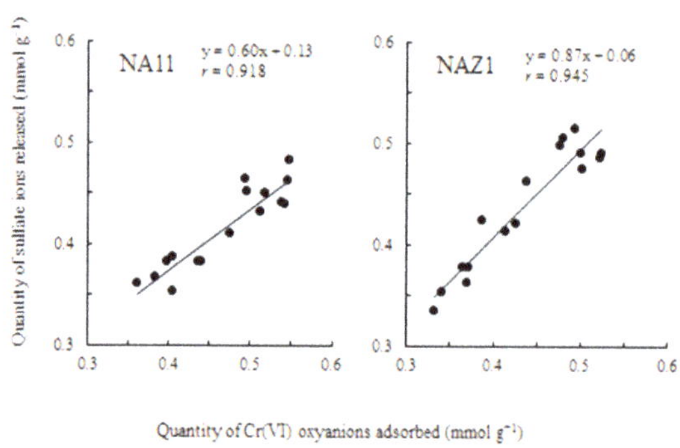

Figure 5. Relationship between quantity of Cr(VI) adsorbed and quantity of sulfate ions released.

Adsorption isotherm models are usually used to describe the interactions between adsorbates and adsorbents. Freundlich and Langmuir isotherms are empirical expressions of the adsorption process attributed to the special micro and macro structures of the adsorbents [37]. Freundlich's isotherm model allows for multilayered adsorption of adsorbates onto adsorbent surfaces [28]. Additionally, the Langmuir isotherm assumes an energetically homogeneous support surface with pervasively identical adsorption sites [38]. The Freundlich model is described by Equation (3) [39]:

$$\log q = \frac{1}{n} \log C_e + \log k, \quad (3)$$

The Langmuir model is described by Equation (4) [40]:

$$1/q = 1/(W_s a C_e) + 1/W_s, \quad (4)$$

where q is the quantity of Cr(VI) adsorbed (mg g^{-1}), W_s is the maximum quantity of adsorbed Cr(VI) (mg g^{-1}), and C_e is the equilibrium concentration (mg L^{-1}). The adsorption capacity and strength of adsorption are k and $1/n$, respectively. Additionally, a is the Langmuir isotherm constant (binding energy) (L mg^{-1}).

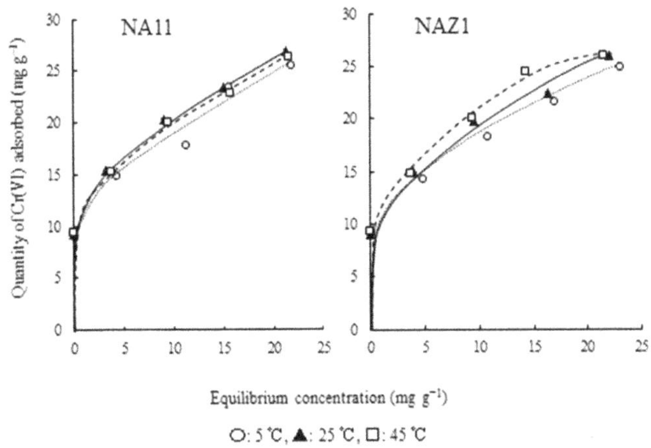

Figure 6. Adsorption isotherms of Cr(VI) using NA11 and NAZ1.

Table 3 showed the Freundlich and Langmuir constants for Cr(VI) adsorption using NA11 and NAZ1. As a result, the regression correlation coefficients (over 0.987) were presented for the Freundlich model, which were higher than those of the Langmuir model (over 0.917). These results indicate that the Freundlich isotherm model was more suitable for describing the adsorption equilibrium of Cr(VI) onto NA11 and NAZ1. The maximum adsorption capability (W_s) of Cr(VI) increased with increasing adsorption temperature from 5 to 45 °C. These phenomena agree with the adsorption behaviors of Cr(VI) using NAZ1. Finally, when the value of $1/n$ is 0.1–0.5, Cr(VI) adsorption using NA11 and NAZ1 easily occurs. On the other hand, when the value of $1/n$ is over 2, adsorption is difficult [41]. In this study, the value of $1/n$ is 0.31–0.34, and then Cr(VI) adsorption from aqueous media using NA11 and NAZ1 were more favorable under our experimental condition.

Table 3. Freundlich and Langmuir constants for the adsorption of Cr(V).

Adsorbents	Temperature (°C)	Freundlich Constants			Langmuir Constants		
		log$k$	$1/n$	r	W_s (mg g^{-1})	a (L mg^{-1})	r
NA11	5	0.96	0.33	0.987	27.4	0.28	0.917
	25	0.99	0.34	0.993	27.9	0.38	0.979
	45	0.98	0.33	0.998	28.5	0.30	0.984
NAZ1	5	0.96	0.31	0.997	28.0	0.22	0.979
	25	0.96	0.33	0.998	28.4	0.27	0.986
	45	0.97	0.34	0.998	30.0	0.26	0.989

3.4. *Adsorption/Desorption Capability of Cr(VI) using NAZ1*

To evaluate the recovery of Cr(VI) adsorbed onto the NAZ1 surface, the adsorption/desorption capability of Cr(VI) using NAZ1 was investigated in this study (Figure 7). As a result, the quantity of Cr(VI) desorbed from NAZ1 using a sodium hydroxide solution increased with increasing concentration of the desorption solution. The recovery percentages of Cr(VI) using 1, 10, and 100 mmol L^{-1} was 29.8%, 72.9%, and 80.1%, respectively, under our experimental conditions. The in-use studies in real samples are needed to elucidate the application of NAZ1 in further experiments.

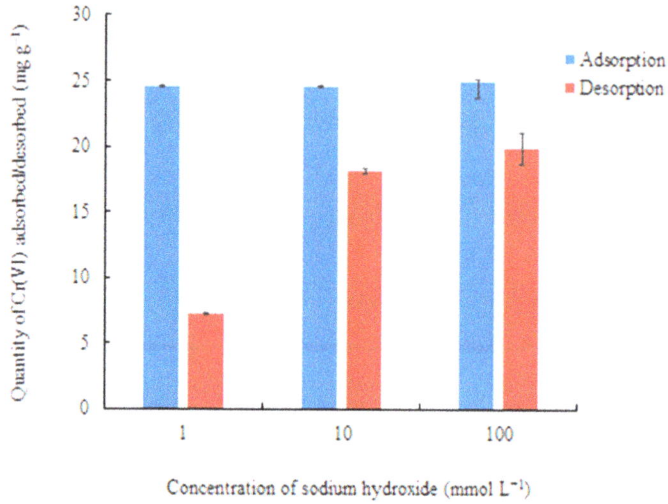

Figure 7. Adsorption/Desorption capability of Cr(VI) using NAZ1.

4. Conclusions

We prepared the NA and NAZ series adsorbents for Cr(VI) removal from aqueous media. Our results showed that NAZ1 was more suitable for Cr(VI) removal than NA11 under our experimental conditions. The optimal pH condition was approximately from 7.0 to 9.0 for removal of Cr(VI) from aqueous media. The kinetic experiments indicated that the adsorption process could reach equilibrium in approximately 20 h, and the kinetic data were accurately described by pseudo-second-order model (Correlation coefficient was 0.999 for NA11 and NAZ1). From the results of elemental distribution, binding energy analysis, and relationship between quantity of Cr(VI) adsorbed and quantity of sulfate ion released from adsorbents, adsorption mechanism related to the adsorbent surface properties (surface hydroxyl groups), and ion exchanges in this study. The adsorption isotherm data agreed with well with the Freundlich model (Correlation coefficient was over 0.987) compared to Langmuir model (Correlation coefficient was over 0.917). Additionally, the recovery percentage of Cr(VI) from NAZ1 using 100 mmol L^{-1} was approximately 80%. These findings are useful for the removal of Cr(VI) from aqueous media.

Author Contributions: Conceptualization, F.O. and N.K.; investigation, N.N., A.T., C.S., and T.N.; resources, M.T. and M.O.; writing—original draft preparation, F.O. and A.T.; writing—review and editing, F.O. and N.K.; project administration, N.K. All authors have read and agreed to the published version of the manuscript.

Funding: This research was supported in part by Kurita Water and Environment Foundation (20A003).

Institutional Review Board Statement: Not applicable.

Informed Consent Statement: Not applicable.

Data Availability Statement: MDPI is committed to supporting open scientific exchange and enabling our authors to achieve best practices in sharing and archiving research data. We encourage all authors of articles published in MDPI journals to share their research data. More details in section "MDPI Research Data Policies" at https://www.mdpi.com/ethics.

Conflicts of Interest: The authors declare no conflict of interest. The funders had no role in the design of the study; in the collection, analyses, or interpretation of data; in the writing of the manuscript, or in the decision to publish the results.

Abbreviations

$1/n$	adsorption strength
a	the Langmuir isotherm constant (binding energy) (L mg^{-1})
C_e	the equilibrium concentration (mg L^{-1})
IARC	the International Agency for Research on Cancer
k	adsorption capacity
k_1	the pseudo-first-order (h^{-1}) rate constant
k_2	the pseudo-second-order (g mg^{-1} h^{-1}) rate constants
NA11	nickel-aluminum type hydroxide (the Ni^{2+}:Al^{3+} molar ratios of 1.0:1.0)
NA12	nickel-aluminum type hydroxide (the Ni^{2+}:Al^{3+} molar ratios of 1.0:2.0)
NAZ1	nickel-aluminum-zirconium type hydroxide (the Ni^{2+}:Al^{3+}:Zr^{4+} molar ratios of 0.9:1.0:0.09)
NAZ2	nickel-aluminum-zirconium type hydroxide (the Ni^{2+}:Al^{3+}:Zr^{4+} molar ratios of 0.9:2.0:0.09)
q_e	the quantity of Cr(VI) adsorbed (mg g^{-1}) at equilibrium
q_t	the quantity of Cr(VI) adsorbed (mg g^{-1}) at given time t
SGDs	Sustainable Development Goals
USEPA	U.S. Environmental Protection Agency
WHO	World Health Organization
W_s	the maximum quantity of adsorbed Cr(VI) (mg g^{-1})

References

1. Liu, X.J.; Zeng, H.Y.; Xu, S.; Chen, C.R.; Zhang, Z.Q.; Du, Z.J. Metal oxides as dual-functional adsorbents/catalysts for Cu^{2+}/Cr(VI) adsorption and, ethyl orange oxidation catalysis. *J. Taiwan Inst. Chem. Eng.* **2016**, *60*, 414–422. [CrossRef]
2. Li, Y.; Wang, J.; Li, Z.; Liu, Q.; Liu, J.; Liu, L.; Zhang, Z.; Yu, J. Ultrasound assisted synthesis of Ca-Al hydrotalcite for U(VI) and Cr(VI) adsorption. *Chem. Eng. J.* **2013**, *218*, 295–302. [CrossRef]
3. Dlugosz, A.; Rembacz, K.P.; Pruss, A.; Durlak, M.; Lembas-Bogaczyk, J. Influence of chromium on the natural antioxidant barrier. *Polym. J. Environ. Stud.* **2012**, *21*, 331–335.
4. Rao, M.V.; Jhala, D.D.; Patel, A.; Chettiar, S.S. Cytogenetic alteration induced by nickel and chromium in human blood cultures and its amelioration by curcumin. *Int. J. Hum. Genet.* **2008**, *8*, 301–306. [CrossRef]
5. Zhu, Y.; Jin, Y.; Chang, K.; Chen, Z.; Li, X.; Wu, X.; Jin, C.; Ye, F.; Shen, R.; Dong, W.; et al. Use of molybdenum disulfide nanosheets embellished nanoiron for effective capture of chromium(VI) ions from aqueous solution. *J. Mol. Liq.* **2018**, *259*, 376–383. [CrossRef]
6. Cao, E.; Duan, W.; Yi, L.; Wang, A.; Zheng, Y. Poly(*m*-phenylenediamine) functionalized *Calotropis gigantea* fiber for coupled adsorption reduction for Cr(VI). *J. Mol. Liq.* **2017**, *240*, 225–232. [CrossRef]
7. Ogata, F.; Nagai, N.; Itami, R.; Nakamura, T.; Kawasaki, N. Potential of virgin and calcined wheat bran biomass for the removal of chromium(VI) ion from a synthetic aqueous solution. *J. Environ. Chem. Eng.* **2020**, *8*, 103710. [CrossRef]
8. Ogata, F.; Ueta, E.; Kawasaki, N. Characteristics of a novel adsorbent Fe-Mg-type hydrotalcite and its adsorption capability of As(III) and Cr(VI) from aqueous solution. *J. Ind. Eng. Chem.* **2018**, *59*, 56–63. [CrossRef]
9. Juang, R.S.; Shiau, R.C. Metal removal from aqueous solutions using chitosan-enhanced membrane filtration. *J. Membr. Sci.* **2000**, *165*, 159–167. [CrossRef]
10. Baral, S.S.; Das, S.N.; Rath, P.; Chaudhury, G.R. Chromium (VI) removal by calcined bauxite. *Biochem. Eng. J.* **2007**, *34*, 69–75. [CrossRef]
11. Harry, I.D.; Saha, B.; Cumming, I.W. Effect of electrochemical reduction and oxidation of a viscose rayon based activated carbon cloth for Cr(VI) sorption from aqueous solution. *Ind. Eng. Chem. Res.* **2008**, *47*, 6734–6741. [CrossRef]
12. Cappelletti, G.; Bianchi, C.L.; Ardizzone, S. Nano-titania assisted photoreduction of Cr(VI): The role of the different TiO_2 polymorphs. *Appl. Catal. B Environ.* **2008**, *78*, 193–201. [CrossRef]
13. Ogata, F.; Kawasaki, N. Adsorption of As(III) from aqueous solutions by novel Fe-Mg type hydrotalcite. *Chem. Pharm. Bull.* **2015**, *63*, 1040–1046. [CrossRef]
14. Alvarez-Ayuso, E.; Nugteren, H.W. Purification of chromium(VI) finishing wastewaters using calcined and uncalcined Mg-Al-CO_3-hydrotalcite. *Water Res.* **2005**, *39*, 2535–2542. [CrossRef] [PubMed]
15. Lazaridis, N.K.; Asouhidou, D.D. Kinetics of sorptive removal of chromium(VI) drom aqueous solutions by calcined Mg-Al-CO_3 hydrotalcite. *Water Res.* **2003**, *37*, 2875–2882. [CrossRef]
16. Rodrigues, E.; Almeida, O.; Brasil, H.; Moraes, D.; dos Reis, M.A.L. Adsorption of chromium(VI) on hydrotalcit-hydroxyapatite material doped with carbon nanotubes: Equilibrium, kinetic and thermodynamic study. *Appl. Clay Sci.* **2019**, *172*, 57–64. [CrossRef]
17. Deng, L.; Shi, Z.; Li, B.; Yang, L.; Luo, L.; Yang, X. Adsorption of Cr(VI) and phosphate on Mg-Al hydrotalcite supported kaoline clay prepared by ultrasound-assisted coprecipitation method using batch and fixed-bed systems. *Ind. Eng. Chem. Res.* **2014**, *53*, 7746–7757. [CrossRef]
18. Chen, C.R.; Zeng, H.Y.; Xu, S.; Liu, X.J.; Duan, H.Z.; Han, J. Preparation of mesoporous material from hydrotalcite/carbon composite precursor for chromium(VI) removal. *J. Taiwan Inst. Chem. Eng.* **2017**, *70*, 302–310. [CrossRef]

19. Aregay, G.G.; Jawad, A.; Du, Y.; Shahzad, A.; Chen, Z. Efficient and selective removal of chromium(VI) by sulfide assembled hydrotalcite compounds through concurrent reduction and adsorption processes. *J. Mol. Liq.* **2019**, *294*, 111532. [CrossRef]
20. Xiao, L.; Ma, W.; Han, M.; Cheng, Z. The influence of ferric iron in calcined nano-Mg/Al hydrotalcite on adsorption of Cr(VI) from aqueous solution. *J. Hazard. Mater.* **2011**, *186*, 690–698. [CrossRef]
21. Ogata, F.; Nakamura, T.; Toda, M.; Otani, M.; Kawasaki, N. Evaluation of nickel-aluminum complex hydroxide for adsorption of chromium(VI) ion. *Chem. Pharm. Bull.* **2020**, *68*, 70–76. [CrossRef]
22. Ogata, F.; Nagai, N.; Toda, M.; Otani, M.; Saenjum, C.; Nakamura, T.; Kawasaki, N. Removal of arsenic(III) ion from aqueous media using complex nickel-aluminum and nickel-aluminum-zirconium hydroxides. *Water* **2020**, *12*, 1697. [CrossRef]
23. Ogata, F.; Iijima, S.; Toda, M.; Otani, M.; Nakamura, T.; Kawasaki, N. Characterization and phosphate adsorption capability of novel nickel-aluminum-zirconium complex hydroxide. *Chem. Pharm. Bull.* **2020**, *68*, 292–297. [CrossRef]
24. Long, F.; Gong, J.L.; Zeng, G.M.; Chen, L.; Wang, X.Y.; Deng, J.H.; Niu, Q.Y.; Zhang, H.Y.; Zhang, X.R. Removal of phosphate from aqueous solution by magnetic Fe–Zr binary oxide. *Chem. Eng. J.* **2011**, *171*, 448–455. [CrossRef]
25. Faria, P.C.C.; Orfao, J.J.M.; Pereira, M.F.R. Adsorption of anionic and cationic dyes on activated carbons with different surface chemistries. *Water Res.* **2004**, *38*, 2043–2052. [CrossRef] [PubMed]
26. Chitrakar, R.; Tezuka, S.; Sonoda, A.; Sakane, K.; Ooi, K.; Hirotsu, T. Synthesis and phosphate uptake behavior of Zr^{4+} incorporated MgAl-layered double hydroxides. *J. Colloid Interf. Sci.* **2007**, *313*, 53–63. [CrossRef]
27. Miyauchi, H.; Yamamoto, T.; Chitrakar, R.; Makita, Y.; Wang, Z.; Kawai, J.; Hirotsu, T. Phosphate adsorption site on zirconium ion modified MgAl-layered double hydroxides. *Top. Catal.* **2009**, *52*, 714–723. [CrossRef]
28. Wang, W.; Zhou, J.; Achari, G.; Yu, J.; Cai, W. Cr(VI) removal from aqueous solutions by hydrothermal synthetic layered double hydroxides: Adsorption performance, coexisting anions and regeneration studies. *Colloids Surf. A Physicochem. Eng. Asp.* **2014**, *457*, 33–40. [CrossRef]
29. Lu, Y.; Jiang, B.; Fang, L.; Ling, F.; Gao, J.; Wu, F.; Zhang, X. High performance NiFe layered double hydroxide for methyl orange dye and Cr(VI) adsorption. *Chemosphere* **2016**, *152*, 415–422. [CrossRef]
30. Yue, X.; Liu, W.; Chen, Z.; Lin, Z. Simultaneous removal of Cu(II) and Cr(VI) by Mg-Al-Cl layered double hydroxide and mechanism insight. *J. Environ. Sci.* **2017**, *53*, 16–26. [CrossRef]
31. Lazaridis, N.K.; Pandi, T.A.; Matis, K.A. Chromium(VI) removal from aqueous solutions by Mg-Al-CO_3 hydrotalcite: Sorption-desorption kinetic and equilibrium studies. *Ind. Eng. Chem. Res.* **2004**, *43*, 2209–2215. [CrossRef]
32. Khitous, M.; Salem, Z.; Halliche, D. Effect of interlayer anions on chromium removal using Mg-Al layered double hydroxides: Kinetic, equilibrium and thermodynamic studies. *Chinese J. Chem. Eng.* **2016**, *24*, 433–445. [CrossRef]
33. Li, Y.; Gao, B.; Wu, T.; Sun, D.; Li, X.; Wang, B.; Lu, F. Hexavalent chromium removal from aqueous solution by adsorption on aluminum magnesium mixed hydroxide. *Water Res.* **2009**, *43*, 3067–3075. [CrossRef] [PubMed]
34. Yan, L.G.; Yang, K.; Shan, R.R.; Yu, H.Q.; Du, B. Calcined ZnAl- and Fe_3O_4/ZnAl-layered double hydroxides for efficient removal of Cr(VI) from aqueous solution. *RSC Adv.* **2015**, *5*, 96495. [CrossRef]
35. Enshirah, D.; Silva, N.D.; Sayari, A. Adsorption of copper on amine functionalized SBA-15 prepared by co-condensation: Kinetic properties. *Chem. Eng. J.* **2011**, *166*, 454–459.
36. Demirbas, E.; Dizge, N.; Sulak, M.T.; Kobya, M. Adsorption kinetics and equilibrium of copper from aqueous solutions using hazelnut shell activated carbon. *Chem. Eng. J.* **2009**, *148*, 480–487. [CrossRef]
37. Atkin, P.W. *Physical Chemistry*, 6th ed.; Oxford University Press: Oxford, UK, 1998.
38. Vimonses, V.; Lei, S.; Bo, J.; Chow, C.W.K.; Saint, C. Kinetic study and equilibrium isotherm analysis of congo red adsorption by clay materials. *Chem. Eng. J.* **2009**, *148*, 354–364. [CrossRef]
39. Freundlich, H.M.T. Over the adsorption in solution. *J. Phys. Chem.* **1906**, *57*, 385–471.
40. Langmuir, I. The constitution and fundamental properties of solids and liquids. *J. Am. Chem. Soc.* **1916**, *38*, 2221–2295. [CrossRef]
41. Abe, I.; Hayashi, K.; Kitagawa, M. Studies on the adsorption of surfactants on activated carbons. I. Adsorption of nonionic surfactants. *Yukagaku* **1976**, *25*, 145–150.

water

MDPI

Article

Equilibrium, Kinetics and Thermodynamics of Chromium (VI) Adsorption on Inert Biomasses of *Dioscorea rotundata* and *Elaeis guineensis*

Angel Villabona-Ortíz [1], Ángel González-Delgado [2] and Candelaria Tejada-Tovar [1,*]

[1] Process Design and Biomass Utilization Research Group (IDAB), Chemical Engineering Department, Universidad de Cartagena, Avenida del Consulado St. 30, Cartagena de Indias 130015, Colombia; avillabonao@unicartagena.edu.co

[2] Nanomaterials and Computer Aided Process Engineering Research Group (NIPAC), Chemical Engineering Department, Universidad de Cartagena, Avenida del Consulado St. 30, Cartagena de Indias 130015, Colombia; agonzalezd1@unicartagena.edu.co

* Correspondence: ctejadat@unicartagena.edu.co

Abstract: Adsorption equilibrium and kinetics on lignocellulosic base adsorbents from oil palm bagasse (OPB) and yam peels (YP) were studied for the removal of hexavalent chromium present in aqueous solution, in a batch system, evaluating the effect of temperature, adsorbent dose and particle size on the process. Isotherms were fitted to Langmuir, Freundlich and Dubinin–Radushkevich isothermal models. Kinetic data were adjusted to the pseudo-first-order, pseudo-second-order and Elovich models. Thermodynamic parameters were estimated by the van't Hoff method. From characterization of adsorbents, the presence of a porous surface typical of lignocellulosic materials was found, with hydroxyl, amine and carboxyl functional groups. It was also found that the highest adsorption capacity was obtained at 0.03 g of adsorbent, 55 °C and 0.5 mm, reporting an adsorption capacity of 325.88 and 159 mg/g using OPB and YP, respectively. The equilibrium of adsorption on OPB is described by Langmuir and Freundlich isotherms, while that of YP is described by Dubinin–Radushkevich's model, indicating that the adsorption is given by the ion exchange between the active centers and the metallic ions. A maximum adsorption capacity was obtained of 63.83 mg/g with OPB and 59.16 mg/g using YP, according to the Langmuir model. A kinetic study demonstrated that equilibrium time was 200 min for both materials; kinetic data were described by pseudo-second-order and Elovich models, thus the mechanism of Cr (VI) adsorption onto the evaluated materials is dominated by a chemical reaction. The thermodynamic study determined that the elimination of YP is endothermic, irreversible and not spontaneous, while for OPB it is exothermic, spontaneous at low temperatures and irreversible.

Keywords: bio-adsorption; Langmuir; Freundlich; Dubinin–Radushkevich

Citation: Villabona-Ortíz, A.; González-Delgado, Á.; Tejada-Tovar, C. Equilibrium, Kinetics and Thermodynamics of Chromium (VI) Adsorption on Inert Biomasses of *Dioscorea rotundata* and *Elaeis guineensis*. *Water* **2022**, *14*, 844. https://doi.org/10.3390/w14060844

Academic Editors: Cristina Palet and Julio Bastos-Arrieta

Received: 12 January 2022
Accepted: 3 March 2022
Published: 8 March 2022

1. Introduction

The presence of heavy metal ions in wastewater is a serious problem due to the discharge of contaminated effluents into the environment [1]. Among the pollutants generally discharged to water bodies are heavy metals; these are found in the wastewater from industrial processes involving inks, metallurgy, batteries, oil, mining and tanneries, among others [2]. Chromium appears in aqueous systems in the forms +3 and +6, with the hexavalent chromium being much more toxic than the trivalent [3]; these ions are widely used in the manufacture of inks, industrial dyes and paint pigments, chrome plating, aluminum anodizing and other cleaning, coating and electroplating operations of metals and gold mining, among others [4].

Several removal methods have been used in the capture of hexavalent chromium due to its toxic health effects [5]. These technologies include filtration, chemical precipitation, adsorption, electrodeposition and membrane systems or even ion exchange processes [6]. Among these methods, bio-adsorption is one of the most economically favorable methods; it is technically easier, with high availability of adsorbents and the possibility of reuse [7]. The removal of hexavalent chromium has been studied using biomasses from different origin, such as corn residues [8], coffee pulp [3], kenaf [9], oil palm ash [5] and peels of lime [4], nut [10], orange [11] and plantain [12], obtaining good performance of biomaterials due to the presence of hydroxyl, carboxyl, amino and carbonyl functional groups, due to their lignocellulosic nature [13].

In Colombia, the intensive and extensive planting of African palm for use in oil extraction generates large volumes of waste during the process, including palm bagasse and shells; due to the 60% of increase in oil production from oil palm, about 500,000 tons of waste are generated per year [14]. Although the shell obtained as a by-product is generally used as fuel for boilers and composting, the residual fiber is a rejected product that does not have a use after fruit separation. Thus, the recovery of waste such as waste fiber is very important to mitigate its effects on the environment, and guarantee a sustainable process. Among the various potential applications that could be given to residual palm fiber, its use as an adsorbent in the treatment of water contaminated with heavy metals, among a wide range of applications, seems attractive taking into account the reported removal yields for similar residues [15]. In addition, Colombia was ranked among the 12 countries with the highest yam production worldwide and ranked first in terms of product yield with 28.3 tons per planted hectare, generating after harvesting around 5000 tons of waste per year [16].

In the studies of hexavalent chromium adsorption, several variables have been evaluated such as adsorbent dose, initial concentration, particle size and temperature, which are parameters that affect the availability of active sites of the adsorbent, the diffusive and mass transfer phenomena of the pollutant from the solution to the adsorbent, the exposed surface area and the mechanisms that dominate the adsorption of the pollutant [17]. In this sense, the determination of the adsorption isotherm and thermodynamic parameters (change in Gibbs free energy, enthalpy, and entropy) are necessary for helping to understand the mechanisms and phases involved in chromium adsorption [18]. Thus, the objective of this study was to determine the effect of adsorbent dose, particle size and temperature on the adsorption capacity of chromium (VI) using yam peels (YPs) and oil palm residues (OPBs) in a batch system. Adsorption equilibrium on adsorbents was studied and fitted to Langmuir, Freundlich, and Dubinin–Radushkevich isothermal models. Thermodynamic parameters were estimated by the van't Hoff graphical method.

2. Methodology

2.1. Materials and Reagents

Oil palm waste (OPB) was obtained as a rejection product of the oil extraction process and yam peels as a by-product of post-harvest handling in the department of Bolívar (Colombia). The lignocellulosic material was washed, sun-dried to a constant mass, and subjected to size reduction in a roller mill and size classification in a sieve-type shaker on stainless steel meshes. The 100 mg/L chromium solution used in the adsorption tests was prepared with analytical grade potassium dichromate ($K_2Cr_2O_7$) (Merck Millipore®, Burlington, MA, USA). To adjust the pH we used hydrochloric acid (HCl) and sodium hydroxide (NaOH) 0.5 M solutions to adjust to pH 2 [8]. All the reagents used in the present study were analytical grade.

The adsorbents were characterized by Fourier transform infrared spectroscopy analysis in a Shimadzu (Kyoto, Japan) IRAinfinity-1S spectrophotometer model, with a frequency of 32 scans between 400 and 4000 cm^{-1}, for the purpose of identifying the functional groups that would work as active centers during the process of metal removal. The morphological properties and elemental composition of the biomass were determined using a scanning electron microscope together with an energy dispersive spectroscopy (SEM-EDS) model

JSM-6490LV JEOL Ltd. (JEOL Ltd., Akishima, Japan). All trials were conducted before and after the metal removal process.

2.2. Adsorption Tests

The YH and OPB wastes were washed, sun-dried to remove moisture and reduced in size in a hand-roller mill, and sized in a vibrating sieve machine using stainless steel meshes. The materials were stored in airtight plastic bags to keep them in good condition.

For the experimental development, a continuous factor central composite design with a star-type response surface was selected and is shown in Table 1, the design was made using the software Statgraphics Centurion XVI.II. For the classification of biomaterial sizes, mesh numbers 120, 45, 35, 18 and 16 were used, which select size ranges of 0.1060.125 mm, 0.3–0.355 mm, 0.425–0.5 mm, 1.0–1.18 mm and 1.18–1.4 mm, respectively.

Table 1. Experimental design.

Independent Variables	Units	Range and Levels				
		Lower Level	Low Level	Center Point	High Level	Higher Level
Particle size	mm	0.14	0.355	0.5	1.0	1.22
Adsorbent dose	g	0.003	0.15	0.325	0.5	0.62
Temperature	°C	29.8	40	55	40	80

Adsorption experiments were performed in a batch system using an orbital shaker by contacting 100 mL of solution at pH 2 [19], for 200 rpm per 24 h, at the conditions described in Table 1. The remaining concentration of hexavalent chromium in the solution was determined by ASTM D 1687-02, by adding 48 mL of deionized water, 2 mL of 1,5-diphenylcarbazide solved in acetone and 0.5 mL of Cr (VI) sample in acid conditions; for this purpose, a Biobase spectrophotometer BK-UV1900 was used, setting the wavelength to 540 nm [20]. The spent adsorbent after adsorption tests was removed by filtration, using cellulose membrane microfilters of 0.45 μm. The adsorption capacity was determined according to Equation (1), where q_t (mg/g) is the adsorption capacity of the material, C_o and C_f (mg/L) are the initial and final concentration of chromium, V (L) is the volume of solution and m (g) is the mass of adsorbent. In order to evaluate the reproducibility of the results, the experiments were conducted in triplicate.

$$q_t = \frac{\left(C_o - C_f\right) * V}{m} \tag{1}$$

2.3. Adsorption Equilibrium

An adsorption isotherm describes the equilibrium of the material's adsorption on a surface at a constant temperature; it represents the amount of material bound to the surface as a function of the material present in the gas phase or dissolution, as well as determining the interactions that control the process [21]. The data from the effect of concentration of the solution (25, 50, 75, 100, 125 and 150 mg/L) were adjusted to the models of Langmuir, Freundlich, and Dubinin–Radushkevich.

Freundlich's isotherm (Equation (2)) assumes the formation of multilayers during the adsorption process due to the different activation energies and affinities of the active centers of the adsorbent [22].

$$q_e = k_f C_e^{1/n} \tag{2}$$

where k_f is the Freundlich constant and represents the distribution coefficient (L/g), n represents the adsorption intensity and indicates the heterogeneity of the active sites, q_e is the amount of metal adsorbed at equilibrium (mg/g) and C_e is the residual metal concentration in solution (mg/L) [23].

Langmuir's adsorption isotherm (Equation (3)) quantitatively describes the deposition of ions uniformly on an adsorbing surface as a function of the concentration of the adsorbed material in the liquid with which it is in contact [24].

$$q_e = q_{max}\frac{K_L C_e}{1 + K_L C_e} \tag{3}$$

where q_{max} is the maximum adsorption of the Langmuir model (mg/g) and K_L is the Langmuir constant and can be correlated with the variation of the adsorption area and the porosity of the adsorbent [23].

The Dubinin–Radushkevich isothermal model (Equation (4)) is a semi-empirical equation applied to express the mechanism of adsorption with Gaussian energy distribution on heterogeneous surfaces and is only suitable for an intermediate range of adsorbate concentrations because it exhibits asymptotic behavior. It assumes a multilayer character involving Van Der Waals forces, applicable for physical adsorption processes; it is usually applied to differentiate between physical and chemical adsorption of metal ions [25].

$$q_e = q_{DR} e^{-k_{DR}\varepsilon^2} \tag{4}$$

$$\varepsilon = RT \times ln\left(1 + \frac{1}{C_e}\right) \tag{5}$$

$$E = \frac{1}{\sqrt{2K_{DR}}} \tag{6}$$

where ε^2 is the Polanyi potential, K_{DR} is the Dubinin–Radushkevich constant related to adsorption energy (mol^2/kJ^2), E is the average adsorption energy per molecule of adsorbate required to transfer one mol of ion from solution to the adsorbent surface (kJ/mol), R is the gas constant (8.314 J/mol K) and T is the absolute temperature.

2.4. Kinetic Study

After the experimental design proposed in Table 1, the best conditions were estimated and the kinetic study was carried out, putting 10 mL of solution in contact with the adsorbent, taking 8 samples at different time intervals in 24 h. The experimental kinetics data were modeled to the pseudo-first-order, pseudo-second-order and Elovich models through non-linear adjustment in the Origin Pro 9 program, maximizing R^2; equations are shown in Table 2.

Table 2. Kinetic models evaluated in this work.

Kinetic Model	Equation	Parameters
Pseudo-first-order	$q_t = q_e\left(1 - e^{-k_1 t}\right)$	k_1 (min^{-1}): kinetic constant of pseudo-first-order
Pseudo-second-order	$q_t = \frac{t}{\left(\frac{1}{q_e^2 k_2}\right)+\left(\frac{t}{q_e}\right)}$	k_2 (g/mg min): kinetic constant of pseudo-first-order
Elovich	$q_t = \frac{1}{\beta}\ln(\alpha\beta) + \frac{1}{\beta}\ln(t)$	β (g/mg): exponent indicating the capacity of adsorption
		α (mg/g min): Elovich constant indicating the rate of adsorption

2.5. Estimation of Thermodynamic Parameters

The thermodynamic parameters were estimated using the graphical method based on the Vant't Hoff equation in order to determine the spontaneity, type of adsorption and predict the magnitude of changes on the surface of the adsorbent [26]. The analysis of these parameters will make it possible to estimate the feasibility of the adsorption process and the effect of temperature [27]. We estimated the change in Gibbs' standard free energy (ΔG^o), the standard enthalpy (ΔH^o) and standard entropy (ΔS^o). Adsorption experiments were performed by varying the temperature from 29.8 to 81 °C, considering the conditions established in the design of experiments; then, a graph of $Ln(K_c)$ vs. T^{-1} was plotted and the parameters were calculated using the following equations:

$$lnK_c = \frac{-\Delta H^o}{RT} + \frac{\Delta S^o}{R} \tag{7}$$

$$\Delta G^o = -RT \times lnK_c \tag{8}$$

$$K_c = \frac{C_{ac}}{C_{se}} \tag{9}$$

where R is the constant of the ideal gases (8.314 J/mol K), K_c is the constant of equilibrium, C_{ac} is the concentration of the adsorbate in the equilibrium contained on the surface of the adsorbent, C_{se} is the concentration in solution at equilibrium and T is the absolute temperature. ΔH^o (kJ/mol K) and ΔS^o (kJ/mol) are determined from the slope and the intercept with the y-axis of the Arrhenius graph of $lnKc$ vs. T^{-1}, respectively.

The final disposition of the residues produced during the experimental part of this research was made according the institutional politics of the Universidad de Cartagena-Colombia. Liquid residue products of experimentation were collected by a specialized company hired for this purpose. Solid residues, such as saturated adsorbent, were dried and used as aggregates in mortars, in order to evaluate their lixiviation possibility in structures like concrete, asphalt and brick.

3. Results

3.1. Characterization of OPB and YP

The adsorption process is a complex phenomenon controlled by several mechanisms, which are generally sequential. Among these, the surface chemistry and the pore structure of the adsorbents have a considerable effect on the process, since they establish the transfer and diffusion interactions for the capture of the metal ions [28]. Thus, the adsorbents were characterized by Brunauer–Emmett–Teller (BET) analysis, obtaining that YP presented a surface area of 0.9463 m^2/g, pore volume of 0.005452 m^3/g and pore size of 23.04419 nm; while OPB presented 2.7317 m^2/g, 0.011207 m^3/g and 16.41 nm for surface area, volume and pore size, respectively. It was established that the OPB area is higher than the YP, which could be due to its fibrous structure since it would present higher surface wear, as it is a residue from the extraction of palm oil, leading to a change in the arrangement of cellulose, hemicellulose and lignin structures in the material structure and causing an increase in surface area [29]. Both materials have pore sizes between 2 and 50 nm, which indicates that they are mesoporous materials; this makes them suitable adsorbents for adsorption in the liquid phase, since it facilitates the diffusion of the adsorbent in the adsorbent structure [30].

Figures 1 and 2 show the scanning electron microscopy (SEM) images of the adsorbents under study. YP has an irregular and porous surface, which allows a large interface for heterogeneous biosorption, while OPB has a smooth, fibrous surface, which presents a certain porosity in its structure that is typical of lignocellulosic materials [31].

Table 3 shows the chemical composition in weight % and atomic % for the evaluated adsorbents obtained from the EDS analysis; carbon and oxygen are the ones with the highest presence in all the materials studied, which can be attributed to their organic nature [28]. For YP there is 57.07% C, 39.05% O, 1.65% K and elements in small proportions: Si, P, S, Ca, Fe and Cu. For OPB, 59.07% C, 38.85% O, 1.92% Si, Ca and Cu were obtained.

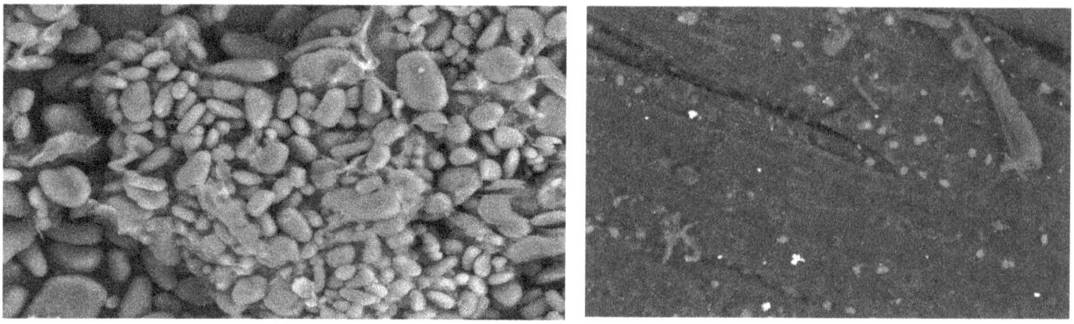

(a) (b)

Figure 1. SEM micrographs of OPB (**a**) before and (**b**) after Cr (VI) adsorption with magnification of ×500.

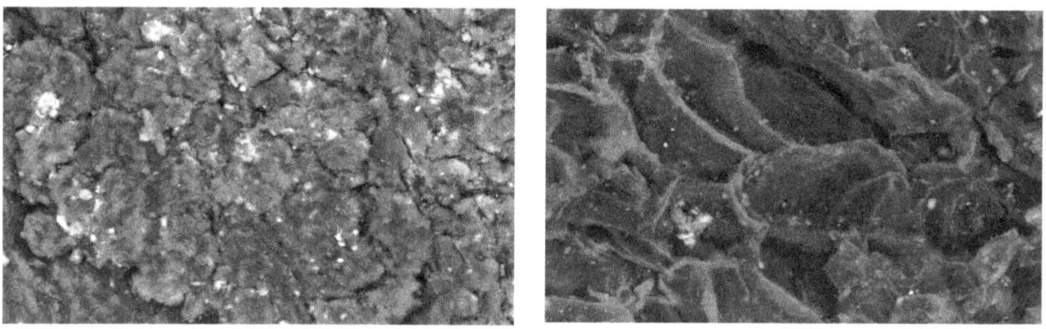

(a) (b)

Figure 2. SEM micrographs of YP (**a**) before and (**b**) after Cr (VI) adsorption with magnification of ×500.

Table 3. EDS composition analysis.

Element	OPB		OPB-Cr (VI)		YP		YP-Cr (VI)	
	Weight %	Atomic %	Weight %	Atomic %	Weight %	Atomic %	Weight %	Atomic %
C	50.90	2.36	48.65	56.49	47.40	57.07	54.50	63.25
O	44.60	2.22	48.82	42.56	43.20	39.05	39.67	34.56
Al					0.75	0.40	0.82	0.42
Si	3.87	0.20	1.12	0.56	2.16	1.11	1.21	0.60
P					0.66	0.31		
S					0.17	0.08		
K			0.20	0.07	4.46	1.65	0.44	0.16
Ca	0.19	0.05	0.25	0.09	0.34	0.12	0.99	0.34
Fe					0.60	0.16		
Cu	0.45	0.1	0.49	0.11	0.27	0.06		
Cl							0.45	0.18
Cr			0.48	0.13			0.88	0.24
Totals	100.00		100.00		100.00		100.00	

From the FTIR analysis, we established which functional groups are present, and which could also be involved in the adsorption of the metal under study. Thus, in Figure 3, the spectrum of the biomaterials evaluated before and after the adsorption of Cr (VI) can be observed. From these, the shifting of the bands and broadening of the peaks after the adsorption process in both materials are observed.

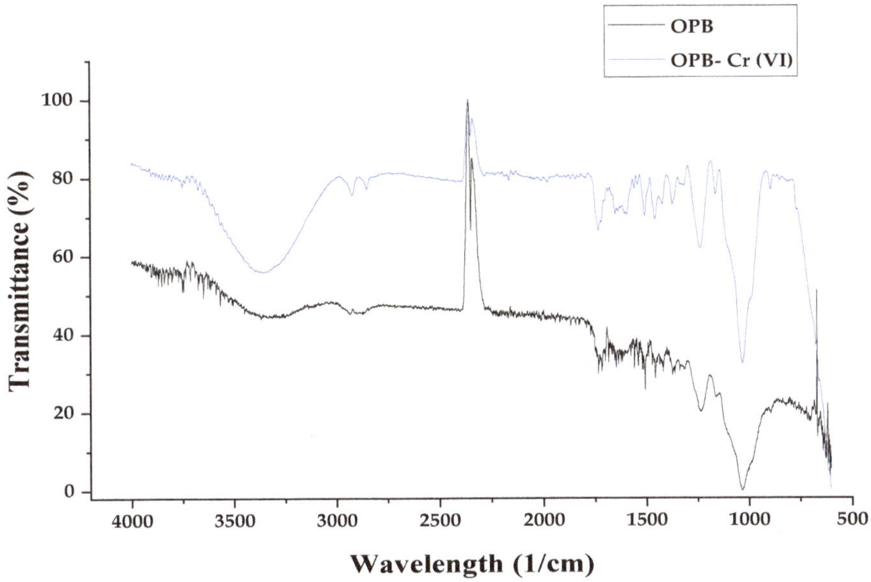

(**a**)

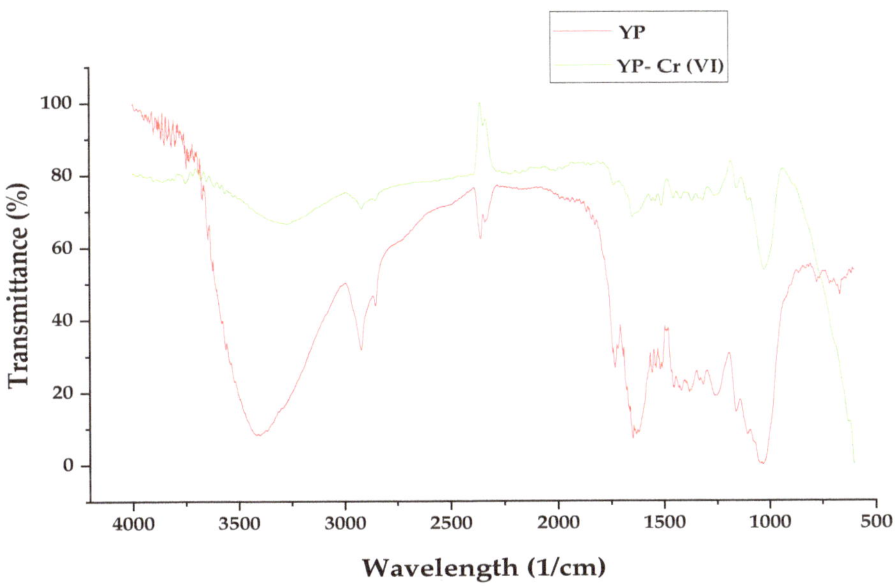

(**b**)

Figure 3. Normalized FTIR spectra for (**a**) OPB and (**b**) YP, before and after Cr (VI) adsorption.

The spectrum in Figure 3 initially shows a broad band at 3400 cm^{-1} that is attributed to the stretching vibrations of the OH bond, while the peaks at 2950 cm^{-1} and 1350 cm^{-1}

denote the C-H vibration [11]. The peak that appears at 1650 cm^{-1} corresponds to the intense signal characteristic of the stretching of the carbonyl group of carboxylic acids, as well as the one that appears at 1400 cm^{-1} related to the torsion of the carboxyl, and at 900 cm^{-1} the peak indicates the out-of-plane twisting of the carboxyl dimer [32]. The stretched band at 1150 cm^{-1} is related to the stretching of the COO- anion, at 1080 cm^{-1} to the symmetric and asymmetric stretching of the C-O group of the ester vibration and at 1030 cm^{-1} to the functional group C-O-C [3]. The presence of hydroxyl groups, carbonyl groups, ethers and aromatic compounds is due to the high content of cellulose and lignin in the biomaterial; it is evidence of the lignocellulosic structure of the biomaterial under study, which contributes to the metal ion adsorption process [33].

3.2. *Effect of the Temperature, Particle Size and Adsorbent Dose*

From the hexavalent chromium adsorption tests using YP and OPB as the adsorbent, we evaluated the effect of temperature, particle size and amount of metal adsorbent by calculating the hexavalent chromium adsorption capacity using Equation (1). Table 4 lists the results of the experiments with their standard deviations (SD), to evaluate the adsorption capacity of Cr (VI) on OPB and YP.

Table 4. Experimental results for adsorption capacity of Cr (VI) onto OPB and YP, evaluating the effect of temperature, particle size and adsorbent dose.

Variable Conditions			Adsorption Capacity (mg/g)	
Temperature (°C)	Particle Size (mm)	Adsorbent Dose (g)	OPB	YP
40.0	0.355	0.15	38.36 ± 0.02	66.67 ± 0.07
70.0	1.0	0.15	15.82 ± 0.03	66.67 ± 0.04
40.0	1.0	0.5	18.02 ± 0.007	19.99 ± 0.04
55.0	0.5	0.62	14.89 ± 0.01	15.96 ± 0.01
55.0	0.14	0.325	7.15 ± 0.07	30.769 ± 0.08
80.0	0.5	0.325	30.77 ± 0.01	30.77 ± 0.01
55.0	1.22	0.325	30.77 ± 0.09	30.77 ± 0.05
29.8	0.5	0.325	25.22 ± 0.06	30.77 ± 0.07
55.0	0.5	0.03	325.88 ± 0.008	159.77 ± 0.001
70.0	0.355	0.15	66.66 ± 0.02	66.67 ± 0.04
40.0	1.0	0.15	34.21 ± 0.02	64.36 ± 0.08
55.0	0.5	0.325	29.48 ± 0.005	29.49 ± 0.006
40.0	0.355	0.5	19.99 ± 0.009	19.99 ± 0.04
70.0	1.0	0.5	4.69 ± 0.0005	19.99 ± 0.03
70.0	0.355	0.5	19.99 ± 0.03	19.99 ± 0.02
55.0	0.5	0.325	29.47 ± 0.002	29.49 ± 0.004

From the results presented in Table 4, we found that at 0.03 g of biomaterial, 0.5 mm of main particle size and 55 °C, the best adsorption was achieved for both biomaterials, reaching 159.77 mg/g when using YP and 325.88 mg/g with OPB. There is evidence of a decrease in adsorption capacity as the amount of biomaterial increases, which is due to the diversity of the biomasses evaluated, so there are a large number of active adsorption sites available due to the presence of the hydroxyl, carbonyl, amine and hydrocarbon groups present in the structure of lignin, hemicellulose, cellulose and pectin [22]. Due to this presence of active adsorption centers, it would appear that the greater the amount of adsorbent, the greater part of these would remain free after the process, thus decreasing the adsorption capacity [34].

The effect of temperature can be explained due to the stimulation of the active adsorption centers with the increase in temperature, as well as the increase in the intra-particle diffusion speed from the solution to the adsorbent, assuming that the diffusion is an endothermic process at the initial temperatures evaluated [35]. It is possible that at higher temperatures (70 and 80 °C) the interaction forces between adsorbate and adsorbent weaken [36]. An increase in temperature would favor the desorption of the sorbate molecules on the structure of the sorbent, since by gaining more energy than the bonds formed between the compound support, the forces of interaction are reduced and cause the release, which is what is not known [4].

In addition, it is evident that the adsorption capacity increases with particle size in the range 0.14 mm to 0.5 mm, and then decreases from 1.0 mm to 1.2 mm. This is explained by the influence of particle size on the surface contact area of the adsorbents, which increases when the particle size is smaller, and in turn, favors the rate of diffusion increased by increasing temperature [36]. However, a very small size would resist diffusive surface phenomena and subsequent mass transfer through the pores [37]. Considering that a smaller particle size increases the contact area between the adsorbent material and the ion, it would be expected that at 0.14 mm the best performance of YP and OPB can be obtained; however, this does not happen. This can be explained because Cr (VI) has an atomic radius of 1.27 Å and an ionic radius of 0.69 Å, which are smaller than the size of the pores of both adsorbents (23.04 nm for YP and 16.41 nm for YP). It has been shown that, as long as the radius of the particles is higher than the penetration depth of the diffusion, the adsorption capacity will increase, since, if the penetration of the contaminant is not sufficient, it would not increase the adsorption capacity [38]. Table 5 shows the results obtained from the analysis of variance (ANOVA), performed with the Statgraphics Centurión XVI.I software for the adsorption capacity of Cr (VI) on OPB and YP.

Table 5. Variance analysis for Cr (VI) adsorption capacity onto OPB and YP.

Factor	OPB			YP		
	Sum of Squares	F-Ratio	*p*-Value	Sum of Squares	F-Ratio	*p*-Value
A: Temperature	2.58	0.00	0.98	0.39	0.00	0.9697
B: Particle size	77.59	0.02	0.91	0.39	0.00	0.9697
C: Adsorbent dose	27,727.1	5.55	0.06	13301.7	53.51	0.0003
AA	1261.07	0.25	0.63	18.41	0.07	0.7946
AB	450.20	0.09	0.77	0.66	0.00	0.9604
AC	67.49	0.01	0.91	0.66	0.00	0.9604
BB	2046.08	0.41	0.55	18.41	0.07	0.7946
BC	177.76	0.04	0.86	0.66	0.00	0.9604
CC	13,858.1	2.77	0.15	3266.42	13.14	0.0110
Total error	29,980.2			1491.49		
Total (corr.)	87,589.2			19569.7		

From the values presented in Table 5, it was established that only for the adsorption of Cr (VI) when using YP does the amount of adsorbent (C) and its interaction CC have a significantly negative effect on the adsorption process. This is explained by the fact that the higher the amount of adsorbent, although there is higher removal efficiency, the material requirement increases to remove the same amount of metal ions. In the same way, when using a higher dose of material, there would be empty adsorption sites, thus reducing the retention capacity of the metal ions per available adsorbent mass [39]. It can be said that the temperature and the particle size do not have a significant influential effect on the adsorption capacity.

3.3. Adsorption Isotherm

The effect of the initial concentration on the adsorption capacity of Cr (VI) was evaluated by varying the concentration of the contaminant in intervals of 25 mg/L to 150 mg/L at pH 2 per 24 h, in order to understand the driving forces involved in the removal [40].

Figure 4 shows the non-linear fitting of experimental data to Langmuir, Freundlich, and Dubinin–Radushkevich models; the fitting parameters are summarized in Table 6, for the non-linearized and linearized forms of the models.

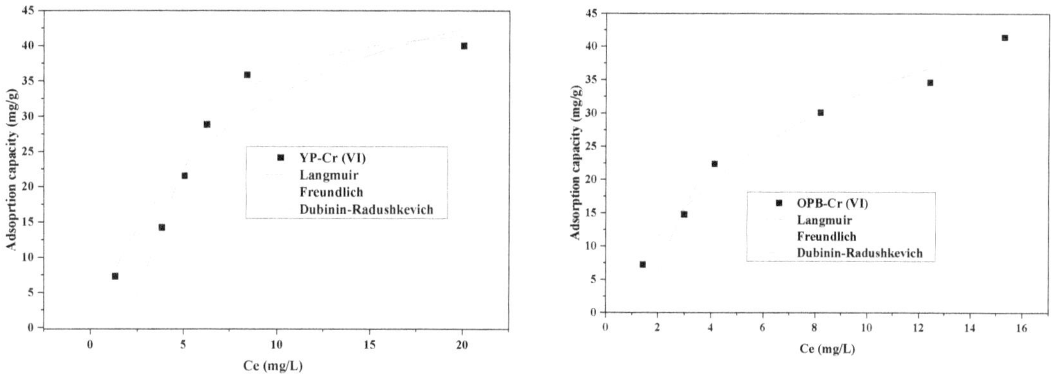

Figure 4. Modeling of hexavalent chromium adsorption isotherms over (**a**) YP and (**b**) OPB; at pH = 2, 200 rpm, 0.03 g of adsorbent dose and 55 °C.

Table 6. Adjustment parameters of hexavalent chromium adsorption isotherms on YP and OPB.

Model	Parameters	Non-Linear		Linear	
		YP	OPB	YP	OPB
Langmuir	q_{max} (mg/g)	59.16	63.83	60.60	68.49
	K_L	0.13	0.11	0.11	0.09
	R^2	0.89	0.98	0.89	0.94
Freundlich	K_f	9.94	8.27	6.74	6.65
	$1/n$	0.49	0.59	1.48	1.44
	n	2.03	1.69	0.68	0.69
	R^2	0.90	0.96	0.90	0.95
Dubinin–Radushkevich	q (mg/g)	43.28	38.39	38.95	32.9
	k_{DR}	2.65×10^{-6}	1.45×10^{-6}	6.86×10^{-6}	1.05×10^{-5}
	E	434.29	587.04	269.95	218.11
	R^2	0.91	0.90	0.85	0.90

The results reveal that Dubinin–Radushkevich's model describes the equilibrium of hexavalent chromium adsorption on OPB, while Langmuir's and Freundlich's models adjust the isotherm of ion adsorption in the study on OPB, because it is established that the physical and chemical adsorption phenomena control the adsorption process; thus, the retention of the contaminant occurs inside and in the internal pores of the adsorbent [41]. Dubinin–Radushkevich's model adjusts the isotherm on YP, with an average energy of adsorption of the ions per sorbate (E) of 434.29 kJ/mol, which is much higher than 8 kJ/mol, indicating that the process is mostly controlled by chemical adsorption with strong interactions between the active sites and the hexavalent chromium [42]. Dubinin–Radushkevich's model presents a good fit for the equilibrium of chromium adsorption using the two adsorbents evaluated because the calculated maximum adsorption capacity (q_{DR}) values are the closest to the experimental ones, with $R^2 \geq 0.9$ in both cases. This means that the adsorbents under study present a heterogeneous structure, which is required for a good fitting by the Dubinin–Radushkevich isotherm [43]. Similar R^2 values to those obtained in the present study when the Dubinin–Radushkevich model was evaluated for removing Cr (VI) were found for *Pleurotus mutilus* (R^2 = 0.93) [44], *Sterculia villosa* (R^2 values between

0.81 and 0.96) [43], activated carbon from Ziziphus jujuba (R^2 between 0.98 and 0.99) [18] and rice husk (R^2 = 0.93) [45].

Table 7 presents a comparison of results from previous studies when using lignocellulosic-based adsorbents for removing Cr (VI). It is observed that in all cases the adsorption process was evaluated at acid pH. In addition, a better performance was obtained for the OPB and YP; this might be due to the contact time between the adsorbate and adsorbent.

Table 7. Maximum adsorption capacity comparison of different adsorbent materials.

Adsorbent	Conditions	q_{max} (mg/g)	Reference
Oil palm fuel ash	pH = 2, 200 rpm, 80 g/L of adsorbent	0.462	[5]
Banana peel dust	pH = 1, 0.4 g of adsorbent dose, 300 rpm, 50 °C	26.46	[12]
Modified lychee peel	50 °C, pH = 2, 130 rpm, 0.08 g of adsorbent and 20 mL of solution	9.76	[17]
Foxtail millet shell	4 g/L of adsorbent dose, 0.25–0.35 mm of particle size, 30 mL of solution, 120 min, 350 rpm, and 25 °C	11.70	[21]
Oil palm bagasse	pH = 2, 200 rpm, 55 °C, 24 h	63.83	This work
Yam peel		59.16	

3.4. Kinetics of Adsorption

The kinetic study is important to determine the behavior of the adsorption process over time. Figure 5 shows the non-linear fit of the kinetic curves obtained for the removal of Cr (VI) on OPB and YP, and it is observed that the equilibrium times were 200 and 400 min for each adsorbent evaluated.

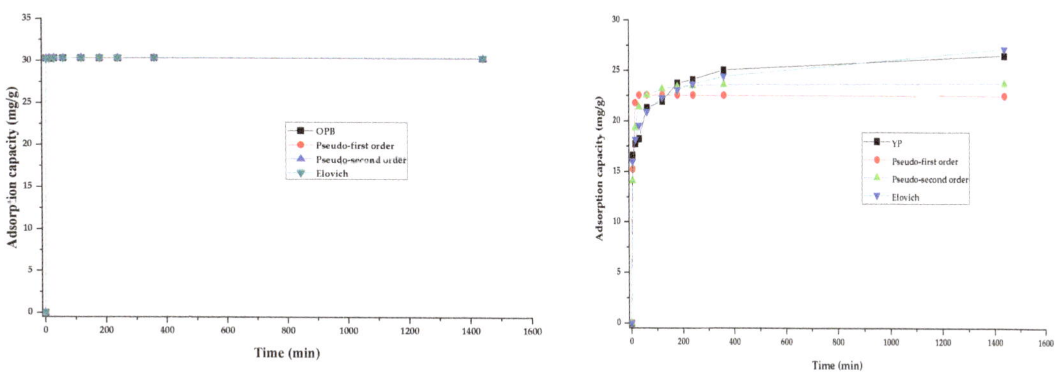

(a) (b)

Figure 5. Non-linear fit to kinetic models of Cr (VI) adsorption on (**a**) OPB and (**b**) YP; at pH = 2, 200 rpm, 0.03 g of sweet adsorbent, 100 mg/L of initial concentration and 55 °C.

For both biomaterials, it is highlighted that about 70% of the equilibrium concentration was reached 120 min after the start of the test, showing that the available sites in the adsorbent probably begin to saturate and therefore the capacity does not change significantly at higher times [45]. In order to study the adsorption mechanisms involved during the Cr (VI) removal process, pseudo-first-order, pseudo-second-order and Elovich linearized kinetic models were used, in their non-linearized forms. The parameters of these models were adjusted by linear regression and the results are shown in Table 8.

Table 8. Adjustment parameters of kinetic models.

Model	Parameters	OPB	YP
Pseudo-first-order	q_e	30.33	22.58
	k_1	1.334	0.23
	SS	0.009	64.93
	R^2	0.9999	0.9359
Pseudo-second-order	q_e	2,499,562.65	3925.03
	k_2	30.34	23.87
	SS	0.006	32.29
	R^2	0.9999	0.9961
Elovich	β	23.27	0.50
	α	1.25×10^{30}	1247.80
	SS	0.02	4.03
	R^2	0.9999	0.9867

For the YP-Cr (VI) system, it is evidenced that the three evaluated models present a good fit of the experimental data with $R^2 > 0.99$ in all cases, indicating that the removal process occurs through physis and chemisorption, due to the heterogeneity of the active centers of the YP [18]. However, the pseudo-second-order model is the one that describes the experimental data of removal of the metal under study on OPB, which indicates that the adsorption occurs by chemical reaction between the functional groups of the biomaterial and the metal ions [46]. From the values of the reaction rate constants of the pseudo-first- (k_1) and pseudo-second-order (k_2) models, it can be said that the initial sorption rate on YP is lower than for OPB, the latter presenting a greater adsorption capacity, which could be due to the fibrous nature of the OPB, as well as its superior surface area [5].

Previous results have been reported regarding the adsorption capacity of different materials such as 13.48 mg/g when using coffee pulp [3], 10.31 mg/g for lime peel [4] and 28 mg/g while using kenaf fiber [9], compared to 31.34 and 23.87 mg/g for OPB and YP, respectively, found in the present study. We note that these results are superior to the others mentioned for adsorbents prepared from residual crude lignocellulosic biomass. From the results obtained when using activated carbon from melon peel in the removal of Cr (VI), it was found that the process can be described by the Elovich equation, so that the limiting step in the adsorption of the metal ions may be the chemical interaction between metal ions and functional groups on the surface of the adsorbent [1].

3.5. Thermodynamic Adsorption Parameters

By calculating the thermodynamic parameters, it is possible to establish the feasibility, the adsorption mechanism, the energetic character of the process (endothermic or exothermic) and the spontaneity of the adsorption; this is done by determining the entropy (ΔS^o), enthalpy (ΔH^o) and the change in Gibbs free energy (ΔG^o), respectively. In Figure 6, we present the van't Hoff equation by graphing to 81 °C, considering the conditions established in the design of experiments; a graph was plotted of $Ln(K_c)$ vs. $1/T$.

The thermodynamic parameters of hexavalent chromium adsorption on OPB and YP are summarized in Table 9. From the positive value of ΔH° when using YP it is established that the processes of chromium removal are endothermic; consequently, energy must be supplied to the system to promote the diffusive phenomena in the solution and inside the pores of the adsorbent [32]. For OPB the process is exothermic and the limiting step is chemical adsorption [47]. The negative value of ΔS° shows that on the surface there was a good bond between the metal and the biomass, thus obtaining a low possibility of reversibility; there was also the probability of some structural changes due to the formation of bonds with functional groups at the interface [42].

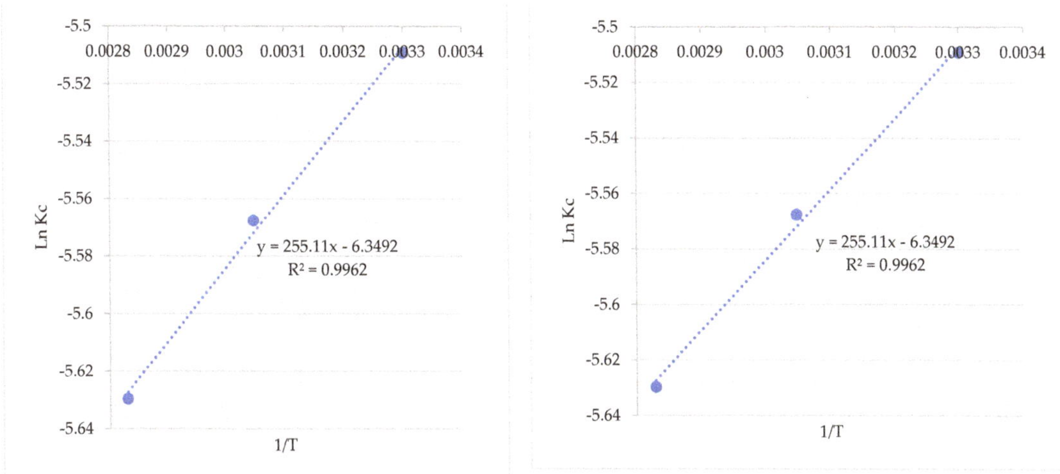

Figure 6. lnK_c vs. $1/T$ plot for thermodynamic parameters calculation.

Table 9. Thermodynamic parameters for the adsorption of hexavalent chromium at pH 2, 200 rpm, 0.325 g of adsorbent dose, 0.5 mm of particle size and 100 mg/L of initial concentration.

Biomass	T (K)	ΔG^o (kJ/mol)	ΔH^o (kJ/mol)	ΔS^o (kJ/mol × K)
OPB	302.95 328.15 353.15	−13.1779 −14.9243 −16.6701	−7.9993	−0.0698
YP	302.95 328.15 353.15	14.1406 15.3020 16.4635	4.3693	−0.029

4. Conclusions

SEM-EDS and FTIR analysis of OPB and YP showed that they present a porous surface typical of lignocellulosic materials, with the presence of hydroxyl, amine and carboxyl functional groups. From the adsorption experiments, the highest adsorption capacity was obtained at 0.03 g of adsorbent, 55 °C and 0.5 mm, reporting 325.88 and 159 mg/g using OPB and YP, respectively. We obtained a maximum adsorption capacity of 63.83 mg/g with OPB and 59.16 mg/g using YP, according to the Langmuir model. The equilibrium of adsorption on OPB was described by the Langmuir and Freundlich models, while YP was described by the Dubinin–Radushkevich model, whose parameters of fitting suggest that adsorption is given by ion exchange between the active centers and the metal ions. A kinetic study demonstrated that equilibrium time was 200 min for both materials; kinetic data were described by pseudo-second-order and Elovich models; thus, the mechanism of Cr (VI) adsorption onto the evaluated materials was dominated by a chemical reaction. The thermodynamic parameters determined that the removal for YP is endothermic, irreversible and non-spontaneous, while for OPB it is exothermic, spontaneous at low temperatures and irreversible. The rapid removal rate and high adsorption capacities obtained show that the adsorbents evaluated are effective in the removal of hexavalent chromium present in aqueous solution.

Author Contributions: C.T.-T., Á.G.-D. and A.V.-O. conceived and designed the paper, and wrote the Introduction and Materials and Methods. C.T.-T. and A.V.-O. wrote the Results and prepared figures and tables. Discussions and Conclusions were the collective work of all authors. The writing—review and editing was performed by Á.G.-D. All authors have read and agreed to the published version of the manuscript.

Funding: This investigation did not have external funding.

Data Availability Statement: The data that support the findings of this study are available on request from the corresponding author.

Acknowledgments: The authors thank the Universidad de Cartagena for the support to develop this research, as well as laboratories, software and time of the research team.

Conflicts of Interest: The authors declare no conflict of interest.

References

1. Manjuladevi, M.; Anitha, R.; Manonmani, S. Kinetic study on adsorption of Cr (VI), Ni (II), Cd (II) and Pb (II) ions from aqueous solutions using activated carbon prepared from Cucumis melo peel. *Appl. Water Sci.* **2018**, *8*, 36. [CrossRef]
2. Feizi, M.; Jalali, M. Removal of heavy metals from aqueous solutions using sunflower, potato, canola and walnut shell residues. *J. Taiwan Inst. Chem. Eng.* **2015**, *54*, 125–136. [CrossRef]
3. Gómez-Aguilar, D.L.; Rodríguez-Miranda, J.P.; Esteban-Muñóz, J.A.; Betancur, J.F. Coffee Pulp: A Sustainable Alternative Removal of Cr (VI) in Wastewaters. *Processes* **2019**, *7*, 403. [CrossRef]
4. Premkumar, P.; Sudha, R. Comparative studies on the removal of chromium (VI) from aqueous solutions using raw and modified Citrus Limettioides peel. *Indian J. Chem. Technol.* **2018**, *25*, 255–265.
5. Nordin, N.; Asmadi, N.A.A.; Manikam, M.K.; Halim, A.A.; Hanafiah, M.M.; Hurairah, S.N. Removal of Hexavalent Chromium from Aqueous Solution by Adsorption on Palm Oil Fuel Ash (POFA). *J. Geosci. Environ. Prot.* **2020**, *8*, 112–127. [CrossRef]
6. Bhanvase, B.A.; Ugwekar, R.P.; Mankar, R.B. (Eds.) *Novel Water Treatment and Separation Methods*; Apple Academic Press: Toronto, NJ, Canada, 2017.
7. Tran, H.N.; Nguyen, D.T.; Le, G.T.; Tomul, F.; Lima, E.C.; Woo, S.H.; Sarmah, A.K.; Nguyen, H.Q.; Nguyen, P.T.; Nguyen, D.D.; et al. Adsorption mechanism of hexavalent chromium onto layered double hydroxides-based adsorbents: A systematic in-depth review. *J. Hazard. Mater.* **2019**, *373*, 258–270. [CrossRef] [PubMed]
8. Núñez-Zarur, J.; Tejada-Tovar, C.; Villabona-Ortíz, A.; Acevedo, D.; Tejada-Tovar, R. Thermodynamics, Kinetics and Equilibrium Adsorption of Cr (VI) and Hg (II) in Aqueous Solution on corn cob (*Zea mays*). *Int. J. ChemTech Res.* **2018**, *11*, 265–280.
9. Omidvar-Borna, M.; Pirsaheb, M.; Vosoughi-Niri, M.; Khosravi-Mashizie, R.; Kakavandi, B.; Zare, M.R.; Asadi, A. Batch and column studies for the adsorption of chromium(VI) on low-cost Hibiscus Cannabinus kenaf, a green adsorbent. *J. Taiwan Inst. Chem. Eng.* **2016**, *68*, 80–89. [CrossRef]
10. Corral-Escárcega, M.C.; Ruiz-Gutiérrez, M.G.; Quintero-Ramos, A.; Meléndez-Pizarro, C.O.; Lardizabal-Gutiérrez, D.; Campos-Venegas, K. Use of biomass-derived from pecan nut husks (*Carya illinoinensis*) for chromium removal from aqueous solutions. column modeling and adsorption kinetics studies. *Rev. Mex. Ing. Quim.* **2017**, *16*, 939–953.
11. Ben Khalifa, E.; Rzig, B.; Chakroun, R.; Nouagui, H.; Hamrouni, B. Application of response surface methodology for chromium removal by adsorption on low-cost biosorbent. *Chemom. Intell. Lab. Syst.* **2019**, *189*, 18–26. [CrossRef]
12. Mondal, N.K.; Samanta, A.; Chakraborty, S.; Shaikh, W.A. Enhanced chromium (VI) removal using banana peel dust: Isotherms, kinetics and thermodynamics study. *Sustain. Water Resour. Manag.* **2017**, *4*, 489–497. [CrossRef]
13. Mahmood-Ul-Hassan, M.; Yasin, M.; Yousra, M.; Ahmad, R.; Sarwar, S. Kinetics, isotherms, and thermodynamic studies of lead, chromium, and cadmium bio-adsorption from aqueous solution onto Picea smithiana sawdust. *Environ. Sci. Pollut. Res.* **2018**, *25*, 12570–12578. [CrossRef] [PubMed]
14. Torres Pérez, C.I.; Quintero Lopez, L.A. Analysis of African palm solid waste, as an alternative to use renewable energy in the department of Cesar. *Ing. USBMed* **2019**, *10*, 8–18. [CrossRef]
15. Souza, P.R.; Dotto, G.L.; Salau, N.P.G. Artificial neural network (ANN) and adaptive neuro-fuzzy interference system (ANFIS) modelling for nickel adsorption onto agro-wastes and commercial activated carbon. *J. Environ. Chem. Eng.* **2018**, *6*, 7152–7160. [CrossRef]
16. Astudillo, P.; Isabel, C.; Barrero, R.; Carlos, A.; Puello, B.; Luis, M. Diagnostic of the main agricultural residues produced in the Bolivar region. *Sci. Agroaliment.* **2015**, *2*, 39–50.
17. Yi, Y.; Lv, J.; Liu, Y.; Wu, G. Synthesis and application of modified Litchi peel for removal of hexavalent chromium from aqueous solutions. *J. Mol. Liq.* **2017**, *225*, 28–33. [CrossRef]
18. Labied, R.; Benturki, O.; Eddine Hamitouche, A.Y.; Donnot, A. Adsorption of hexavalent chromium by activated carbon obtained from a waste lignocellulosic material (*Ziziphus jujuba cores*): Kinetic, equilibrium, and thermodynamic study. *Adsorpt. Sci. Technol.* **2018**, *36*, 1066–1099. [CrossRef]
19. Heffron, J.; Marhefke, M.; Mayer, B.K. Removal of trace metal contaminants from potable water by electrocoagulation. *Sci. Rep.* **2016**, *6*, 28478. [CrossRef]
20. Mädler, S.; Sun, F.; Tat, C.; Sudakova, N.; Drouin, P.; Tooley, R.J.; Reiner, E.J.; Switzer, T.A.; Dyer, R.; Kingston, H.M.S.; et al. Trace-Level Analysis of Hexavalent Chromium in Lake Sediment Samples Using Ion Chromatography Tandem Mass Spectrometry. *J. Environ. Prot.* **2016**, *7*, 422–434. [CrossRef]
21. Peng, S.H.; Wang, R.; Yang, L.Z.; He, L.; He, X.; Liu, X. Biosorption of copper, zinc, cadmium and chromium ions from aqueous solution by natural foxtail millet shell. *Ecotoxicol. Environ. Saf.* **2018**, *165*, 61–69. [CrossRef]

22. Cherdchoo, W.; Nithettham, S.; Charoenpanich, J. Removal of Cr (VI) from synthetic wastewater by adsorption onto coffee ground and mixed waste tea. *Chemosphere* **2019**, *221*, 758–767. [CrossRef] [PubMed]
23. Ayawei, N.; Ebelegi, A.N.; Wankasi, D. Modelling and Interpretation of Adsorption Isotherms. *J. Chem.* **2017**, *2017*. [CrossRef]
24. Rasmey, A.-H.M.; Aboseidah, A.A.; Youssef, A.K. Application of Langmuir and Freundlich Isotherm Models on Biosorption of Pb2+ by Freez-dried Biomass of Pseudomonas aeruginosa. *Egypt. J. Microbiol.* **2018**, *53*, 37–48. [CrossRef]
25. Akpomie, K.G.; Eluke, L.O.; Ajiwe, V.I.E.; Alisa, C.O. Attenuation kinetics and desorption performance of artocarpus altilis seed husk for Co (II), Pb (II) And Zn (II) Ions. *Iran. J. Chem. Chem. Eng.* **2018**, *37*, 171–186.
26. Blanes, P.S.; Bordoni, M.E.; González, J.C.; García, S.I.; Atria, A.M.; Sala, L.F.; Bellú, S.E. Application of soy hull biomass in removal of Cr (VI) from contaminated waters. Kinetic, thermodynamic and continuous sorption studies. *J. Environ. Chem. Eng.* **2016**, *4*, 516–526. [CrossRef]
27. Tran, H.N.; You, S.J.; Chao, H.P. Thermodynamic parameters of cadmium adsorption onto orange peel calculated from various methods: A comparison study. *J. Environ. Chem. Eng.* **2016**, *4*, 2671–2682. [CrossRef]
28. Manirethan, V.; Gupta, N.; Balakrishnan, R.M.; Raval, K. Batch and continuous studies on the removal of heavy metals from aqueous solution using biosynthesised melanin-coated PVDF membranes. *Environ. Sci. Pollut. Res.* **2019**, *27*, 1–15. [CrossRef]
29. Asuquo, E.D.; Martin, A.D. Sorption of cadmium (II) ion from aqueous solution onto sweet potato (*Ipomoea batatas* L.) peel adsorbent: Characterisation, kinetic and isotherm studies. *J. Environ. Chem. Eng.* **2016**, *4*, 4207–4228. [CrossRef]
30. Asuquo, E.; Martin, A.; Nzerem, P.; Siperstein, F.; Fan, X. Adsorption of Cd(II) and Pb(II) ions from aqueous solutions using mesoporous activated carbon adsorbent: Equilibrium, kinetics and characterisation studies. *J. Environ. Chem. Eng.* **2017**, *5*, 679–698. [CrossRef]
31. Pradhan, P.; Arora, A.; Mahajani, S.M. Pilot scale evaluation of fuel pellets production from garden waste biomass. *Energy Sustain. Dev.* **2018**, *43*, 1–14. [CrossRef]
32. Afroze, S.; Sen, T.K. A review on heavy metal ions and dye adsorption from water by agricultural solid waste adsorbents. *Water Air Soil Pollut.* **2018**, *229*, 225. [CrossRef]
33. Vu, X.H.; Nguyen, L.H.; Van, H.T.; Nguyen, D.V.; Nguyen, T.H.; Nguyen, Q.T.; Ha, L.T. Adsorption of Chromium (VI) onto Freshwater Snail Shell-Derived Biosorbent from Aqueous Solutions: Equilibrium, Kinetics, and Thermodynamics. *J. Chem.* **2019**, *2019*, 1–11. [CrossRef]
34. Chakraborty, P.; Show, S.; Ur Rahman, W.; Halder, G. Linearity and non-linearity analysis of isotherms and kinetics for ibuprofen remotion using superheated steam and acid modified biochar. *Process Saf. Environ. Prot.* **2019**, *126*, 193–204. [CrossRef]
35. Wu, Y.; Fan, Y.; Zhang, M.; Ming, Z.; Yang, S.; Arkin, A. Functionalized agricultural biomass as a low-cost adsorbent: Utilization of rice straw incorporated with amine groups for the adsorption of Cr (VI) and Ni (II) from single and binary systems. *Biochem. Eng. J.* **2016**, *105*, 27–35. [CrossRef]
36. Haroon, H.; Ashfaq, T.; Gardazi, S.M.H.; Sherazi, T.A.; Ali, M.; Rashid, N.; Bilal, M. Equilibrium kinetic and thermodynamic studies of Cr (VI) adsorption onto a novel adsorbent of Eucalyptus camaldulensis waste: Batch and column reactors. *Korean J. Chem. Eng.* **2016**, *33*, 2898–2907. [CrossRef]
37. Cai, J.; He, Y.; Yu, X.; Banks, S.W.; Yang, Y.; Zhang, X.; Yu, Y.; Liu, R.; Bridgwater, A.V. Review of physicochemical properties and analytical characterization of lignocellulosic biomass. *Renew. Sustain. Energy Rev.* **2017**, *76*, 309–322. [CrossRef]
38. Pan, L.; Nishimura, Y.; Takaesu, H.; Matsui, Y.; Matsushita, T.; Shirasaki, N. Effects of decreasing activated carbon particle diameter from 30 μm to 140 nm on equilibrium adsorption capacity. *Water Res.* **2017**, *124*, 425–434. [CrossRef] [PubMed]
39. Ibisi, N.; Asoluka, C. Use of agro-waste (*Musa paradisiaca peels*) as a sustainable biosorbent for toxic metal ions removal from contaminated water. *Chem. Int.* **2018**, *4*, 52–59.
40. Villabona-Ortíz, A.; Tejada-Tovar, C.; Ortega-Toro, R. Modelling of the adsorption kinetics of Chromium (VI) using waste biomaterials. *Rev. Mex. Ing. Química* **2020**, *19*, 401–408. [CrossRef]
41. Da'Na, E.; Awad, A. Regeneration of spent activated carbon obtained from home filtration system and applying it for heavy metals adsorption. *J. Environ. Chem. Eng.* **2017**, *5*, 3091–3099. [CrossRef]
42. Ajmani, A.; Shahnaz, T.; Subbiah, S.; Narayanasamy, S. Hexavalent chromium adsorption on virgin, biochar, and chemically modified carbons prepared from Phanera vahlii fruit biomass: Equilibrium, kinetics, and thermodynamics approach. *Environ. Sci. Pollut. Res.* **2019**, *26*, 32137–32150. [CrossRef] [PubMed]
43. Patra, C.; Medisetti, R.M.N.; Pakshirajan, K.; Narayanasamy, S. Assessment of raw, acid-modified and chelated biomass for sequestration of hexavalent chromium from aqueous solution using Sterculia villosa Roxb. shells. *Environ. Sci. Pollut. Res.* **2019**, *26*, 23625–23637. [CrossRef] [PubMed]
44. Alouache, A.; Selatnia, A.; Sayah, H.E.; Khodja, M.; Moussous, S.; Daoud, N. Biosorption of hexavalent chromium and Congo red dye onto Pleurotus mutilus biomass in aqueous solutions. *Int. J. Environ. Sci. Technol.* **2021**, 1–16. [CrossRef]
45. Khalil, U.; Shakoor, M.B.; Ali, S.; Ahmad, S.R.; Rizwan, M.; Alsahli, A.A.; Alyemeni, M.N. Selective removal of hexavalent chromium from wastewater by rice husk: Kinetic, isotherm and spectroscopic investigation. *Water* **2021**, *13*, 263. [CrossRef]
46. Alorabi, A.Q.; Alharthi, F.A.; Azizi, M.; Al-Zaqri, N.; El-Marghany, A.; Abdelshafeek, K.A. Removal of lead(Ii) from synthetic wastewater by lavandula pubescens decne biosorbent: Insight into composition–adsorption relationship. *Appl. Sci.* **2020**, *10*, 7450. [CrossRef]
47. Amro, A.N.; Abhary, M.K.; Shaikh, M.M.; Ali, S. Removal of lead and cadmium ions from aqueous solution by adsorption on a low-cost Phragmites biomass. *Processes* **2019**, *7*, 406. [CrossRef]

water

Article

Agro-Industrial Waste as Potential Heavy Metal Adsorbents and Subsequent Safe Disposal of Spent Adsorbents

Daiana Simón [1], Cristina Palet [2], Agustín Costas [3] and Adrián Cristóbal [1,*]

1 Instituto de Investigaciones en Ciencia y Tecnología de Materiales (INTEMA), Universidad Nacional de Mar del Plata (UNMdP)–Consejo Nacional de Investigaciones Científicas y Técnicas (CONICET), Mar del Plata 7600, Argentina
2 GTS–UAB Research Group, Department of Chemistry, Faculty of Science, Edifici C-Nord, Universitat Autònoma de Barcelona, Carrer dels Til·lers, 08193 Bellaterra, Spain
3 Centro Biotecnológico Fares Taie, Mar del Plata 7600, Argentina
* Correspondence: acristobal@fi.mdp.edu.ar

Abstract: Water pollution is an environmental problem that affects the ecosystem and living beings. Adsorption is one of the best technologies for the removal of heavy metals. Since waste recovery is the basis of the Circular Economy, agro-industrial waste is emerging as low-cost adsorbents for these pollutants from wastewater. Residues of pine sawdust, sunflower seed hulls and corn residues mix were evaluated as adsorbents of synthetic aqueous solutions of Ni(II), Zn(II) and Cd(II). These residues were characterized to determine their structure and composition, and to understand the adsorption mechanism. Adsorption efficiencies and capacities for the adsorbents and adsorbates were determined and compared. From the obtained results, it is possible to affirm that all biomasses used are good alternatives to the synthetic materials, with adsorption efficiencies greater than 50%. The order of adsorption was Cd > Zn > Ni. At the concentration range checked, adsorption efficiencies decreased in sawdust when a mixture of all metals together was considered (as present in real sewage). Finally, the heavy metals were immobilized, with efficiencies over 88.5%, in clay ceramics (as brick's precursors). This procedure would help to minimize the contamination that could be generated by the disposal of spent adsorbents, rarely explored in the literature.

Keywords: heavy metals; agro-industrial waste; adsorption; wastewater treatment; contaminant immobilization

Citation: Simón, D.; Palet, C.; Costas, A.; Cristóbal, A. Agro-Industrial Waste as Potential Heavy Metal Adsorbents and Subsequent Safe Disposal of Spent Adsorbents. *Water* **2022**, *14*, 3298. https://doi.org/10.3390/w14203298

Academic Editor: Laura Bulgariu

Received: 23 September 2022
Accepted: 14 October 2022
Published: 19 October 2022

1. Introduction

Water pollution is a serious world environmental problem mainly caused by the climate change, rapid urbanization and advance of industrialization [1,2]. Heavy metals are among most released pollutants or contaminants into the water and are not biodegradable; therefore, they accumulate in living organisms entering in the food chain, also through the consumption of water and other contaminated products, producing corresponding pollution biomagnification [3,4]. These metals are an environmental and public health concern, not only because of their persistence and concentration that influence exposure, but also because of their toxicity, and their mobility in the environment that determines their bioavailability, which is given by the type of compound or metabolite that each metal can form, and also by the characteristics of each the specific environment [5,6].

Nickel, zinc and cadmium are common and relevant heavy metals in the environment [4,7]. Electroplating, metallurgical and batteries industries are some of the anthropogenic sources of nickel, zinc and cadmium contamination [8,9]. Also, nickel and zinc can easily leach due to mineral weathering [10]. Ni(II) and Zn(II) are essential elements and, in low concentrations, they are necessary for the metabolic development of humans, plants or animals. However, these elements can be toxic and harmful to health effects when exposure/assimilation exceeds the upper limit of the physiologically required range [11–13].

For example, high exposure to nickel can cause cancer, dry cough and lung problems, dermatitis, nausea, gastrointestinal and kidney problems in humans, and high exposure to zinc can cause fever, vomiting, anemia, and skin problems in humans [7,8]. Cd(II) is a highly toxic metal, even at very low concentrations, and is a non-essential element because it has no known benefit to human health or other living beings [6,10]. Cadmium is a human carcinogen as established by The International Agency for Research on Cancer (IARC) and can cause kidney problems, hypertension, stomach irritation, among others, and its chronic exposure can lead to the development of "Itai-Itai" disease [11,14].

In the province of Buenos Aires, Argentina, the permitted maximum discharge limits of these heavy metals in sewers, surface water or stormwater conduits and the open sea vary from 2 to 3 mg/L for Ni, 2 to 5 mg/L for Zn and 0.1 to 0.5 mg/L for Cd [15]. Industrial effluents generally can contain concentrations of heavy metals above the maximum permissible limits; therefore, industries must treat their effluents before discharging them into the environment [13,14]. There are conventional technologies for treating wastewater and minimizing heavy metal pollution (e.g., chemical precipitation, coagulation/flocculation, membrane filtration, electrochemical technologies, ion exchange), however they can be expensive, and can generate by-products or sludge, involving complicated procedures [8,16]. Adsorption is considered one of the best options for heavy metal removal due to its flexibility in operation and design, low energy consumption, minimization of sludge and by-products, possibility of regenerating adsorbents, and high removal efficiency even at a very low metal concentrations [17,18]. Activated carbon (AC) is the most used and recognized heavy metal adsorbent but it is expensive due to its preparation process and the impossibility of its regeneration, which limits its use at large-scale application [19,20].

For developing countries, the application and development of heavy metal removal technologies represents a challenge [20]. In recent years, agro-industrial residues have emerged as low-cost adsorbents, also for heavy metals, due to its availability and abundance, allowing to apply processes under the bases of the Circular Economy (which corresponds to the recovery and reuse of wastes) [7,10]. Every year, worldwide, tons of waste are produced from the agro-industrial sector that are stored in the open air and disposal in landfills, causing negative environmental impacts due to leachates and gases, following with the CO_2 generation with their burning [21]. Literature examples of agro-industrial residues used as heavy metal adsorbents are: cow dung [5], potato peel [22], cucumber peel [23], groundnut husk [24], eggshells [25], pine and modified pine [26], rice and rapeseed [27], coffee husk and lignin [28], among others. All the plant-based wastes are made up of hemicellulose, cellulose and lignin, and has a wide variety of functional groups (e.g., aldehydes and ketones, carboxyl groups, phenolics, hydroxyls, methyls, ethers, amides, aminos, etc.) that can interact with pollutants through various mechanisms [7,8].

As reported by the national government in 2020, in Argentina agro-industrial is really important and constitutes the 25% of the manufacturing industry and represents the 40% of exports. Among the agro-industrial residues, sawdust constitutes from 9 to 15% of the forest biomass discarded by sawmills and comes mainly from pine and eucalyptus plantations [29]. Sunflower crop is one of the most important in Argentina with a production of 3.5 million tons of seeds per year, obtaining 50% by weight of discarded hulls per seed [30]. On another hand, the corn production extends over a large area of the country and generates a great volume of biomass when compared to others such as wheat or barley [31].

This paper focuses the study of adsorption processes of Ni(II), Zn(II) and Cd(II) by using pine sawdust, sunflower seeds hulls and corn residues mix as adsorbents. A comparison of these three agro-industrial wastes as adsorbents is here presented, by checking firstly each individual heavy metal adsorption process, and secondly the influence of the mixture of these three heavy metals on their adsorption (as they are present in real sewage all together), by corresponding adsorption experiments. Later, such biomass residues containing heavy metals are immobilized in clay ceramics (as brick's precursors), to here propose an environmentally safe way to dispose the spent adsorbents together with adsorbates (heavy metals). This procedure would help to minimize the secondary contamination that could be generated by the disposal of spent adsorbents, which is rarely explored in the adsorption literature and is fundamental for the real application of the adsorption from low-cost materials.

2. Materials and Methods

2.1. Chemicals and Reagents

All reagents used were of analytical grade. Stock solutions of concentration 1000 mg/L of individual heavy metals ions, Ni(II), Zn(II) and Cd(II), were prepared dissolving adequate amounts of $Ni(NO_3)_2.6H_2O$, $Zn(NO_3)_2.6H_2O$ and $Cd(NO_3)_2.4H_2O$ (all from Panreac, Castellar del Vallès, Spain), respectively. From these solutions the corresponding dilutions used in the adsorption experiments were prepared and the pH (Omega 300 pH meter, Crison Instruments, S.A, Barcelona, Spain) was adjusted with HNO_3 70% (Panreac, Spain). Solutions were prepared with Milli-Q water.

2.2. Biomass

The biomasses of sawdust, sunflower and corn were selected according to the reasons already mentioned. The pine (*Pinus elliottii*) sawdust residues were provided by a sawmill in the province of Corrientes, Argentina, and corresponded to the main cutting process of the wood, before any addition. Sunflower seed hulls (*Helianthus annuus*) were provided by a company located in the province of Santa Fe, Argentina, dedicated to the oilseed market, and were obtained from the processing of sunflower grains. The corn residues mix (*Zea mays var. saccharata*) were kindly provided by the National Institute of Agricultural Technology (INTA), and corresponded to the harvest stage.

The development of the adsorbents included the collection of the biomass, grinding with a knife mill (IKA A10) and sieving to a particle size of less than 1 mm to promote adsorption. The waste did not receive additional processing (chemical or thermal treatment) to make it as friendly as possible to the environment and reduce costs. Figure 1 shows the macroscopic appearance of the used waste, after grinding and sieving.

Figure 1. Ground and sieved residues (**A**) pine sawdust, (**B**) sunflower seed hulls and (**C**) corn residues mix.

2.3. Biomass Characterization

Physicochemical properties of adsorbents contribute to the process of adsorption of contaminants. Characteristics of potential adsorbents were determined from a number of techniques, that included the Brunauer-Emmett-Teller (BET) (Micromeritics Accusorb, model 2100), Scanning Electron Microscopy (SEM) (ZEISS EVO® MA 10 at the UAB Microscopy Service and FEI ESEM Quanta 200), Energy-Dispersive X-ray Spectroscopy (EDS) (Oxford SDD X-Act, software: AZTecOne), Attenuated Total Reflectance–Fourier Transform Infrared Spectroscopy (ATR-FTIR) (Nicolet 6700, Thermo Electron Corp. equipment, Waltham, MA, USA), X-ray Fluorescence (XRF) (PW4024 Minipal2 PANalytical X-ray spectrometer with copper anode and operating conditions nitrogen flow, voltage 20 kV, current 5 mA and time 100 s), and Differential Thermal Analysis (DTA) and Thermogravimetric

Analysis (TGA) (Shimadzu TGA-50 and Shimadzu DTA-50 instruments, with TA-50 WSI analyzer and operating conditions air, heating rate of 10 °C/min to 1000 °C and approximately 20 mg of mass). The mineral content of biomass it was determined following the guidelines of the ASTM E1755-01 standard [32]. In addition, the possible changes produced in the biomasses after the adsorption experiments were analyzed.

2.4. Batch Adsorption Experiments

The removal of heavy metals by means of the aforementioned agro-industrial residues was carried out under batch adsorption experiments at room temperature. A volume of 10 mL of a mono-metal solution, of Ni(II), Zn(II) or Cd(II) of 0.18 mmol/L, or a multi-metal solution of all together with a concentration of 0.18 mmol/L for each heavy metal was placed in contact with 0.1 g of adsorbent in tubes. The pH of the solutions was initially adjusted to 4–5 following previous literature [26,33]. The system was stirred at 40 rpm in a rotary mixer (CE 2000 ABT-4, SBS Instruments SA) for 24 h to ensure that equilibrium was reached. The liquid phase was filtered through 0.22 μm filters (Millex-GS, Millipore, Burlington, MA, USA). The metal concentration in the aqueous solution (not adsorbed) was determined by Inductively Coupled Plasma-Mass Spectrometry (ICP-MS) (XSERIES 2 ICP-MS, Thermo Scientific, Waltham, MA, USA) from the Autonomous University of Barcelona. The adsorption of each heavy metal was expressed as adsorption percentage (A%), calculated from Equation (1), and the adsorption capacity of each adsorbent (q_e) was calculated from Equation (2):

$$A\%\ (\%) = \frac{(C_0 - C_e)}{C_0} \times 100 \quad (1)$$

$$q_e\left(\frac{mmol}{g}\right) = \frac{(C_0 - C_e) \times V}{m} \quad (2)$$

where C_0 and C_e (mmol/L) are the initial and equilibrium concentrations of heavy metal in solution, respectively, V (L) is the volume of the heavy metal solution, and m (g) is the mass of the adsorbent [34,35]. Adsorption experiments are prepared by duplicate and the average results are reported.

2.5. Spent Adsorbents Disposal in Clay Ceramics

Safe disposal of spent adsorbents is necessary to minimize secondary contamination, especially if large-scale adsorption technology implementation is considered. Pine sawdust, sunflower seed hulls and corn residues mix, after being used as adsorbents, were added to the clay, and clay ceramics were prepared with the aim of immobilizing adsorbed heavy metals.

The amount of residue contaminated with Ni(II), Zn(II) and Cd(II) that was used in the clay ceramics corresponded to 20% in volume with respect to the volume of clay, in accordance with what was observed by the authors in previous studies [36]. For this reason, adsorption experiments were previously carried out scaling 20 times the amount of adsorbent (2 g) and 20 times the moles of metal ion (3.6×10^{-5}). As a consequence of the increase in adsorbent mass, the solution volume had to be increased to 40 mL so that the liquid covers the entire surface of the biomass.

The methodology used in the preparation process of the clay ceramics is detailed in Figure 2 and it was designed considering the characteristics of the raw materials, the pressure and firing conditions used in the brick factory. The clay was provided by a local brick factory and is the same that the manufacturer uses in the hollow brick production process. Firing of clay ceramic was carried out in an electric oven INDEF 332.

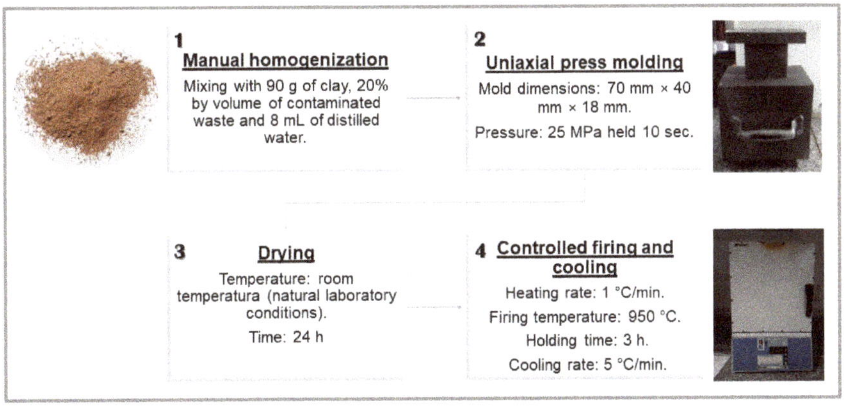

Figure 2. Clay ceramics preparation process.

It is essential to evaluate the mobility of heavy metals present in manufactured clay ceramics to determine the feasibility of immobilize these contaminants in the ceramic structure. Leaching tests were carried out based on EPA method 1311 [37], which is a method accepted by Argentine laws for hazardous waste.

Clay ceramics prepared from spent adsorbents were crushed and sieved to a particle size of less than 9.5 mm. In beakers, crushed clay ceramics were mixed with leaching solution in a 1:20 solid-liquid ratio. According to the alkalinity of the ceramic, an extraction fluid of pH 4.93 ± 0.05 was prepared from 5.7 mL of acetic acid, 64.3 mL of sodium hydroxide 1 mol/L and completing with distilled water up to 1 L. The covered beakers were shaken at 100 rpm in an orbital shaker (SK-0330-Pro) for 22 h. The mixtures were then filtered through filter paper washed with 1 mol/L nitric acid and rinsed with distilled water. The TCLP extracts of the solid phases were acidified with concentrated nitric acid until pH < 2 and stored refrigerated at 4 °C. Finally, the extracts were analyzed by Atomic Absorption Spectrophotometry (AAS) (Shimadzu 6800 with flame) from the Fares Taie Biotechnological Center. The results of the leaching tests were compared with the estimated initial mass of heavy metals in the clay ceramics. In this way, the retention efficiency was calculated according to Yilmaz et al. [38].

3. Results and Discussion

3.1. Biomass Characterization

The phenomena that occur in an adsorbent are related to its specific surface and, therefore, to the total volume of pores and their dimensions, that influence the interaction with the adsorbate and the obtained adsorption efficiency. Table 1 shows the results obtained from the BET analysis for pine sawdust, sunflower seed hulls and corn residues mix. All biomasses showed the presence of mesopores and the surface area values are in agreement with those reported in the literature for adsorbents of lignocellulosic origin [34,39]. Corn biomass presented a higher surface area, total pore volume, and mean pore size comparing with the other two biomass residues (pine sawdust and sunflower seed hulls). Bilal et al. [7] reported that the adsorption of contaminants increases with the increase in the surface area of the adsorbent, since the adsorption process is a surface phenomenon.

Table 1. BET analysis results of surface area, total pore volume and mean pore size for biomass residues.

Biomass	Surface Area (m^2/g)	Total Pore Volume (cm^3/g)	Mean Pore Size (nm)
Pine sawdust	1.1	0.003	9.4
Sunflower seed hulls	0.7	0.0009	4.9
Corn residues mix	1.5	0.006	14.9

SEM images allow to obtain information about the morphological characteristics (texture, topography and surface characteristics) of the adsorbents. So, the SEM images of the three studied agro-industrial wastes are presented in Figure 3. Biomass analyzed particles showed an elongated shape and a fibrous microstructure. An irregular and rough surface with cavities can be observed in all cases, which forms a network of holes and fibers. These characteristics can facilitate the adsorption of heavy metals [40]. No appreciable changes were detected in the morphology of the adsorbents related to the interaction with metal ions, as reported by Zhang et al. [33], after the adsorption of heavy metals. The combination of SEM with EDS detector and analyzer system allowed to obtain the distribution of heavy metals on the biomasses after adsorption by mapping. As shown in Figure 4, the distribution, and therefore adsorption, of nickel, zinc and cadmium in the adsorbents was across the entire surface.

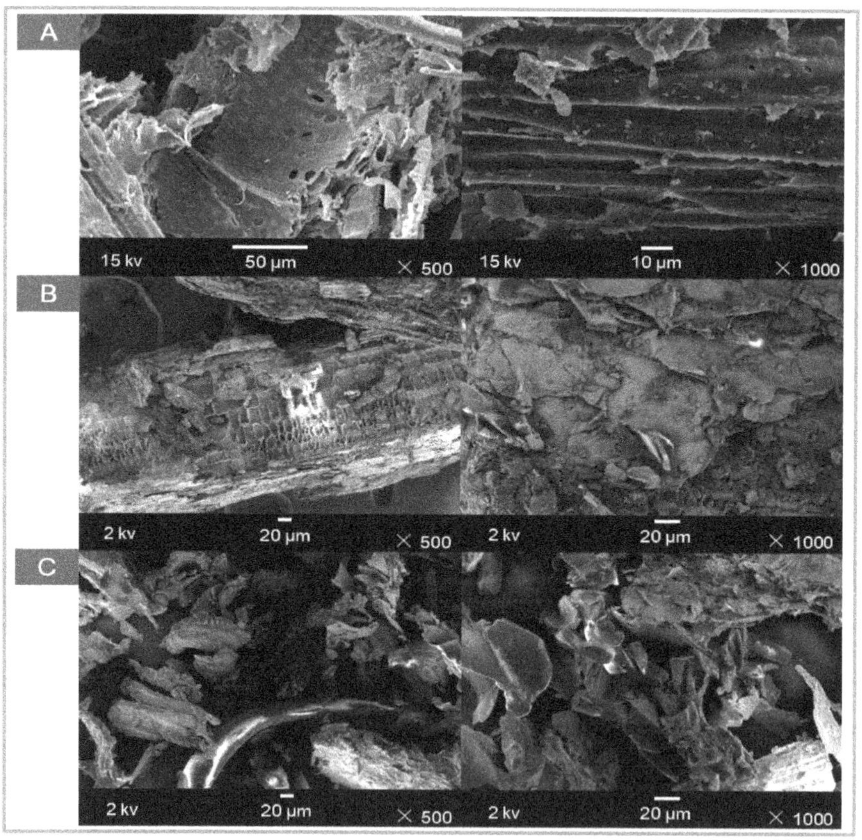

Figure 3. SEM images of: (A) pine sawdust, (B) sunflower seed hulls, (C) corn residues mix.

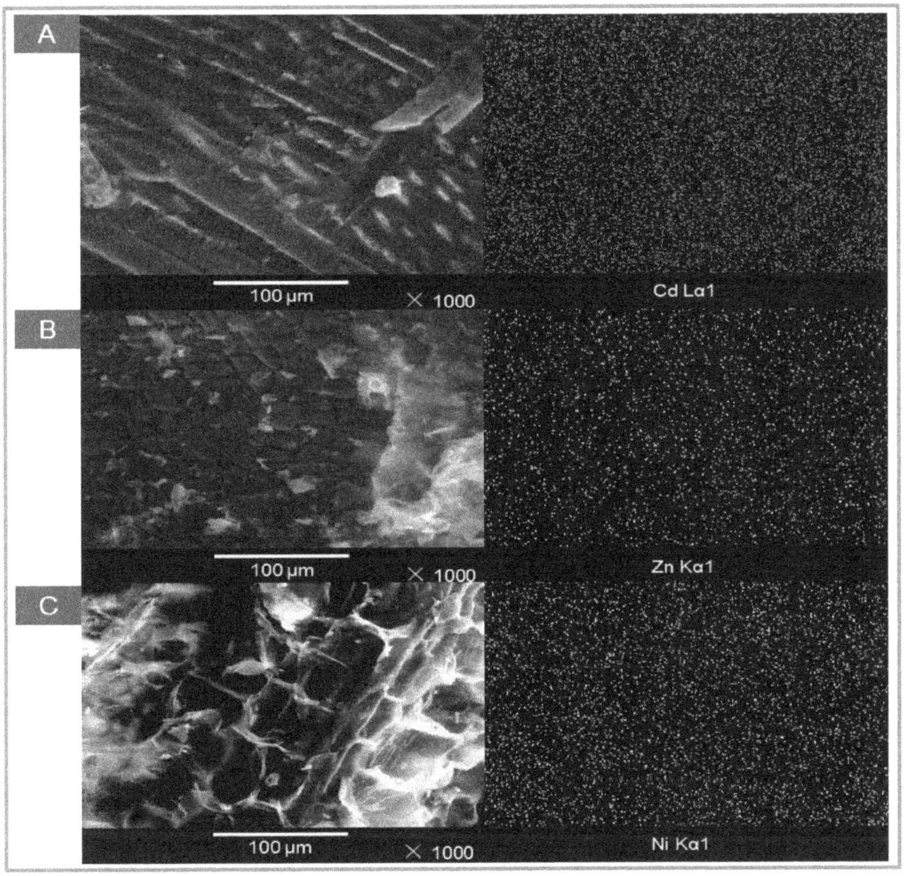

Figure 4. Elemental distribution obtained by SEM-EDS by mapping analysis for (**A**) pine sawdust, (**B**) sunflower seed hulls, (**C**) corn residues mix, after adsorption of Cd(II), Zn(II) and Ni(II), respectively.

ATR-FTIR made it possible to determine the presence of functional groups in biomass responsible for the metal adsorption mechanism, for example, either by electrostatic forces or complexation. ATR-FTIR spectra for the biomass residues studied here are shown in Figure 5. The large number of IR bands was associated with the typical complex nature of agro-industrial biomasses [41–43]. The assignment of the main IR bands at the respective approximate wavelengths are summarized in Table 2. The great similarity between the ATR-FTIR spectra of sawdust, sunflower and corn was due to the fact that the composition of these three biomasses is based on cellulose and lignin.

The presence of numerous functional groups in biomass facilitates the adsorption of heavy metals [7]. A comparison of the ATR-FTIR spectra of the biomasses before and after the adsorption of the heavy metals is also shown in Figure 5, being both spectra were very similar in each biomass case. However, slight differences were observed, such as a shift of the band at 1603–1624 cm^{-1} in the three biomasses, and shift of the band at 1224–1238 cm^{-1} in the pine sawdust and corn residues, after the contact with heavy metals. These results may be indicative that carboxyl, alcohol, phenol, amide, and other functional groups could provide possible adsorption sites for the retention of the studied heavy metals, and were similar to those found in previous works in the literature [44,45].

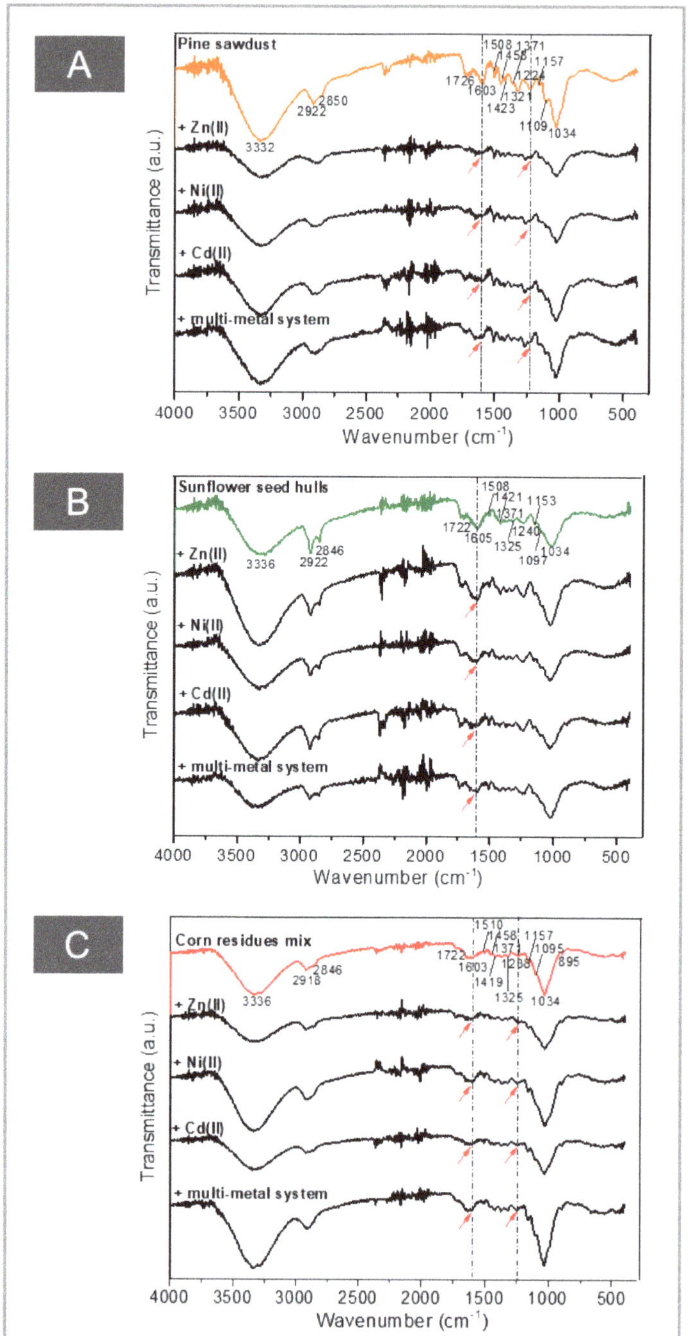

Figure 5. ATR-FTIR spectra of: (**A**) pine sawdust, (**B**) sunflower seed hulls, (**C**) corn residues mix, before and after the adsorption process.

Table 2. Assignment of the main bands obtained by ATR-FTIR for biomass residues.

Wavenumber (cm^{-1})	Assignment
3332–3336	O-H stretching of carboxylic acids and alcohols/phenols, and N-H of amino and amide groups
2918–2922	Asymmetric C-H stretching of CH_3
2846–2850	Symmetric C-H stretching of CH_2 and stretching of methoxy groups
1722–1726	C-O stretching of carbonyl, C=O of acetyl, carboxyl, aldehydes and aromatic/conjugated esters
1603–1624	COO- stretching of carboxyl groups, C=C of the aromatic ring, and C=O, C-N, C-N-H stretching of amides, and O-H bending
1508–1510	C=C stretching of aromatic ring, N-H bending
1458	C=C aromatic, C=O stretching and symmetric bending of C-H, O-H
1419–1437	O-H bending of acids, vibrations of aromatic rings and bending of CH_2 and aromatic functional groups such as C=C and C=O
1371	Asymmetric C-H bending of CH_3, CH_2
1321–1325	O-H bending of phenol group, C–N groups
1224–1240	C-O stretching of phenols and carboxylic acids, and alkyl aryl ether bonds
1153–1157	Asymmetric stretching of the C-O-C pyranose backbone
1095–1109	C-OH and C-H stretching
1034–1036	C-O stretching in carboxyl group, C-O-C, dialkyl ether, C-H of aromatics, and C=C and C-C-O
874–895	Changes in aromatic structures such as C–H stretching of aromatics

The XRF analysis for the biomasses before and after the adsorption of heavy metals is presented in Figure 6. The Cr peaks come from the tube used as the source of the equipment (anode). Signals corresponding to Cl, K, Ca, Mn and Fe were observed, although the intensities varied according to the residue. The presence in biomasses of Ni, Zn and Cd was observed after adsorption (Cd signals were detected in the Ca and K energy zone), associated with the decrease in the intensities of elements such as Ca and K, mainly.

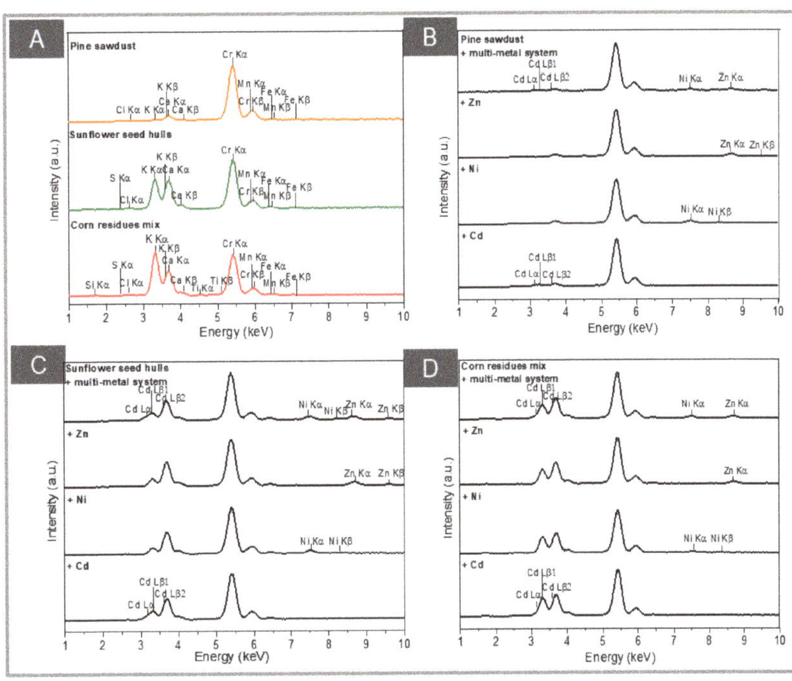

Figure 6. XRF of: (**A**) biomasses before the adsorption process, and after adsorption process with (**B**) pine sawdust, (**C**) sunflower seed hulls, (**D**) corn with adsorbed heavy metals.

The percentage of ash obtained was 0.2%, 2.0% and 10.2% for pine sawdust, sunflower seed hulls and corn residues mix, respectively. The XRD patterns of the biomasses (Figure S1 of the supplementary material) evidenced the presence of a significant amount of amorphous phase, in agreement with the mentioned results. The XRF equipment used does not allow the measurement of elements lighter than Na, and H, C and O are part of hemicellulose, cellulose and lignin as the main components of the biomasses. For that reason, the composition of the corresponding ashes was also analyzed by XRF (results collected in Figure 7). Differences in mineral content and composition were observed. Probably potassium, calcium, magnesium and phosphorus are involved in the adsorption of heavy metals through ion exchange mechanism, as reported previously [46].

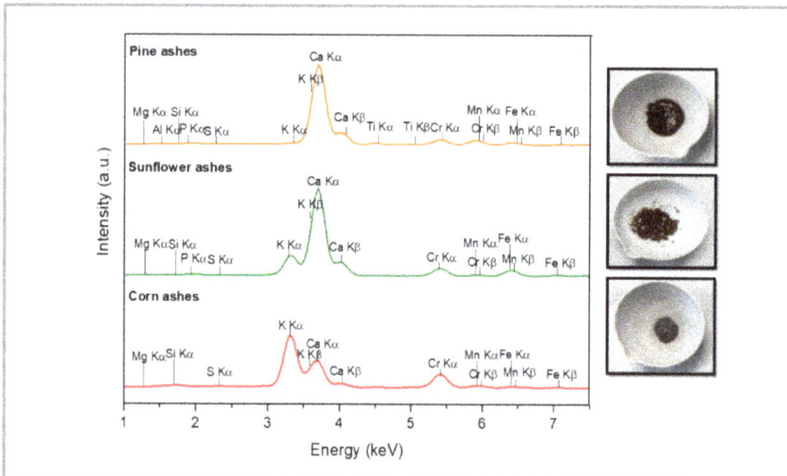

Figure 7. XRF of pine ashes, sunflower ashes and corn ashes.

Biomasses of plant origin are mainly composed of lignin and cellulose, as mentioned above, together with hemicellulose, low molecular weight compounds, lipids, proteins, starch, water, etc. [7]. The DTA-TGA profiles were obtained for the three biomasses here selected, and provided a description of the thermal behavior and an estimated percentage composition of biomasses, which was in agreement with the literature [47,48]. As seen in Figure 8, TGA analysis, the total weight loss was 96% for pine sawdust, 97% for sunflower seed hulls, and 86% for corn residues mix, and was divided into three stages. The first weight loss stage (up to 230 °C) was 7% for pine sawdust, 12% for sunflower seed shells and 10% for corn residues mix. This stage was related to the loss of moisture, which was characterized by an endothermic peak at 52 °C in pine sawdust and at 60 °C in sunflower seed hulls. The degradation of low molecular weight compounds was identified with an exothermic peak at 263 °C, and was observed in sawdust. The second weight loss stage was 48% for sawdust (up to 311 °C), 56% for sunflower (up to 297 °C) and 50% for corn (up to 338 °C). This stage could be related to the degradation of hemicellulose and pectin into volatile compounds of lower molecular weight. Finally, the third weight loss stage was 41% (up to 500 °C), 29% (up to 460 °C) and 26% (up to 508 °C) for biomasses of pine sawdust, sunflower seed hulls and corn, respectively, and it could be related to the degradation of cellulose and lignin into CO_2, H_2O and ashes. From the TGA analysis, the final residue corresponds to the mineral content and was higher for corn than for sawdust and sunflower, in accordance with the results obtained following the guidelines of the ASTM E1755-01 standard. The two large exothermic peaks in the three DTA curves were assigned to the decomposition of the biopolymers present in the biomass: such as hemicellulose, cellulose and lignin. Since these biopolymers are closely related in the biomass structure, the thermal degradation of each biopolymer cannot be clearly defined independently, probably being hemicellulose responsible of the first peak and cellulose and lignin together of the second peak [30].

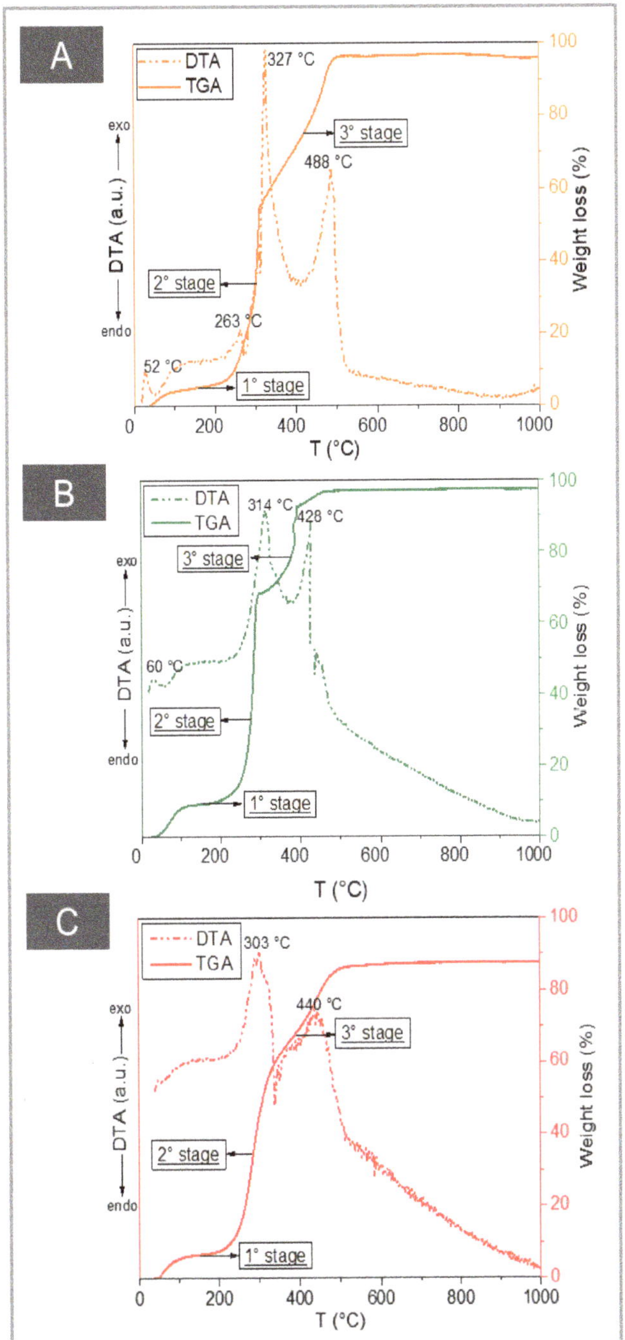

Figure 8. DTA-TGA of: (**A**) pine sawdust, (**B**) sunflower seed hulls, (**C**) corn residues mix.

3.2. Adsorption Process Characterization

Biomasses of pine sawdust, sunflower seed hulls and corn residues mix were evaluated as adsorbents of mono-metal aqueous solutions of Ni(II), Zn(II) and Cd(II). The results obtained are shown in Figure 9.

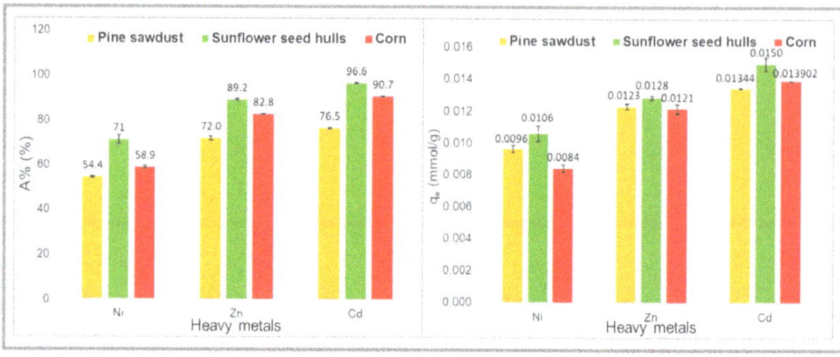

Figure 9. Comparison of the adsorption of mono-metal solutions of Ni(II), Zn(II) and Cd(II) on pine sawdust, sunflower seed hulls and corn residues mix.

According to the results of adsorption percentage (A%) and adsorption capacity (q_e) obtained, Ni(II) presented a lower adsorption, compared to Zn(II) and Cd(II), in all the biomasses studied. The adsorption follows the order: Ni(II) < Zn(II) < Cd(II), as shown in Figure 9, for sawdust, sunflower and corn, as it was also reported for other waste of lignocellulosic origin, such as coffee residues, rice husks, cocoa husks, and paper manufacturing wastes [49,50]. This behavior is related to the different characteristics and affinity of the metal ions for adsorbent adsorption sites [51]. Comparing the values of the hydration energies (Ni(II): −2106 kJ/mol, Zn(II): −2044 kJ/mol and Cd(II): −1806 kJ/mol), related to hydrolysis of metal ions, the nickel ion has a higher hydration energy than the zinc and cadmium ions and, therefore, it is less easy to lose its water molecules from its coordination sphere, which would prevent it from being adsorbed by the adsorbent through complexation or ion-exchange mechanisms. According to Mahmood-ul-Hassan et al. [52] and Qu et al. [53], smaller ions are more hydrated than larger ones, which could hinder adsorption.

The higher surface area, pore volume and mineral fraction in corn residues could have positively contributed to the adsorption of heavy metals. However, the differences in the adsorption results are not too significant between the three biomasses.

The obtained results are of the same order as results reported in the literature for batch adsorption experiments of Ni(II), Zn(II) and Cd(II) using sawdust, sunflower and corn residues (Table S1, supplementary material). However, the differences between the results of adsorption percentage (A%) are not only due to the characteristics of the biomass adsorbent but were also due to the concentration of contaminant, dosage of the adsorbent, pH of the solution, temperature, contact time, among others, that are factors that can affect the adsorption process [7]. Some of these factors were evaluated by the authors in previous work on the adsorption of heavy metals on adsorbent materials of plant origin [26,46,54,55].

The determination of the adsorption of heavy metals of a mixing metal solutions represents a closer situation to a real effluent. The competition between the metal ions for the adsorption sites occurs due to the saturation of the adsorption sites of the adsorbent whose dosage remains fixed to that of the individual systems [7]. Figure 10 compares the results of A% for the adsorption of mono-metal aqueous solutions of Ni(II), Zn(II) and Cd(II), and the multi-metal aqueous solution made up of heavy metals mentioned above with a concentration of 0.18 mmo/L of each of them. At this initial concentration, the results of A% for each of the heavy metals were similar when the adsorption was carried out separately and when it was carried out within the mixture, on sunflower seed hulls and corn residues mix. However, the adsorption of heavy metals in pine sawdust was lower for the multi-metal aqueous solution than for each metal separately, being the nickel ion the most affected by the competition with the other two heavy metals for the adsorption sites of the adsorbent, as expected (as Ni was the less adsorbed, as seen in Figure 9). Zhao et al. [46], also reported a lower performance of sawdust as an adsorbent when comparing the results of A% obtained for the adsorption of a mixture consisting of Cr(III), Cd(II), Cu(II) and Pb(II), on poplar sawdust and two other agricultural residues.

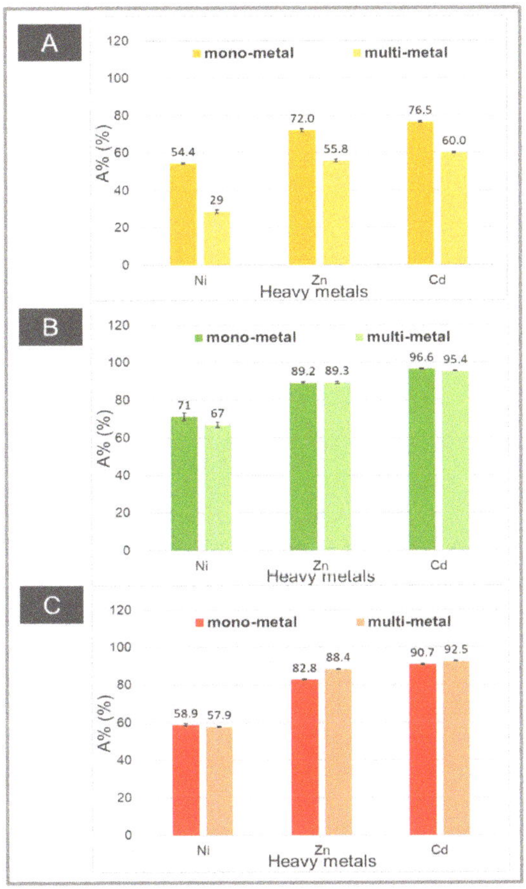

Figure 10. Comparison of the adsorption percentage (A%) of Ni(II), Zn(II) and Cd(II) in mono-metal and multi-metal solutions on: (**A**) pine sawdust, (**B**) sunflower seed hulls, (**C**) corn residues mix.

3.3. Spent Adsorbent Disposal

There is limited information available in the literature on the toxic effects of spent adsorbents, and their regeneration decreases their performance and generate new contaminant materials. The safe disposal of used and/or spent adsorbents is nowadays raising as a need to consider it for a more sustainable processes that can help to preserve the environment [56]. Based on the previous experience of the authors [54,55], the local production of ceramics for bricks construction is presented as a possible alternative for the safe disposal of spent adsorbents that would contribute to the real applicability of them as metal adsorbents.

In Figure 11, the adsorption percentages obtained from the experiments using 0.1 g of adsorbent and 1.8×10^{-6} moles of each of the heavy metals in the multi-metal system are compared with those results obtained by increasing the residue mass by 20 times to 2 g and moles of adsorbate to 3.6×10^{-5} moles. As can be seen, by maintaining the adsorbent/adsorbate ratio constant, the adsorption percentage remained constant. These results were considered in the preparation of clay ceramics with 20% by volume of contaminated biomass with respect to the volume of clay.

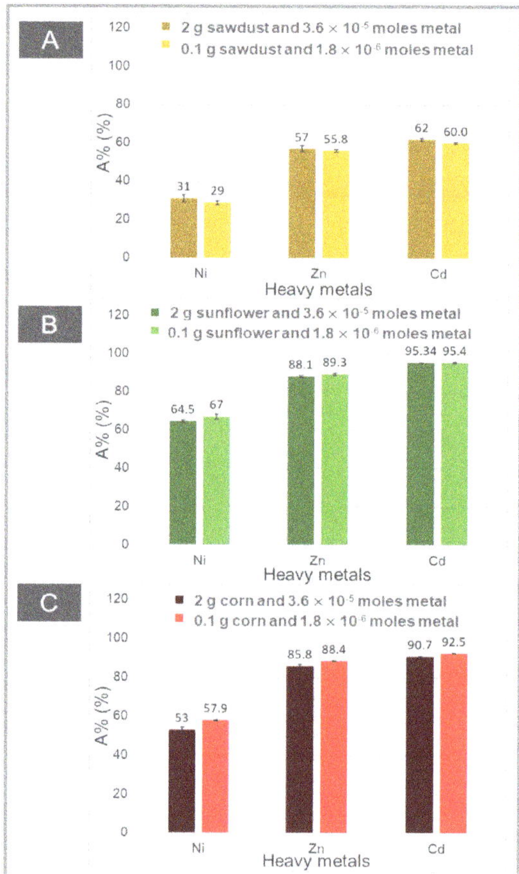

Figure 11. Comparison of the percentage of adsorption (A%) when increasing 20 times the mass of adsorbent and moles of adsorbate in the multi-metal system of heavy metals for: (**A**) pine sawdust, (**B**) sunflower seed hulls, (**C**) corn residues mix.

The macroscopic appearance of the clay ceramics is shown in Figure 12. All the samples presented a reddish color due to the Fe content of the natural clay and a porous surface according to the TGA and DTA results of the included lignocellulosic residues. As can be seen in Figure 8, at the ceramic firing temperature (950 °C), the added biomass burned out creating pores and releasing gases in the clay ceramics matrix.

Leaching tests based on EPA Method 1311 were performed to determine the possible leaching levels of Ni(II), Zn(II) and Cd(II) from the clay ceramics prepared with the addition of each of the spent adsorbents. Heavy metal concentrations were not detected in the TCLP extracts of the ceramic matrices because they were below the detection limits of the equipment used (Ni(II) < 0.05 mg/L, Zn(II) < 0.02 mg/L and Cd(II) < 0.05 mg/L, by AAS). For this reason, they were lower than the permissible limits of Argentina (nickel 5 mg/L and cadmium 1 mg/L, zinc not reported) [57].

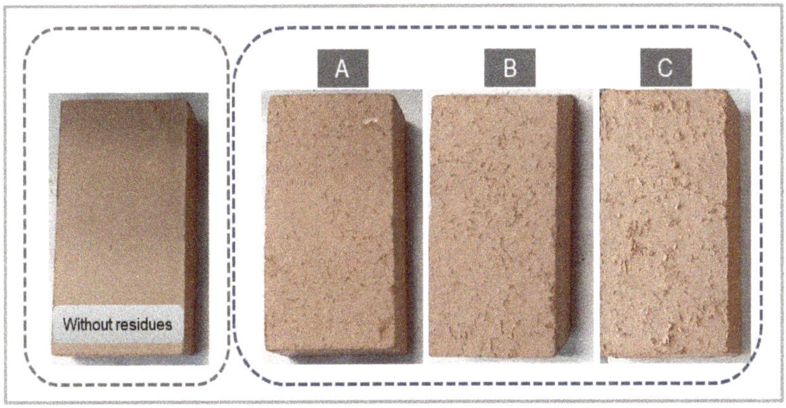

Figure 12. Macroscopic appearance of clay ceramics including spent biomass of: (**A**) pine sawdust, (**B**) sunflower seed hulls, (**C**) corn residues mix, and comparison with a sample without residue addition.

Table 3 shows the retention efficiency calculated for each one clay ceramics matrices. It was calculated from the mass of each of the heavy metals added in the ceramic (included in the spent adsorbent) and in the TCLP extract obtained for each tested ceramic. Based on these results, with heavy metal retentions above 88.5% in all cases, we can propose such clay ceramics prepared with added spent adsorbents with potential use in construction, and useful for the stabilization and immobilization of heavy metals together with the corresponding spent adsorbents. At firing temperatures in the followed leaching tests (EPA Method 1311), the organic residues can burn out and the heavy metals would be able to form stable phases with the clay minerals, which would decrease their bioavailability [58].

Table 3. Heavy metal retention efficiency for clay ceramics prepared from spent adsorbent with the mixture of heavy metals.

Clay Ceramics	Retention Efficiency (%)		
	Ni(II)	Zn(II)	Cd(II)
Pine sawdust	>88.5	>98.2	>97.1
Sunflower seed hulls	>95.9	>99.1	>98.5
Corn residues mix	>93.4	>98.7	>97.9

According to Mohajeran et al. [59], the particle size of the sample determines the contact surface with the leaching solution and therefore influences the concentration of heavy metals detected. This fact is important because in the leaching tests the clay ceramics were used crushed, but in practice the clay bricks will be used whole, so the leaching may be even lower, even there is still no legislation that imposes a test and limits on the leaching of heavy metals in construction materials [60].

4. Conclusions

One of the challenges of these times is the minimization of waste generated by agro-industrial activities and/or the reuse of this waste in applications that improve the quality of life. In this sense, considering that there is still no universal process to remove heavy metals from wastewater and effluents, adsorption from agroindustry residues is emerging as a simple, low-cost and efficient alternative.

Agro-industrial residues such as pine sawdust, sunflower seed hulls and corn residues mix, without any additional treatment, are characterized. These results are correlated with the performance of these materials as adsorbents of heavy metals such as Ni(II), Zn(II) and

Cd(II). In addition, SEM-EDS and XRF confirmed the presence of these heavy metals in the residues after the adsorption process.

Batch adsorption experiments from aqueous mono-metal solutions of nickel, zinc and cadmium ions, with concentrations of 0.18 mmol/L, with all three biomass residues selected showed promising results, with adsorption percentages greater than 50%. Ni(II) presented the lowest adsorption percentages and adsorption capacities compared to Zn(II) and Cd(II), possibly due to the higher hydration energy that could hinder its accessibility to the adsorbent. At this concentration, in a multi-metal solution, the decrease in adsorption due to the competition of heavy metals for the limited adsorption sites of pine sawdust determined that it is necessary to study multi-component systems to evaluate the actual performance of the adsorption process in practice.

Sawdust, sunflower and corn residues could be used as an alternative to traditional synthetic materials to remove heavy metals from wastewater given their properties, low cost and availability. However, more research is needed for the scale-up and possible commercial application of these agro-industrial residues as adsorbents of toxic metals from industrial wastewater.

Furthermore, a solution to the safe disposal of such biomass adsorbents after the adsorption process (containing heavy metals as pollutants) was proposed in this research work. The stabilization of these spent adsorbents in clay ceramics with possible use in construction is presented as an alternative for the immobilization of Ni(II), Zn(II), Cd(II) and their mixture. The heavy metal leaching tests of the ceramic matrices prepared with added spent adsorbents confirmed an effective immobilization for all heavy metals, whose concentrations were found to be below the permissible limits. The clay ceramics showed retention efficiencies over 88.5%.

Supplementary Materials: The following supporting information can be downloaded at: https://www.mdpi.com/article/10.3390/w14203298/s1, Figure S1: XRD patterns of pine sawdust, sunflower seed hulls, and corn residues mix; Table S1: Main results of literature studies about the adsorption of Ni(II), Zn(II) and Cd(II) on sawdust, sunflower and corn. References [44,52,61–67] are cited in the "supplementary materials".

Author Contributions: Conceptualization, C.P. and A.C. (Adrián Cristóbal); methodology, D.S., C.P. and A.C. (Adrián Cristóbal); validation, D.S., A.C. (Agustín Costas) and A.C. (Adrián Cristóbal); investigation, D.S. and A.C. (Adrián Cristóbal); resources, C.P., A.C. (Agustín Costas) and A.C. (Adrián Cristóbal); writing—original draft preparation, D.S., C.P. and A.C. (Adrián Cristóbal); writing—review and editing, D.S., C.P. and A.C. (Adrián Cristóbal); project administration, C.P. and A.C. (Adrián Cristóbal); funding acquisition, C.P. and A.C. (Adrián Cristóbal). All authors have read and agreed to the published version of the manuscript.

Funding: This research was funded by grants from Spanish research projects (CTM2015-65414-C2-1-R and AGL2015-70393-R), National Scientific and Technical Research Council (PIP 2021-2023 GI 11220200100739CO), National University of Mar del Plata (UNMdP 15/G577, 2020-2021) and H2020-MSCA-RISE-2014 Research Executive Agency (UE) (GA645024 (NANOREMOVAS)).

Institutional Review Board Statement: Not applicable.

Informed Consent Statement: Not applicable.

Data Availability Statement: Not applicable.

Acknowledgments: All the authors are grateful to the UAB Microscopy Service (*Servei de Microscòpia Electrònica* from UAB, Catalunya, Spain) for the SEM analysis and the *Centro Biotecnológico Fares Taie* (Mar del Plata, Argentina) for the AAS analysis of the TCLP extracts.

Conflicts of Interest: The authors declare no conflict of interest.

References

1. Uddin, M.K. A Review on the Adsorption of Heavy Metals by Clay Minerals, with Special Focus on the Past Decade. *Chem. Eng. J.* **2017**, *308*, 438–462. [CrossRef]
2. Hokkanen, S.; Bhatnagar, A.; Sillanpää, M. A Review on Modification Methods to Cellulose-Based Adsorbents to Improve Adsorption Capacity. *Water Res.* **2016**, *91*, 156–173. [CrossRef] [PubMed]
3. Tounsadi, H.; Khalidi, A.; Machrouhi, A.; Farnane, M.; Elmoubarki, R.; Elhalil, A.; Sadiq, M.; Barka, N. Highly Efficient Activated Carbon from *Glebionis coronaria* L. Biomass: Optimization of Preparation Conditions and Heavy Metals Removal Using Experimental Design Approach. *J. Environ. Chem. Eng.* **2016**, *4*, 4549–4564. [CrossRef]
4. Ali, H.; Khan, E.; Ilahi, I. Environmental Chemistry and Ecotoxicology of Hazardous Heavy Metals: Environmental Persistence, Toxicity, and Bioaccumulation. *J. Chem.* **2019**, *2019*, 6730305. [CrossRef]
5. Ojedokun, A.T.; Bello, O.S. Sequestering Heavy Metals from Wastewater Using Cow Dung. *Water Resour. Ind.* **2016**, *13*, 7–13. [CrossRef]
6. Rahman, Z.; Singh, V.P. The Relative Impact of Toxic Heavy Metals (THMs) (Arsenic (As), Cadmium (Cd), Chromium (Cr)(VI), Mercury (Hg), and Lead (Pb)) on the Total Environment: An Overview. *Environ. Monit. Assess.* **2019**, *191*, 419. [CrossRef]
7. Bilal, M.; Ihsanullah, I.; Younas, M.; Ul Hassan Shah, M. Recent Advances in Applications of Low-Cost Adsorbents for the Removal of Heavy Metals from Water: A Critical Review. *Sep. Purif. Technol.* **2022**, *278*, 119510. [CrossRef]
8. Zamora-Ledezma, C.; Negrete-Bolagay, D.; Figueroa, F.; Zamora-Ledezma, E.; Ni, M.; Alexis, F.; Guerrero, V.H. Heavy Metal Water Pollution: A Fresh Look about Hazards, Novel and Conventional Remediation Methods. *Environ. Technol. Innov.* **2021**, *22*, 101504. [CrossRef]
9. Tanong, K.; Tran, L.-H.; Mercier, G.; Blais, J.-F. Recovery of Zn(II), Mn(II), Cd(II) and Ni(II) from the Unsorted Spent Batteries Using Solvent Extraction, Electrodeposition and Precipitation Methods. *J. Clean. Prod.* **2017**, *148*, 233–244. [CrossRef]
10. Elgarahy, A.M.; Elwakeel, K.Z.; Mohammad, S.H.; Elshoubaky, G.A. A Critical Review of Biosorption of Dyes, Heavy Metals and Metalloids from Wastewater as an Efficient and Green Process. *Clean. Eng. Technol.* **2021**, *4*, 100209. [CrossRef]
11. Engwa, G.A.; Ferdinand, P.U.; Nwalo, F.N.; Unachukwu, M.N. Mechanism and Health Effects of Heavy Metal Toxicity in Humans. In *Poisoning in the Modern World—New Tricks for an Old Dog?* Karcioglu, O., Arslan, B., Eds.; Intechopen: London, UK, 2019; pp, 1–23.
12. Anastopoulos, I.; Pashalidis, I.; Hosseini-Bandegharaei, A.; Giannakoudakis, D.A.; Robalds, A.; Usman, M.; Escudero, L.B.; Zhou, Y.; Colmenares, J.C.; Núñez-Delgado, A.; et al. Agricultural Biomass/Waste as Adsorbents for Toxic Metal Decontamination of Aqueous Solutions. *J. Mol. Liq.* **2019**, *295*, 111684. [CrossRef]
13. Soliman, N.K.; Moustafa, A.F. Industrial Solid Waste for Heavy Metals Adsorption Features and Challenges; a Review. *J. Mater. Res. Technol.* **2020**, *9*, 10235–10253. [CrossRef]
14. Sall, M.L.; Diaw, A.K.D.; Gningue-Sall, D.; Efremova Aaron, S.; Aaron, J.J. Toxic Heavy Metals: Impact on the Environment and Human Health, and Treatment with Conducting Organic Polymers, a Review. *Environ. Sci. Pollut. Res.* **2020**, *27*, 29927–29942. [CrossRef] [PubMed]
15. RESOLUCION N° 336/2003. Autoridad del Agua de Provincia de Buenos Aires. ANEXO II Parámetros de Calidad de Las Descargas Límites Admisibles. Available online: https://normas.gba.gob.ar/documentos/0P3k4eSA.html (accessed on 22 September 2022).
16. Bolisetty, S.; Peydayesh, M.; Mezzenga, R. Sustainable Technologies for Water Purification from Heavy Metals: Review and Analysis. *Chem. Soc. Rev.* **2019**, *48*, 463–487. [CrossRef] [PubMed]
17. Fiyadh, S.S.; AlSaadi, M.A.; Jaafar, W.Z.; AlOmar, M.K.; Fayaed, S.S.; Mohd, N.S.; Hin, L.S.; El-Shafie, A. Review on Heavy Metal Adsorption Processes by Carbon Nanotubes. *J. Clean. Prod.* **2019**, *230*, 783–793. [CrossRef]
18. Rebello, S.; Sivaprasad, M.S.; Anoopkumar, A.N.; Jayakrishnan, L.; Aneesh, E.M.; Narisetty, V.; Sindhu, R.; Binod, P.; Pugazhendhi, A.; Pandey, A. Cleaner Technologies to Combat Heavy Metal Toxicity. *J. Environ. Manag.* **2021**, *296*, 113231. [CrossRef]
19. Danish, M.; Ahmad, T. A Review on Utilization of Wood Biomass as a Sustainable Precursor for Activated Carbon Production and Application. *Renew. Sustain. Energy Rev.* **2018**, *87*, 1–21. [CrossRef]
20. Chai, W.S.; Cheun, J.Y.; Kumar, P.S.; Mubashir, M.; Majeed, Z.; Banat, F.; Ho, S.H.; Show, P.L. A Review on Conventional and Novel Materials towards Heavy Metal Adsorption in Wastewater Treatment Application. *J. Clean. Prod.* **2021**, *296*, 126589. [CrossRef]
21. Omran, B.A.; Baek, K.H. Valorization of Agro-Industrial Biowaste to Green Nanomaterials for Wastewater Treatment: Approaching Green Chemistry and Circular Economy Principles. *J. Environ. Manag.* **2022**, *311*, 114806. [CrossRef]
22. Guechi, E.K.; Hamdaoui, O. Evaluation of Potato Peel as a Novel Adsorbent for the Removal of Cu(II) from Aqueous Solutions: Equilibrium, Kinetic, and Thermodynamic Studies. *Desalin. Water Treat.* **2016**, *57*, 10677–10688. [CrossRef]
23. Basu, M.; Guha, A.K.; Ray, L. Adsorption of Lead on Cucumber Peel. *J. Clean. Prod.* **2017**, *151*, 603–615. [CrossRef]
24. Gupta, N.; Sen, R. Kinetic and Equilibrium Modelling of Cu(II) Adsorption from Aqueous Solution by Chemically Modified Groundnut Husk (*Arachis hypogaea*). *J. Environ. Chem. Eng.* **2017**, *5*, 4274–4281. [CrossRef]
25. Tizo, M.S.; Blanco, L.A.V.; Cagas, A.C.Q.; Dela Cruz, B.R.B.; Encoy, J.C.; Gunting, J.V.; Arazo, R.O.; Mabayo, V.I.F. Efficiency of Calcium Carbonate from Eggshells as an Adsorbent for Cadmium Removal in Aqueous Solution. *Sustain. Environ. Res.* **2018**, *28*, 326–332. [CrossRef]
26. Zhao, J.; Boada, R.; Cibin, G.; Palet, C. Enhancement of Selective Adsorption of Cr Species via Modification of Pine Biomass. *Sci. Total Environ.* **2020**, *756*, 143816. [CrossRef] [PubMed]

27. Shen, X.; Zhao, J.; Bonet-Garcia, N.; Villagrasa, E.; Solé, A.; Liao, X.; Palet, C. Insights of Microorganisms Role in Rice and Rapeseed Wastes as Potential Sorbents for Metal Removal. *Int. J. Environ. Sci. Technol.* **2022**. [CrossRef]
28. Guevara-Bernal, D.F.; Cáceres Ortíz, M.Y.C.; Gutiérrez Cifuentes, J.A.; Bastos-Arrieta, J.; Palet, C.; Candela, A.M. Coffee Husk and Lignin Revalorization: Modification with Ag Nanoparticles for Heavy Metals Removal and Antifungal Assays. *Water* **2022**, *14*, 1796. [CrossRef]
29. Clauser, N.M.; Felissia, F.E.; Area, M.C.; Vallejos, M.E. Design of Nano and Micro Fibrillated Cellulose Production Processes from Forest Industrial Wastes in a Multiproduct Biorefinery. *Chem. Eng. Res. Des.* **2021**, *167*, 1–14. [CrossRef]
30. Casoni, A.I.; Gutierrez, V.S.; Volpe, M.A. Conversion of Sunflower Seed Hulls, Waste from Edible Oil Production, into Valuable Products. *J. Environ. Chem. Eng.* **2019**, *7*, 102893. [CrossRef]
31. Menéndez, J.E.; Hilbert, J.A. *Cuantificación y Uso de Biomasa de Residuos de Cultivos En Argentina Para Bioenergía;* Hilbert, J.A., Ed.; Ediciones INTA Publicaciones Periódicas; Instituto de Ingeniería Rural: Buenos Aires, Argentina, 2013; ISBN 9789875214484.
32. *ASTM E1755-01;* Standard Test Method for Ash in Biomass. ASTM International: West Conshohocken, PA, USA, 2020.
33. Zhang, H.; Carrillo, F.; López-Mesas, M.; Palet, C. Valorization of Keratin Biofibers for Removing Heavy Metals from Aqueous Solutions. *Text. Res. J.* **2019**, *89*, 1153–1165. [CrossRef]
34. Tran, H.N.; You, S.-J.; Hosseini-Bandegharaei, A.; Chao, H.-P. Mistakes and Inconsistencies Regarding Adsorption of Contaminants from Aqueous Solutions: A Critical Review. *Water Res.* **2017**, *120*, 88–116. [CrossRef]
35. Birniwa, A.H.; Mahmud, H.N.M.E.; Abdullahi, S.S.; Habibu, S.; Jagaba, A.H.; Ibrahim, M.N.M.; Ahmad, A.; Alshammari, M.B.; Parveen, T.; Umar, K. Adsorption Behavior of Methylene Blue Cationic Dye in Aqueous Solution Using Polypyrrole-Polyethylenimine Nano-Adsorbent. *Polymers* **2022**, *14*, 3362. [CrossRef] [PubMed]
36. Simón, D.; Gass, S.; Palet, C.; Cristóbal, A. Disposal of Wooden Wastes Used as Heavy Metal Adsorbents as Components of Buildings Bricks. *J. Build. Eng.* **2021**, *40*, 102371. [CrossRef]
37. Environment Protection Agency. *SW-846 Test Method 1311 Toxicity Characteristic Leaching Procedure;* Environment Protection Agency: Washington, DC, USA, 1992.
38. Yilmaz, O.; Ünlü, K.; Cokca, E. Solidification/Stabilization of Hazardous Wastes Containing Metals and Organic Contaminants. *J. Environ. Eng.* **2003**, *129*, 366–376. [CrossRef]
39. Fayoud, N.; Tahiri, S.; Alami Younssi, S.; Albizane, A.; Gallart-Mateu, D.; Cervera, M.L.; de la Guardia, M. Kinetic, Isotherm and Thermodynamic Studies of the Adsorption of Methylene Blue Dye onto Agro-Based Cellulosic Materials. *Desalin. Water Treat.* **2015**, *57*, 16611–16625. [CrossRef]
40. Najam, R.; Andrabi, S.M.A. Removal of Cu(II), Zn(II) and Cd(II) Ions from Aqueous Solutions by Adsorption on Walnut Shell-Equilibrium and Thermodynamic Studies: Treatment of Effluents from Electroplating Industry. *Desalin. Water Treat.* **2016**, *57*, 27363–27373. [CrossRef]
41. Özgenç, Ö.; Durmaz, S.; Boyaci, I.H.; Eksi-Kocak, H. Determination of Chemical Changes in Heat-Treated Wood Using ATR-FTIR and FT Raman Spectrometry. *Spectrochim. Acta—Part A Mol. Biomol. Spectrosc.* **2017**, *171*, 395–400. [CrossRef]
42. Stanković, S.; Šoštarić, T.; Bugarčić, M.; Janićijević, A.; Pantović-Spajić, K.; Lopičić, Z. Adsorption of Cu (II) Ions from Synthetic Solution by Sunflower Seed Husk. *Acta Period. Technol.* **2019**, *50*, 268–277. [CrossRef]
43. Anastopoulos, I.; Karamesouti, M.; Mitropoulos, A.C.; Kyzas, G.Z. A Review for Coffee Adsorbents. *J. Mol. Liq.* **2017**, *229*, 555–565. [CrossRef]
44. Sulaiman, O.; Ghani, N.S.; Rafatullah, M.; Hashim, R.; Ahmad, A. Sorption Equilibrium and Thermodynamic Studies of Zinc (II) Ions from Aqueous Solutions by Bamboo Sawdust. *J. Dispers. Sci. Technol.* **2011**, *32*, 583–590. [CrossRef]
45. Zhang, H.; Tang, Y.; Cai, D.; Liu, X.; Wang, X.; Huang, Q.; Yu, Z. Hexavalent Chromium Removal from Aqueous Solution by Algal Bloom Residue Derived Activated Carbon: Equilibrium and Kinetic Studies. *J. Hazard. Mater.* **2010**, *181*, 801–808. [CrossRef]
46. Zhao, J.J.; Shen, X.J.; Domene, X.; Alcañiz, J.M.; Liao, X.; Palet, C. Comparison of Biochars Derived from Different Types of Feedstock and Their Potential for Heavy Metal Removal in Multiple-Metal Solutions. *Sci. Rep.* **2019**, *9*, 9869. [CrossRef] [PubMed]
47. Georgiev, A.; Yoleva, A.; Djambazov, S. Effect of Wheat Straw and Sunflower Seeds Husks as Pore Forming Agents on the Properties of Porous Clay Bricks. *J. Chem. Technol. Metall.* **2017**, *52*, 885–891.
48. Khan, E.U.; Khushnood, R.A.; Baloch, W.L. Spalling Sensitivity and Mechanical Response of an Ecofriendly Sawdust High Strength Concrete at Elevated Temperatures. *Constr. Build. Mater.* **2020**, *258*, 119656. [CrossRef]
49. Joseph, L.; Jun, B.-M.; Flora, J.R.V.; Park, C.M.; Yoon, Y. Removal of Heavy Metals from Water Sources in the Developing World Using Low-Cost Materials: A Review. *Chemosphere* **2019**, *229*, 142–159. [CrossRef]
50. Ahmed, J.K.; Ahmaruzzaman, M.; Ahmed, M.J.K.; Ahmaruzzaman, M. A Review on Potential Usage of Industrial Waste Materials for Binding Heavy Metal Ions from Aqueous Solutions. *J. Water Process Eng.* **2016**, *10*, 39–47. [CrossRef]
51. Petrella, A.; Spasiano, D.; Acquafredda, P.; De Vietro, N.; Ranieri, E.; Cosma, P.; Rizzi, V.; Petruzzelli, V.; Petruzzelli, D. Heavy Metals Retention (Pb(II), Cd(II), Ni(II)) from Single and Multimetal Solutions by Natural Biosorbents from the Olive Oil Milling Operations. *Process Saf. Environ. Prot.* **2018**, *114*, 79–90. [CrossRef]
52. Mahmood-ul-Hassan, M.; Suthor, V.; Rafique, E.; Yasin, M. Removal of Cd, Cr, and Pb from Aqueous Solution by Unmodified and Modified Agricultural Wastes. *Environ. Monit. Assess.* **2015**, *187*, 19. [CrossRef]
53. Qu, J.; Tian, X.; Jiang, Z.; Cao, B.; Akindolie, M.S.; Hu, Q.; Feng, C.; Feng, Y.; Meng, X.; Zhang, Y. Multi-Component Adsorption of Pb(II), Cd(II) and Ni(II) onto Microwave-Functionalized Cellulose: Kinetics, Isotherms, Thermodynamics, Mechanisms and Application for Electroplating Wastewater Purification. *J. Hazard. Mater.* **2020**, *387*, 121718. [CrossRef]

54. Simón, D.; Quaranta, N.; Medici, S.; Costas, A.; Cristóbal, A. Immobilization of Zn(II) Ions from Contaminated Biomass Using Ceramic Matrices. *J. Hazard. Mater.* **2019**, *373*, 687–697. [CrossRef]
55. Simón, D.; Quaranta, N.; Gass, S.; Procaccini, R.; Cristóbal, A. Ceramic Bricks Containing Ni Ions from Contaminated Biomass Used as Adsorbent. *Sustain. Environ. Res.* **2020**, *3*, 26. [CrossRef]
56. Rial, J.B.; Ferreira, M.L. Potential Applications of Spent Adsorbents and Catalysts: Re-valorization of Waste. *Sci. Total Environ.* **2022**, *823*, 153370. [CrossRef] [PubMed]
57. Decreto 831/93. Boletín Nacional. Residuos Peligrosos. Reglamentación de La Ley 24051. Generación, Transporte y Disposición de Residuos Peligrosos 1993. Available online: http://servicios.infoleg.gob.ar/infolegInternet/anexos/10000-14999/12830/norma.htm (accessed on 22 September 2022).
58. Luo, L.; Li, K.; Weng, F.; Liu, C.; Yang, S. Preparation, Characteristics and Mechanisms of the Composite Sintered Bricks Produced from Shale, Sewage Sludge, Coal Gangue Powder and Iron Ore Tailings. *Constr. Build. Mater.* **2020**, *232*, 117250. [CrossRef]
59. Mohajerani, A.; Kadir, A.A.; Larobina, L. A Practical Proposal for Solving the World's Cigarette Butt Problem: Recycling in Fired Clay Bricks. *Waste Manag.* **2016**, *52*, 228–244. [CrossRef] [PubMed]
60. Arroyo, F.; Luna-Galiano, Y.; Leiva, C.; Vilches, L.F.; Fernández-Pereira, C. Environmental Risks and Mechanical Evaluation of Recycling Red Mud in Bricks. *Environ. Res.* **2020**, *186*, 109537. [CrossRef] [PubMed]
61. Mtibe, A.; Linganiso, L.Z.; Mathew, A.P.; Oksman, K.; John, M.J.; Anandjiwala, R.D. A Comparative Study on Properties of Micro and Nanopapers Produced from Cellulose and Cellulose Nanofibres. *Carbohydr. Polym.* **2015**, *118*, 1–8. [CrossRef]
62. Demcak, S.; Balintova, M.; Hurakova, M.; Frontasyeva, M.V.; Zinicovscaia, I.; Yushin, N. Utilization of Poplar Wood Sawdust for Heavy Metals Removal from Model Solutions. *Nov. Biotechnol. Chim.* **2017**, *16*, 26–31. [CrossRef]
63. Mahmood-Ul-Hassan, M.; Yasin, M.; Yousra, M.; Ahmad, R.; Sarwar, S. Kinetics, Isotherms, and Thermodynamic Studies of Lead, Chromium, and Cadmium Bio-Adsorption from Aqueous Solution onto Picea Smithiana Sawdust. *Environ. Sci. Pollut. Res.* **2018**, *25*, 12570–12578. [CrossRef]
64. Abdulhussein, S.A.; Al wared, A.I. Single and Binary Adsorption of Cu(II) and Ni(II) Ions from Aqueous Solutions by Sunflower Seed Husk. *Assoc. Arab Univ. J. Eng. Sci.* **2019**, *26*, 35–43. [CrossRef]
65. Jalali, M.; Aboulghazi, F. Sunflower Stalk, an Agricultural Waste, as an Adsorbent for the Removal of Lead and Cadmium from Aqueous Solutions. *J. Mater. Cycles Waste Manag.* **2013**, *15*, 548–555. [CrossRef]
66. Saeed, A.A.H.; Harun, N.Y.; Sufian, S.; Bin Aznan, M.F. Effect of Adsorption Parameter on the Removal of Nickel (II) by Low-Cost Adsorbent Extracted from Corn Cob. *Int. J. Adv. Res. Eng. Technol.* **2020**, *11*, 981–989. [CrossRef]
67. Melese, T.; Chala, K.; Ayele, Y.; Abdisa, M. Preparation, Characterization of Raw Corncob Adsorbent for Removal of Heavy Metal Ions from Aqueous Solution Using Batch Method. *African J. Pure Appl. Chem.* **2020**, *14*, 81–90. [CrossRef]

water

MDPI

Article

Elimination of Chromium (VI) and Nickel (II) Ions in a Packed Column Using Oil Palm Bagasse and Yam Peels

Angel Villabona-Ortíz [1], Candelaria Tejada-Tovar [1,*] and Ángel Darío González-Delgado [2,*]

[1] Process Design and Biomass Utilization Research Group (IDAB), Chemical Engineering Department, Universidad de Cartagena, Avenida del Consulado St. 30, Cartagena de Indias 130015, Colombia; avillabonao@unicartagena.edu.co

[2] Nanomaterials and Computer Aided Process Engineering Research Group (NIPAC), Chemical Engineering Department, Universidad de Cartagena, Avenida del Consulado St. 30, Cartagena de Indias 130015, Colombia

* Correspondence: ctejadat@unicartagena.edu.co (C.T.-T.); agonzalezd1@unicartagena.edu.co (Á.D.G.-D.)

Abstract: The single-component adsorption of chromium (VI) and nickel (II) on oil palm bagasse (OPB) and yam peels (YP) in a packed bed column was explored and improved using a central 2^2-star T composite design. The temperature, bed height, and particle size were evaluated, and the optimized response variable was the removal efficiency. The remaining concentration of heavy metals in solution was determined by Ultraviolet–Visible and Atomic Absorption Spectroscopy. It was found that bioadsorbents have a porous structure, with the presence of functional groups such as hydroxyl, carboxyl, and amino, which favor adsorption processes, and that the adsorption mechanisms controlling the process is cation exchange, precipitation, and complexation on the exposed surface of the biomaterials. In the adsorption trials, removal percentages higher than 87% were obtained in all cases, showing better results in the removal of Cr(VI), and that particle size is the most influential factor. Maximum Cr(VI) capacities of 111.45 mg g^{-1} and 50.12 mg g^{-1} were achieved on OPB and YP, respectively, while for nickel values of 103.49 mg g^{-1} and 30.04 mg g^{-1} were obtained. From the adjustment of the breakthrough curve to the models, it was determined that the model best able to adjust the data was the Thomas model, and the thermodynamic parameters of Cr(VI) and Ni(II) removal suggest that the process on YP is endothermic, while on OPB it is exothermic. In both biomaterials, the process is controlled by spontaneous chemisorption with a great affinity of the active centers for the ions.

Keywords: continuous system; metal ions; adsorption mechanisms

Citation: Villabona-Ortíz, A.; Tejada-Tovar, C.; González-Delgado, Á.D. Elimination of Chromium (VI) and Nickel (II) Ions in a Packed Column Using Oil Palm Bagasse and Yam Peels. *Water* **2022**, *14*, 1240. https://doi.org/10.3390/w14081240

Academic Editors: Cristina Palet and Julio Bastos-Arrieta

Received: 7 January 2022
Accepted: 8 April 2022
Published: 12 April 2022

1. Introduction

The release of industrial sewage polluted with toxic heavy metals is a global environmental problem, because metals like chromium, mercury, cobalt, nickel, lead, copper, and arsenic [1] can have serious impacts on human, animal, and aquatic life, as they have a very stable nature and a reduced tendency toward the biodegradation of their compounds [2]. Among them, chromium and nickel are especially important, due to their multiple applications at the industrial level. Chrome compounds exist in water and soil, because of the industrial activities of tanneries, electronic manufacturing, pigments, fertilizers, textiles, and photography. Furthermore, when chromium comes into contact with water, it is toxic for living beings, because it is subjected to oxidation (III) and (VI) [3]. For this reason, Cr(VI) is considered a hazardous pollutant; normally, it occurs in chromate (CrO_4^{2-}) or dichromate ($Cr_2O_7^{2-}$) forms, and effortlessly trespasses biological barriers, while also being carcinogenic [4]. On the other hand, nickel is applied in manufacturing activities such as metallization, and in paint, electroplating, powder, batteries, and alloys, among other things, as well as in everyday products such as cosmetics, clothing, and electronic devices [5]. Ni(II) is toxic, and in high concentrations can cause lung, skin, cardiovascular, and gastrointestinal diseases, and even cancer [6].

Several methods have been implemented in the elimination of heavy metals from effluent, including precipitation, electro-flocculation, ionic exchange, solvent extraction, and osmosis; these methods have several disadvantages, including toxic sludge production, the need for chemical additives, energy requirements, and the generation of secondary contaminants [7]. For this reason, bioadsorption is presented as a viable, efficient, and non-destructive method for removing heavy metals from waste to address environmental problems [8]. Although adsorption has been widely used in the removal of heavy metals from water systems, this implementation has most commonly been used in batch systems. Among these, metal–organic frameworks have been used [9]. Additionally, nanomaterials and nanotechnologies have been developed that can be used to generate innovative materials that are able to provide water treatment plants with extraordinary properties, increasing cost efficiency [10].

Most of the research on Cr(VI) and NI(II) bioadsorption has been performed using batch processes, because they are easy to apply on a small laboratory scale, but challenging to use on a massive scale, especially when the volume of industrial effluent is huge [11]. In addition, data from batch systems may not apply to continuous fixed-bed column operating conditions, where residence time is not necessary for achieving equilibrium [12]. Therefore, it is necessary to determine the real applicability of biosorbents in the continuous mode [13]. However, fixed-bed columns are preferred because of their high efficiency, easy method of operation, and ability to be expanded beyond laboratory scale to produce higher quality effluent [14]. Nevertheless, the use of eucalyptus in the elimination of Cr(VI) and Ni(II) in the packed-bed system has been reported [15], as well as palm residues [16], plantain waste [17], vetch [18], alharma [19] and kenaf [20], where the effects of temperature, pH, bed height and particle size were determined, presenting high percentages of removal.

Thus, in the present work, the performance of yam peel (YP) and oil palm bagasse (OPB) in the elimination of Cr(VI) and Ni(II) present in solution in a continuous system was evaluated, and the effects of temperature variation, particle size, and bed height were determined. The bioadsorbents were characterized using the proximal chemical analysis, FTIR (Fourier transform–infrared spectroscopy), SEM (scanning electron microscopy) and EDS (energy-dispersive X-ray spectroscopy) techniques.

2. Materials and Methods

2.1. Preparation of the Biomass

Oil palm bagasse (OPB) was obtained in Bolivar, Colombia, as a rejected by-product of the palm oil extraction process. Yam peel (YP) was obtained as post-harvest residues. Biomasses were selected in their best condition, washed with deionized water, sun-dried until a constant mass, and reduced in size in a blade mill; size classification was performed in a shaker-type sieve with stainless steel screens, selecting the particle diameters according to the design of the experiments.

2.2. Characterization of the Bioadsorbent

Proximal chemical analysis was performed using analytical methods to determine the content of pectin (thermogravimetry), lignin (acid digestion), cellulose (photolysis), hemicellulose (thermogravimetry digestion), carbon (AOAC 949.14), hydrogen (AOAC 949.14), nitrogen (AOAC 949.14-Kjeldahl), sulfur (948.13) and ash (nephelometry digestion). The functional groups and structural change in the biomass were determined by FTIR analysis of the bioadsorbents before and after adsorption using a Shimadzu IRAinfinity-1S spectrophotometer with a frequency of 32 scans between 600 and 4000 cm^{-1}. The surface properties of the biomasses were studied by SEM-EDS analysis using a scanning electron microscope coupled to an energy dispersive spectrophotometer model JSM-6490LV JEOL Ltd. (Tokyo, Japan).

2.3. *Column Adsorption Tests*

Adsorption tests were performed in packed-bed system following an experimental design created in Statgraphics Centurion XVI®, with the central type composed of a 2^3+rotating star on the response surface; there were 5 levels of temperature variation (30, 40, 55, 70 and 80 °C), particle size (0.1, 0.355, 0.5, 1 and 1.2 mm) and bed height (6.1, 30, 65, 100 and 124 mm), for a total of 16 experiments for each biomass–metal system. The experiments were conducted in an acrylic cylinder with height of 15 cm and internal diameter of 4.1 cm. The solution at 100 mg L^{-1} was feed from a tank located at the top of the equipment, flowing into the column by gravity at a flowrate of 0.75 mL/s. To prepare the synthetic solutions, nickel sulfate ($NiSO_4$) and Potassium Dichromate ($K_2Cr_2O_7$) were used (Merck Millipore; Burlington, MA, USA) and dissolved in deionized water using pH of 2 and 6 for Cr(VI) and Ni(II), respectively.

The biomass packaging was filled with the biomaterials under study, and the conditions of temperature, bed height, and particle size were configured according to the proposed design of experiments. The samples were taken 5 min after starting the assays; the final concentration of Cr(VI) was determined by UV-Vis spectrophotometry at 540 nm in a Biobase spectrophotometer using phenolphthalein as an indicator, in accordance with ASTM D1687-17 [3]. The residual nickel (II) was determined using an atomic absorption spectrophotometer, Buck Scientific model 210 VGP, at 228.8 nm [5]. Adsorption efficiency was determined using Equation (1):

$$\%E = \frac{C_0 - C_i}{C_0} \times 100 \quad (1)$$

where C_0 is the initial concentration of the studied metal and C_i is the concentration of the studied metal after the adsorption process [21].

The effects of temperature, particle size, and bed height were studied using the Statgraphics Centurion software through the estimated response surface and the optimization of such response using the Surface Response Method (SRM), which made it possible to maximize the efficiency of the removal, determining the best conditions under which to construct the breakage curve.

2.4. *Breaktrhough Curve*

The breakage curve or determining the time of service of the bed and the saturation of the biomaterial was constructed by means of continuous experiments under the optimum conditions obtained. The breakage curves for Cr(VI) and Ni(II) adsorption were adjusted to the models described below in OriginPro®.

2.4.1. Thomas Model

This model obeys Langmuir's adsorption and considers the axial dispersion to be insignificant in the bed, because the driving force follows second-order kinetics with a reversible reaction given by Equation (2).

$$\frac{C_0}{C} = \frac{1}{1 + exp\left[\frac{K_{th}}{Q}\left(q_0 X - C_0 V\right)\right]} \quad (2)$$

where K_{th} is Thomas' constant (mL min^{-1} mg^{-1}); q_o is the maximum concentration of solute in the solid phase (mg g^{-1}); X is the amount of adsorbent in column (g), Q is the rate of flow (mL min^{-1}), and V (L) is the volume of the effluent at the time of operation.

2.4.2. Yoon–Nelson Model

Yoon–Nelson model. This model relates the dimensionless parameter C/C_0 throughout the operating time by means of two parameters, K_{YN} and τ, which correspond to the proportionality constant in min^{-1} and the time after which 50% of the initial adsorbate is retained, respectively. This is shown in Equation (3).

$$\frac{C}{C_0} = \frac{exp\ (K_{YN}t - \tau K_{YN})}{1 + exp\ (K_{YN}t - \tau K_{YN})} \quad (3)$$

2.4.3. Dose–Response (DR) Model

The DR model can be represented by Equation (4).

$$\frac{C}{C_0} = 1 - \frac{1}{1 + \left[\frac{C_0 \times Qt}{q_0 X}\right]^a} \quad (4)$$

where a is the constant of the model; q_0 is the maximum solute concentration in the solid phase (mg g^{-1}), X is the amount of adsorbent in the column (g), and Q is the rate of flow (mL min^{-1}).

2.4.4. Adams–Bohart Model

This model is implemented to describe the initial part of the breakage curve above the breakage or saturation points, assuming that the adsorption velocity is proportional to the adsorption capacity and the concentration of the adsorbed species. This model is described according to Equation (5).

$$\frac{C}{C_0} = \frac{exp(K_{AB}C_0 t)}{exp\left(\frac{K_{AB}\ N_0 L}{v}\right) - 1 + exp(K_{AB}C_0 t)} \quad (5)$$

where K_{AB} is the kinetic constant of Adams–Bohart in (L/mg.min), N_0 is the maximum volumetric adsorption capacity in mg L^{-1}, v is the linear flow rate in (cm/min), L is the depth of the column bed in cm, C_0 is the internal concentration mg L^{-1}, C_t is the concentration of the effluent, and t is the residence time of the dissolution in the column.

2.4.5. Adsorption Capacity of the Column

The adsorption capacity of the column was determined using Equation (6).

$$q_j = \frac{C_j^f Q}{1000 m_s} \int_0^{t_s} \left(1 - \frac{C_j^{out}}{C_j^f}\right) dt \quad (6)$$

where q_j is the concentration of the ion in the adsorbent (mmol/gram), C_{jf} is the feed concentration of the j-ion in the liquid phase (mmol/L), Q is the volumetric flow rate of the solution flowing through the column (cm^3/min), m_s is the mass of the biomass with which the tower is packed (g), C_j^{out} is the j-ion output concentration in the liquid phase (mmol/L), and t_s is the time at which the column is saturated (min) [22].

2.5. Thermodynamic Parameters

To establish in a general way the type of adsorption of the metals using OPB and YP, the equilibrium constant K_c and the values of adsorption enthalpy (ΔH°), adsorption entropy (ΔS°), and Gibbs' energy (ΔG°) were determined, making it possible to establish the favorability of the process and the effect that the temperature has on it [23]. For this, Van't Hoff's graphical method was used, according to the following equations:

$$\Delta G = \Delta H - T\Delta S \quad (7)$$

$$G = -RT \times lnk_c \quad (8)$$

$$lnK_c = \frac{-\Delta H}{RT} + \frac{\Delta S}{R} \quad (9)$$

3. Results

3.1. Characterization of Bioadsorbents

The bromatological analysis of OPB and YP is shown in Figure 1, where it can be seen that the compounds with the greatest presence in both biomaterials are lignin and cellulose, compounds which stand out for having hydroxyl, carboxyl, and phenolic functional groups in their structure, and which are important in the adsorption of metallic ions [24].

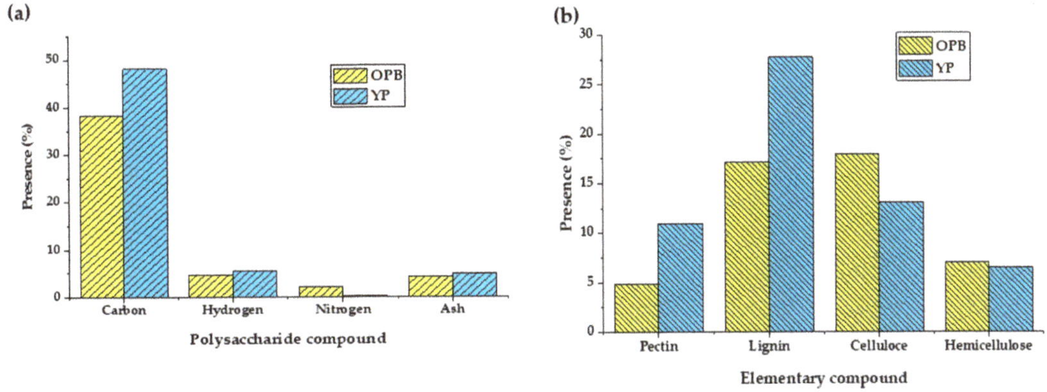

Figure 1. Bromatological analysis of OPB and YP: (**a**) polysaccharide compounds and (**b**) elementary compounds [25,26].

The most abundant element in the two bioadsorbents is carbon, which is typical of lignocellulosic materials due to the presence of lignin, cellulose, and hemicellulose. Therefore, a high efficiency of Cr(VI) and Ni(II) adsorption is expected, because these compounds are known to possess a large number of OH^- groups [27], carboxyl, and phenol, which can promote the adsorption of heavy metals [28]. Similarly, it can be observed that the lignin content in YP and OPB is high, and therefore the removal efficiency may be higher in these materials [29]. The ash content is higher than 4% in both biomaterials, indicating that the biomass contains in its structure oxides of silica, aluminum, iron, calcium, magnesium, titanium, sodium, and potassium [30].

The FTIR spectrum was obtained for OPB and YP before and after the Cr(VI) and Ni(II) adsorption processes (Figures 2 and 3). For both adsorbents, the evident stretching between frequencies of 3500 and 3800 cm^{-1} corresponds to the hydroxyl (O-H) functional group [31], while the presence of primary and secondary amines is indicated by the stretching of the N-H bond between 3200 and 3450 cm^{-1} [32]. Additionally, stretching and deformation of the C-H group and carboxylic acid can be observed between the frequencies of 2700 and 3000 cm^{-1}, as well as the stretching of carbon dioxide O=C=O close to the band at 2400 cm^{-1} [33]. The vibrations between 900 and 1800 cm^{-1} are attributed to alkenes, aromatics, and carboxylic acids [19]. Around the band between 1600 and 1650 cm^{-1}, a stretch of the carbonyl group of pectin, cellulose, and hemicellulose carboxylic acids can be observed, while between 900 and 1200 cm^{-1}, vibrations corresponding to the C-O group of alcohols and phenols are evident [28]. The presence of these multiple adsorption peaks confirms the heterogeneous nature and structural complexity of the bioadsorbents under study.

In the FTIR spectrum of the OPB and the YP following Cr(VI) and Ni(II) adsorption, changes in the functional groups of the biomasses can be observed, and the difference in the vibrations is remarkable. This is attributed to the incorporation of the metals under study onto the surface of the biomass, whereby they are joined to different functional groups on it [34]. The change in the intensity of the bands of the groups OH, CH, CO, NH, and C=C evidences the union of the ions; active and present functional groups in the biomass are key to the mechanism of adsorption of cationic pollutants [35]. Between 900 and 1800 cm^{-1}, the presence of aromatic rings in the lignin and stretching of the C-O groups can be observed [4]. The vibrations of the bands between 700 and 1300 cm^{-1} present inorganic groups [36].

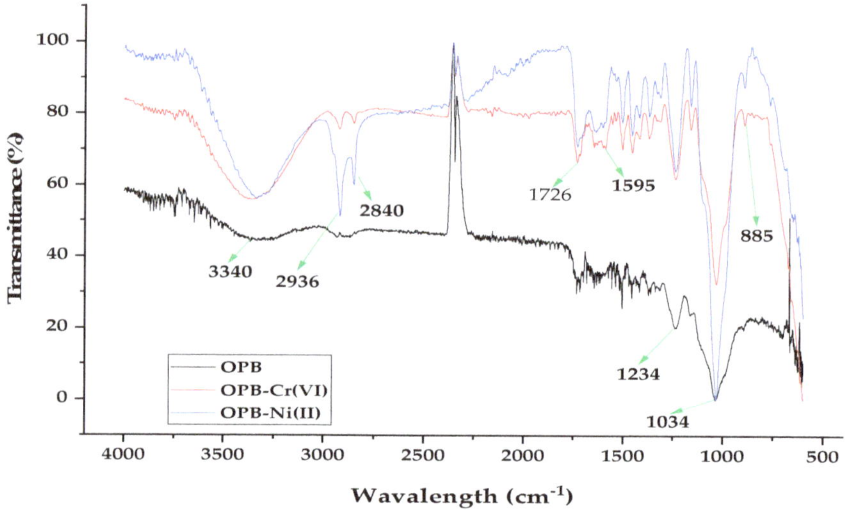

Figure 2. OPB FTIR before and after Cr(VI) and Ni(II) adsorption.

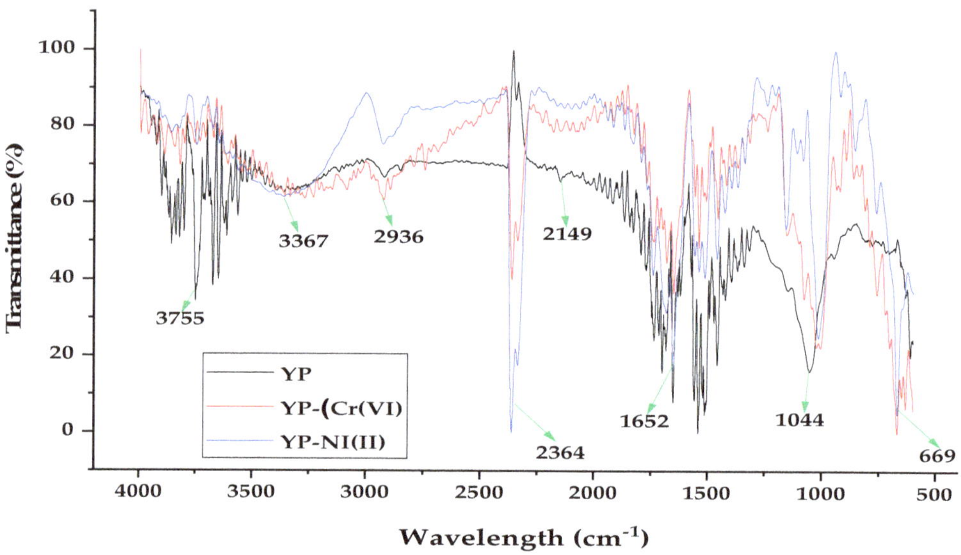

Figure 3. FTIR analysis of yam peels before and after removal of Cr(VI) and Ni(II).

In Figure 4 and Table 1, it can be seen that YP exhibits an irregular surface with the presence of mesopores, while the morphology of OPB is fibrous, with the presence of porosity and cavities [26]. Carbon, calcium, and oxygen are present in both materials, with YP exhibiting the most diverse structure, with traces of multiple elements such as potassium, silicon, aluminum, phosphorus, and iron, among others [37]. The presence of these elements is related to the functional groups present in lignocellulosic materials (cellulose, hemicellulose, lignin, and pectin) [4] and its ability to capture cations due to electrostatic forces; therefore, white particles are observed on the micrographs [38]. It is evident that the pores of the biomaterials are covered by the metals following adsorption, and so they are softened when they are covered by Cr(VI) and Ni(II) [39].

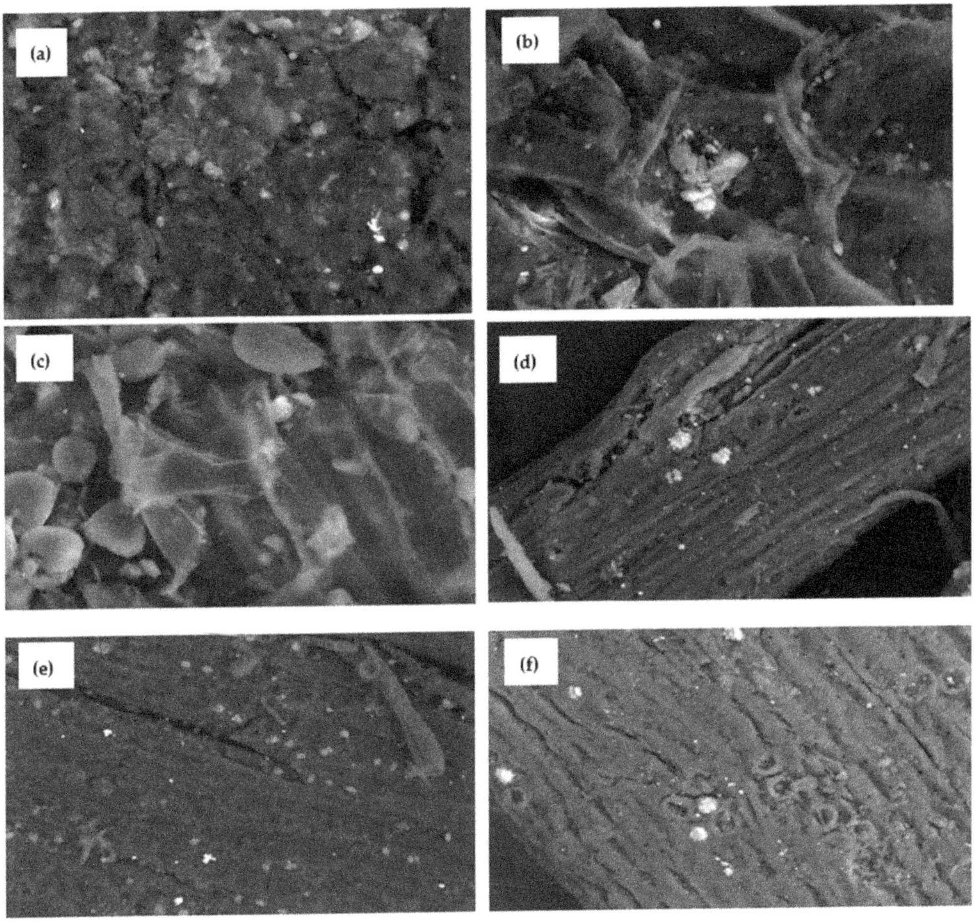

Figure 4. SEM micrograph of OPB (**a**) before the adsorption process, and (**b**) yam after the Cr(VI) adsorption process and (**c**) after the Ni(II) adsorption process; and YP (**d**) before the adsorption process, and (**e**) yam after the Cr(VI) adsorption process and (**f**) after the Ni(II) adsorption process; SEM micrographs have magnification ×500.

With respect to SEM and EDS after adsorption, the presence of Cr(VI) was observed on the basis of the characteristic high-intensity peaks at 1, 5.6, and 5.8 keV, while the appearance of Ni(II) was found on the basis of the peaks at 1.4, 7.8 and 8.2 keV. Due to the structural changes reflected in the FTIR (Figures 1 and 2) and the EDS spectra after adsorption, as well as the increase in the number of white particles precipitated onto the surface of the two adsorbents, it can be established that the mechanism of adsorption of both metals in the biomaterials is an exchange between the studied cations and the disponible centers of the material [40], which promotes the formation of microcomplexes and precipitation [41] when Cr(VI) and Ni(II) are retained at the active centers [42]. In the EDS after Cr(VI) adsorption, a decrease in Ca intensity of 0.2 keV was observed in OPB, while the intensity of O increased at 0.6 keV, and K appeared at 3.6 keV; meanwhile, YP presented a decrease in the peaks of Ca at 0.6 keV, Fe at 1.1 keV, and K at 3.6 keV, while P disappeared at 2.0 keV and S at 2.1 keV. Following the removal of Ni(II), OPB presented an increase in the Fe peak at 0.8 keV, and the appearance of Al at 1.8 keV, S at 2.2 keV, Fe at 6.6, and 6.9 keV, while in YO presented the disappearance of P at 2.0 keV, Al at 1.8 keV, S at 2.1 keV, Fe at 0.7 keV, 6.5 keV, and 7 7 keV, and Cu at 0.9 keV, 8 keV, and 8.6 keV; there was also a decrease in the peaks for K at 3.4 keV and Ca at 0.2 keV and 3.8 keV; the disappearance of these compounds in the structure of the biomaterials was due to the formation of links with the heavy metals, and their entry into the lignocellulosic structure is shown in the FTIR spectra (Figures 2 and 3), corroborating that ion exchange is the mechanism of adsorption [28]. The above occurred along with slight variations in the intensity of the peaks marking the presence of Ca, while a new peak of Cr(VI) and NI(II) was observed with the C, N, K, Na, Mg, Al, silicon, chlorine and Ca groups contained on the surface, confirming the adsorption of the heavy metals under study [39].

Table 1. EDS compositional analysis.

Element	OPB		OPB—Cr(VI)		YP		YP—Cr(VI)	
	Weight %	Atomic %	Weight %	Atomic %	Weight %	Atomic %	Weight %	Atomic %
C	50.90	59.07	48.65	56.49	47.40	57.07	54.50	63.25
O	44.60	38.85	48.82	42.56	43.20	39.05	39.67	34.56
Al					0.75	0.40	0.82	0.42
Si	3.87	1.92	1.12	0.56	2.16	1.11	1.21	0.60
P					0.66	0.31		
S					0.17	0.08		
K			0.20	0.07	4.46	1.65	0.44	0.16
Ca	0.19	0.07	0.25	0.09	0.34	0.12	0.99	0.34
Fe					0.60	0.16	0.59	0.15
Cu	0.45	0.10	0.49	0.11	0.27	0.06	0.45	0.10
Cl							0.45	0.18
Cr			0.48	0.13			0.88	0.24
Totals	100.00	100.00	100.00	100.00	100.00	100.00	100.00	100.00

3.2. *Adsorption Tests*

Tables 2 and 3 summarize the results of Cr(VI) and Ni(II) adsorption on OPB and YP under the evaluated conditions. For the removal of Cr(VI) and Ni (II) using OPB, the results were very similar under all of the evaluated operating conditions; therefore, a deeper statistical analysis was necessary in order to be able to draw any conclusions (see Section 3.3). This can be explained by the fact that OPB, as a fibrous material, ensures the tortuosity of the bed, regardless of the bed height.

On the other hand, when using YP, the best conditions, found on the basis of optimization analysis, for Cr(VI) removal were a particle size of 0.355, a temperature of 55 °C and a bed height of 89 mm, while for Ni(II), they were a particle size of 1 mm, a temperature of 70 °C, and a bed height of 123 mm (Table 3).

OPB exhibits a high efficiency of Ni(II) and Cr(VI) removal, with percentages higher than 87%; this is due to the heterogeneity of the biomass due to its lignocellulosic character, which ensures the presence of hydroxyl, carboxyl, amino, unsaturated hydrocarbon, and phenol groups, which have a direct influence on the removal of these ions [43]. Despite the good performance of the bioadsorbent for the two metals, there is evidence of better behavior with respect to Cr(IV), which can be explained by their ionic radius: Cr(VI) has an ionic radius of 0.69, while Ni(II) has an ionic radius of 0.78, and the lower ionic radius increases the diffusion of the metal in solution and on potential adsorption sites [44]. Regarding the evaluated variables, an increase in temperature from 30 °C to 55 °C decreases the retention of both metals on OPB and YP, while at temperatures greater than 55 °C, increasing temperature results in a slight increase in removal efficiency. This may be due to the fact that the increase in temperature contributes to an increase in the speed of the ions within the solution and the biomass, counterbalancing the thermodynamic effect, which in turn causes a reversal of the trend whereby the temperature continues to increase [45].

Considering the high degree of removal achieved by the biomass, the effect of this variable is considered to be insignificant, since it does not constitute a determining aspect for the process, suggesting that energy changes in the system do not have an important effect on the mass transfer mechanisms that allow the diffusion of ions on the biomass. However, variation in temperature is important, since it constitutes a critical variable when scaling the process up. Therefore, it is considered that the best conditions for carrying out the removal exist at room temperature. The influence of bed height presents a positive behavior when using the two adsorbents, removal efficiency increasing with increasing bed height, which can be explained by the availability of a greater number of link sites for adsorption and the increased residence time. Therefore, the diffusion of the ions is more effective, since the contact between the phases is more intimate [17].

Table 2. Cr(VI) and Ni(II) removal efficiency over OPB in a continuous system.

Temperature (°C)	Particle Size (mm)	Bed Height (mm)	OPB	
			Cr(VI)	Ni(II)
70	0.355	30.0	100.00	99.86
40	0.355	100.0	99.97	99.53
55	0.500	6.13	99.98	89.85
70	1.000	30.0	99.99	87.24
40	1.000	30.0	100.00	90.37
40	0.355	30.0	100.00	98.93
70	0.355	100.0	99.98	99.57
30	0.500	65.0	100.00	98.33
40	1.000	100.0	99.97	97.16
70	1.000	100.0	99.99	96.77
55	0.500	123.9	99.99	99.76
55	0.500	65.0	99.99	96.80
80	0.500	65.0	100.00	96.95
55	0.500	65.0	99.99	96.80
55	0.135	65.0	99.98	99.76
55	1.219	65.0	99.99	91.18
55	0.500	65.0	99.99	96.80

Table 3. Cr(VI) and Ni(II) removal efficiency over YP in a continuous system.

Temperature (°C)	Particle Size (mm)	Bed Height (mm)	YP	
			Cr(VI)	Ni(II)
70	0.355	30.0	96.26	96.65
40	0.355	100.0	87.30	87.4
55	0.500	6.13	65.30	91.40
70	1.000	30.0	97.43	90.20
40	1.000	30.0	98.30	96.72
40	0.355	30.0	96.95	95.22
70	0.355	100.0	87.01	87.0
30	0.500	65.0	87.70	94.46
40	1.000	100.0	88.24	94.07
70	1.000	100.0	91.10	95.91
55	0.500	123.9	92.30	98.74
55	0.500	65.0	94.50	98.17
80	0.500	65.0	94.88	98.82
55	0.500	65.0	97.32	97.25
55	0.135	65.0	87.1	87.2
55	1.219	65.0	98.01	91.66
55	0.500	65.0	94.50	98.15

On the other hand, the increase in particle size positively affects the influence of Cr(VI) removal when OPB is used, which is decisive for establishing the hydrodynamics of the adsorption column, and thus the prevention of overflow [46]. Meanwhile, removing Ni(II) using OPB as well as using YP to remove both metals is favored by decreased particle diameter due to increased area of contact exposed to the smaller particle size; this, in turn, favors the diffusion velocity [47]. However, in practice, for packed-bed systems in which the flow is continuous, very small particle sizes are inadequate, because the close spacing between the bed particles leaves less space for the transit of the fluid, affecting the tortuosity of the bed. In fact, while executing the experiment, it was noticed that the smaller particle sizes caused an obstruction of the fluid on the bed, making its transit difficult; this represents a difficulty, since the ideal is to treat as much of the solution as possible without decreasing the quality of the removal achieved during operation. In this case, sacrificing a small amount of the optimal particle size can lead to better operation of the system, without sacrificing much of the effectiveness of the operation.

Table 4 shows the q_{max} data of the parameters reported for the removal of Cr(VI) and Ni(II) with different evaluated adsorbents, showing that the results obtained in the present study lie in the average range for bioadsorbents of lignocellulosic origin.

Table 4. q_{max} of Cr(VI) and Ni(II) in a fixed bed using various adsorbents.

Metal	Adsorbent	Conditions	q_{max} (mg/g)	Reference
Cr(VI)	Pecan nut husk	0.85 mm particle size, pH = 6.2	180.75	[4]
	Eucalyptus	pH = 2, 15 cm bed height, 50 mg/L initial concentration	381.82	[15]
	Palm Oil Fuel Ash	100 mg/L initial concentration, 36.4 g adsorbent, pH = 2.	0.41	[16]
	Biosynthesized melanin-coated PVDF membranes	0.5 mL/min, 3 mg/L, 10 mg adsorbent	9.29	[28]
	Plantain starch residues	0.75 mL/s, 100 mg/L initial concentration, pH = 2, 68 °C, 81.49 mm bed height	29.85	[17]
	OPB YP	0.75 mL/s, 100 mg/L initial concentration, pH = 2	115.45 50.12	This work
Ni (II)	Plantain starch residues Yam starch residues	0.75 mL/s, 100 mg/L initial concentration, pH = 6	28.01 22.08	[22]
	Fenton modified with *Hydrilla verticillate*	Bed height = 25 cm, 10 mL/min, initial concentration 5 mg/L	87.18	[48]
	Carboxylated sugarcane bagasse	25 °C, pH = 5.5, 0.5 g adsorbent	1020	[49]
	OPB YP	0.75 mL/s, 100 mg/L initial concentration, pH = 6, 55 °C	103.49 30.04	This work

3.3. Application of the Surface Response Method (SRM)

SRM was applied to the experimental data using the Statgraphics Centurion software, and a mathematical expression of adsorption efficiency was obtained that fits the results of the experimental results presented in Tables 2 and 3; in this way, it was possible to vary the parameters of the factors temperature, particle size, and bed height by means of Equation (10) for Cr(VI) and Equation (11) for Ni(II), using OPB as the adsorbent.

$$\begin{aligned}\%R = {} & 100.04838 - 0.0014138539\,A - 0.046518846\,B \\ & -0.00033061127\,C + 0.0000076169902\,A^2 \\ & -0.00010335917\,A\,B + 0.0000071428571\,A\,C \\ & +0.041973067\,B^2 + 0.000022148394\,B\,C + 3.8888414E \\ & -7\,A^2\end{aligned} \tag{10}$$

$$\begin{aligned}\%R = {} & 105.71338 - 0.14294739\,A - 9.4454119\,B - 0.007324277\,C \\ & +0.00153546\,A^2 - 0.11583979\,A\,B \\ & +0.00044214286\,A\,C - 3.7284854\,B^2 + 0.17744186\,B\,C \\ & -0.00052249029\,C^2\end{aligned} \tag{11}$$

where $\%R$ is the percentage removal, A is the temperature in °C, B is the particle size in mm, and C is the bed height in mm. Using these equations, graphs of the estimated response surface were obtained (Figure 5), which described the influence of the evaluated variables on the processes of Cr(VI) and Ni(II) adsorption on OPB. The concave downwards shape indicates the adverse effect of increasing temperature and particle size at different bed heights. The optimal conditions for maximizing the removal efficiency of the process were obtained as follows: temperature of 40 °C, particle size of 1 mm size, and bed height of 30 cm for Cr(VI), while for Ni(II) these values were a temperature of 80 °C, particle size of 0.212 mm, and a bed height of 50.3 mm. The breakage curve of the two metals on OPB was constructed under these optimal conditions.

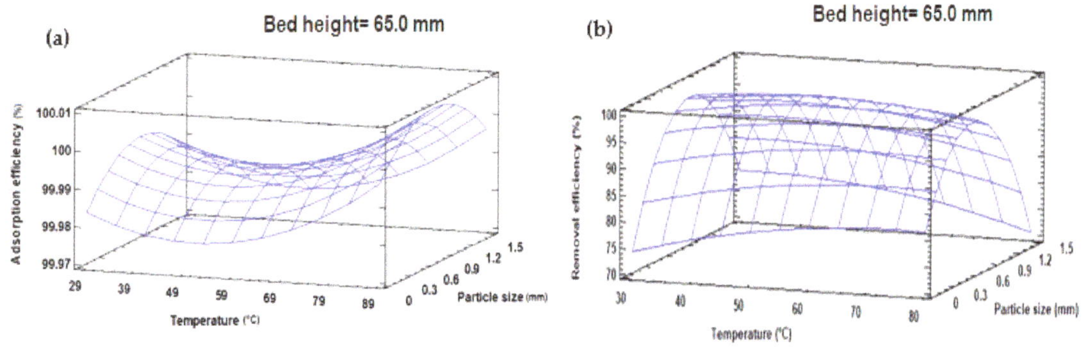

Figure 5. Estimated response area for the removal efficiency of (**a**) CrVI) and (**b**) Ni(II) on OPB.

Previous studies on Cr(VI) and Ni(II) adsorption on Fenton *Hydrilla verticilata* (FMB)-modified dry biomass in industrial waste, in which the effect of bed height, flow rate, influent metal ion concentration, and particle size were evaluated, concluded that the values most favorable for the removal of heavy metals were the highest bed height (25 cm); a lower flow rate (10 mL min^{-1}), and a lower concentration of influential metal (5 mg L^{-1}), in order to ensure optimal tortuosity and hydrodynamic conditions with the aim of maximizing bed removal [48].

Regarding the application of SRM to the results obtained for the adsorption of Cr(VI) and Ni(II) on YP, Equations (12) and (13) were obtained for the two metals, respectively, where $\%R$ is the percentage of removal, A is the temperature in °C, B is particle size in mm, and C is the bed height in mm.

$$\begin{gathered}\%R = 77.51 - 0.0065A - 2.622B\,B + 0.45C - 0.0018\,A^2 + 0.18\,A\,B \\ +0.0019\,A\,C + 0.426\,B^2 - 0.0165\,B\,C - 0.0039C^2\end{gathered} \quad (12)$$

$$\begin{gathered}\%R = 77.67 + 0.438A + 17.45B - 0.0228C - 0.0035A^2 + 0.11AB \\ -0.000826AC - 32.23B^2 + 0.338BC - 0.0011B^2\end{gathered} \quad (13)$$

From these equations, the estimated response surface was obtained (Figure 6), and the significant positive effect of increasing particle size on the quantity of Cr(VI) elimination was observed, while the concave shape of the scheme for nickel shows that intermediate values of the considered variables favored its removal. In addition, a temperature of 62 °C, a particle size of 1.22, and a bed height of 89.3 mm were determined to be the optimal conditions for the removal of Cr(VI), while for Ni(II) these parameters were a temperature of 70 °C, a particle size of 1 mm, and a bed height of 123.8 mm. The breakage curve was constructed under these conditions.

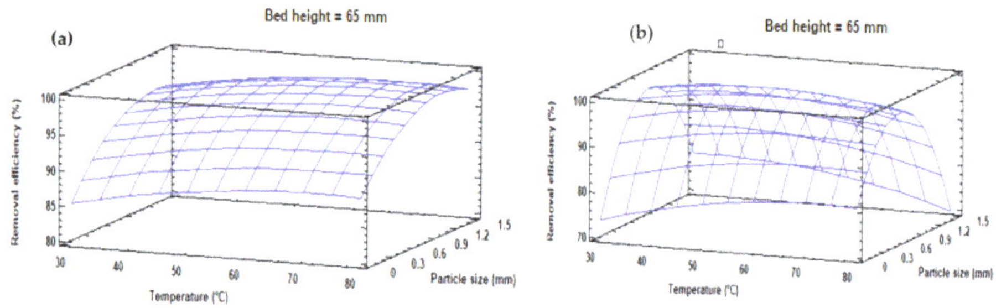

Figure 6. Estimated response area for the removal efficiency of (**a**) CrVI) and (**b**) Ni(II) on YP.

The optimization of the response variable was obtained using the RSM methodology, as shown in Table 5. The experiments for analyzing the breakage curve were performed under these conditions.

Table 5. Optimization of removal efficiency of Cr(VI) and Ni(II) over OPB and YP.

Factor	OPB		YP	
	Cr(VI)	Ni(II)	Cr(VI)	Ni(II)
Temperature (°C)	30	80	55	70
Particle size (mm)	0.5	0.212	0.355	1
Bed height (mm)	65	50.3	89	123

3.4. Breakthrough Curve

The breakage curve shows the performance of a fixed-bed column in terms of the amount of metal that can be retained, expressed in terms of a standardized concentration (C/Ci), as a function of time or effluent volume for a fixed bed height [50]. For this study, the breakpoint was established as the point at which the concentration at the end of the column reached 0.1% of the C_i, while the concentration at which 95% was reached was established as the column saturation point [20]. Figure 7 shows the Cr(VI) and Ni(II) breakage curves on OPB and YP, which were made at the best conditions found and at 100 mg L^{-1}.

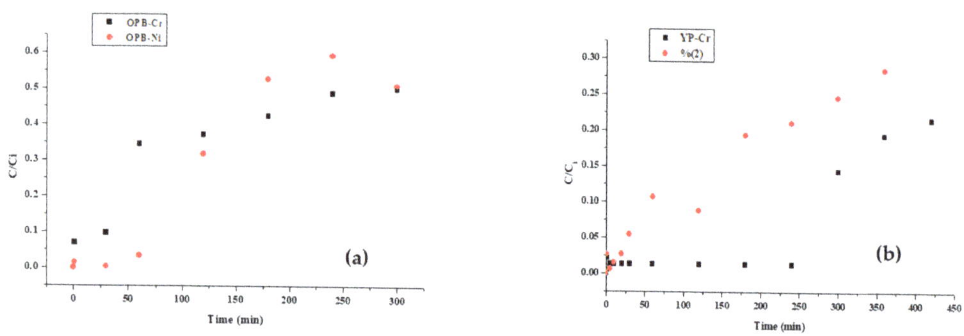

Figure 7. Breakage curve for the adsorption of Cr(VI) and Ni(II) on (**a**) OPB and (**b**) YP.

The experimental data obtained from experiments were analyzed to evaluate the effect on the C_f/C_i vs. time plot, and to stablish the q (mg/g) of the bed. It was evidenced that the highest adsorption of metal ions occurred at the beginning of the tests, due to the high disposal of active sites in the unsaturated bioadsorbent [31].

As a result, metal ions were retained around or inside the active sites, while the effluent from the end of the bed was almost raw without the presence of the cations. As the bed operation time passed, its capability to capture ions progressively diminished due to the occupation of the existing active sites; therefore, the concentration of the effluent increased until the outlet concentration was 30% of the inlet concentration [21]. However, the total saturation of the adsorbent was not reached in the column, so the breakage curve exhibits an upward trend after the contact time. Applying Equation (3), the maximum adsorption capacity of the column was determined for each adsorbent–metal system, indicating that the highest adsorption capacities were obtained for Cr(VI), with 111.45 mg g^{-1} and 50.12 mg g^{-1} on OPB and YP, respectively; meanwhile, for Ni(II), values of 103.49 mg g^{-1} and 30.04 mg g^{-1} were obtained. With respect to the values obtained for the removal of Ni(II) on OPB (*Elaeis guineensis*) in stages, a q_e of 103.3 mg g^{-1} biomass was reported for bagasse, which is very close to the value obtained in the present study [51], which is superior to those reported by [22] using plantain, yam and OPB residues, respectively. Likewise, the results of Cr(VI) removal using OPB and YP are superior to those obtained in other studies, where 29.85 mg g^{-1} on plantain residues was achieved [17]. The experimental adsorption capacity for the removal process of both metals in the column was higher for palm bagasse; this is attributed to its highly porous nature and the typical fibrous surface of lignocellulosic materials [52].

On the other hand, designing a continuous adsorption system requires knowing mainly its adsorption profile (concentration–time), so there are different theoretical models that seek to describe the behavior of a column, estimating some kinetic parameters [21]. Breakage curve data were fitted to the Dose–Response, Yoon–Nelson, Adams–Bohart, and Thomas models. The fitting parameters are summarized in Table 6.

Table 6. Adjustment parameters for Cr(VI) and Ni(II) adsorption on OPB and YP using non-linear regression.

Model	Parameter	OPB-Cr	OPB-Ni	YP-Cr	YP-Ni
Thomas	K_{Th} (mL min^{-1} mg^{-1})	0.06	0.077	0.09	0.06
	q_{Th} (mg g^{-1})	58.85	27.69	47.58	23.94
	R^2	0.66	0.55	0.87	0.84
Dose-Response	$q_{D\text{-}R}$ (mg g^{-1})	62.18	24.74	54.21	54.93
	a	0.67	1.29	2.85	0.84
	R^2	0.91	0.86	0.92	0.95
Yoon–Nelson	$K_{Y\text{-}N}$ (min^{-1})	0.00616	0.01	0.008615	0.00609
	τ (min)	255.9752	269.36	533.67	486.79
	R^2	0.66	0.55	0.87	0.8391
Adams–Bohart	$K_{A\text{-}B}$ (L min^{-1} mg^{-1})	3.622×10^{-5}	4.09×10^{-5}	7.43×10^{-5}	4.98×10^{-5}
	N_o (mg L^{-1})	39388.88	15013.42	13281.04	9302.27
	R^2	0.53	0.21	0.85	0.81

From the data presented in Table 6, it can be stated that, according to their respective square correlation coefficients, the experimental Cr(VI) and Ni(II) removal data on YP and OPB best fit the DR model, with R^2 exceeding 90%. Considering the industrial implementations of adsorption, the established advance time defines the operational limits of the column; however, in the present study, such saturation was not reached. Rather, it was established that all the biomasses could be used during the study time, since they present a high heavy metal adsorption capacity, as reported when Fenton-modified *Hydrilla verticillata* biomass was used as Cr(VI) and Ni(II) adsorbents [48]. Despite the R^2 value reported for the DR model, it was observed that the q reported by this model when using OPB to remove the two metals was much lower than that obtained experimentally. Similarly, the theoretical q obtained by the models was much higher than the experimental q with YP; in this sense, the model that comes closest to the experimental values is the Thomas model, so it was assumed that the mechanism of adsorption was Langmuir-type adsorption followed by chemical sorption of a pseudo-second-order monolayer, which is evident in the precipitation and micro complex formation evidenced in the SEM micrographs reported in Figures 3 and 4 [36]. The prediction of the breakage curve by the Thomas model has previously been reported by Srivastava et al. [36], Nordin et al. [16], and Xavier [49] using red beans impregnated with nanoparticles of magnetite, OPB and carboxylated OPB resin. It has been reported, when using mangosteen peel for Cd(II), that the good fit of the data using the Thomas model suggests that external and internal diffusion will not be the limiting step [6].

3.5. Thermodynamic Parameters

It is established on the basis of the positive sign of ΔH° that the processes of removal of Cr(VI) on OPB and YP is of an endothermic nature, such that the energy must be proportional to the system in order to boost diffusive phenomena (Table 7) [22]. It also follows that the limiting step of chromium adsorption on OPB and YP is the chemical adsorption [53]. In addition, the positive values of ΔG° for Cr(VI) removal onto the two evaluated adsorbents suggest that the system is non-spontaneous, and the increase in the module with increasing temperature indicates that the process becomes energetically more favorable [54]. The positive value of ΔS° exposes the high affinity of Cr(VI) ions with biomaterials and also the possibility of some structural modifications due to the development of links with functional groups at the interface, as well as high randomness at the solid–solution interface, as shown by FTIR spectra and SEM-EDS analysis; it is also inferred that the process is reversible.

Table 7. Thermodynamic parameters for chromium (VI) and nickel (II) adsorption at bed height of 65 mm.

Biomass	Cr (VI)					Ni (II)		
	T (K)	q_e (mg/g)	ΔG° (kJ/mol)	ΔH° (kJ/mol)	ΔS° (kJ/mol × K)	ΔG° (kJ/mol)	ΔH° (kJ/mol)	ΔS° (kJ/mol × K)
OPB	303.15 328.15 353.15	0.79 0.78 0.78	30.19 32.68 35.17	3.46	28.81	0.79 −84.29 0.78 −91.25 0.78 −98.20	−107.32	−25.91
YP	303.15 328.15 353.15	2.70 2.91 2.92	20.16 21.80 23.45	18.65	6.59	−75.54 −81.79 −88.03	−19.65	−24.98

For Ni(II), the ΔH° values establish that the removal process possesses an exothermic nature and occurs by physisorption [53]. ΔG° values indicate the spontaneity of the process and that it becomes energetically more favorable, evolving by itself. On the basis of the negative magnitude of ΔS°, it can be said that the bond between the metal and the biomass is strong, with a low possibility of desorption [54].

The difference found in the adsorption mechanism regarding the thermodynamic parameters may be in accordance with the chemical nature of Cr(VI) and Ni(II), and its influence over the chemical interactions with the active centers in the adsorbent; thus, the different valences, atomic ratios and the ionic energies between the heavy metals cause the energetic requirements for each case to be specific. When using eucalyptus as an adsorbent of Cr(VI), equal results were obtained [15], as well as when using nanocellulose modified with polypyrrole [55], in addition to when using pea peels [18] and composites of Polyaniline and Iron Oxide [56] for removing Ni(II) ions.

4. Conclusions

The monocomponent elimination of Cr(VI) and Ni(II) on OPB and YP was performed. The morphological and physical characterizations showed a porous, heterogeneous structure with high presence of functional groups that favor heavy metal adsorption processes. The SEM-EDS results suggest that the adsorption mechanism that controls the process is cation exchange with the active sites of the adsorbent, as well as precipitation and complexing on the exposed surface of the biomaterials. From the adsorption essays, removal efficiencies above 80% were obtained in all cases, presenting better performance in Cr(VI) removal due to its diffusion capacity through the bioadsorbent pores. Statistical analysis shows that particle size was the most significant factor in the adsorption processes.

Mximum capacities of Cr(VI) of 111.45 mg g^{-1} and 50.12 mg g^{-1} were achieved on OPB and YP, respectively; while for nickel, capacities of 103.49 mg g^{-1} and 30.04 mg g^{-1} were obtained. The Thomas model fitted very well to the experimental breakage curves, so it is assumed that monolayer adsorption was followed by chemisorption. Thermodynamic study suggested an endothermic process that was non-spontaneous, with high affinity between Cr(VI) and the adsorbents; meanwhile, Ni(II) adsorption onto OPB and YP exhibited an exothermic and spontaneous nature.

Author Contributions: Conceptualization, C.T.-T., A.V.-O. and Á.D.G.-D.; methodology, C.T.-T.; software, A.V.-O.; validation, C.T.-T., A.V.-O. and Á.D.G.-D.; formal analysis; investigation; resources, C.T.-T.; data curation, Á.D.G.-D. writing—original draft preparation; writing—review and editing, C.T.-T.; visualization, A.V.-O.; supervision, C.T.-T.; project administration, A.V.-O. and Á.D.G.-D.; funding acquisition, C.T.-T. All authors have read and agreed to the published version of the manuscript.

Funding: This research received no external funding.

Data Availability Statement: The data that support the findings of this study are available from the corresponding author, upon reasonable request.

Acknowledgments: The authors thank the Universidad de Cartagena for the support to develop this research, laboratories, software, and time of the research team.

Conflicts of Interest: The authors declare no conflict of interest.

References

1. Kulbir, S.; Abdullahi, W.S.; Chhotu, R. Removal of Heavy Metals by Adsorption using Agricultural based Residue: A Review. *Res. J. Chem. Environ.* **2018**, *22*, 65–74.
2. De Gisi, S.; Lofrano, G.; Grassi, M.; Notarnicola, M. Characteristics and adsorption capacities of low-cost sorbents for wastewater treatment: A review. *Sustain. Mater. Technol.* **2016**, *9*, 10–40. [CrossRef]
3. Villabona-Ortíz, A.; De Cartagena, U.; Tejada-Tovar, C.N.; Ortega-Toro, R. Modelling of the Adsorption Kinetics Of Chromium (VI) Using Waste Biomaterials. *Rev. Mex. Ing. Química* **2020**, *19*, 401–408. [CrossRef]
4. Corral-Escárcega, M.C.; Ruiz-Gutiérrez, M.G.; Quintero-Ramos, A.; Meléndez-Pizarro, C.O.; Lardizabal-Gutiérrez, D.; Campos-Venegas, K. Use of biomass-derived from pecan nut husks (Carya illinoinensis) for chromium removal from aqueous solutions. column modeling and adsorption kinetics studies. *Rev. Mex. Ing. Quim.* **2017**, *16*, 939–953.
5. Manjuladevi, M.; Anitha, R.; Manonmani, S. Kinetic study on adsorption of Cr(VI), Ni(II), Cd(II) and Pb(II) ions from aqueous solutions using activated carbon prepared from Cucumis melo peel. *Appl. Water Sci.* **2018**, *8*, 36. [CrossRef]
6. Chowdhury, Z.; Zain, S.M.; Rashid, A.; Rafique, R.F.; Khalid, K. Breakthrough Curve Analysis for Column Dynamics Sorption of Mn(II) Ions from Wastewater by UsingMangostana garciniaPeel-Based Granular-Activated Carbon. *J. Chem.* **2013**, *2013*, 1–8. [CrossRef]
7. Bhanvase, B.A.; Ugwekar, R.P.; Mankar, R.B. *Novel Water Treatment and Separation Methods: Simulation of Chemical Processes*; CRC Press: Boca Raton, FL, USA, 2017.
8. Sahu, O.; Singh, N. Significance of bioadsorption process on textile industry wastewater. In *The Impact and Prospects of Green Chemistry for Textile Technology*; Woodhead Publishing: Sawston, UK, 2019; pp. 367–416.
9. Hu, Q.; Xu, L.; Fu, K.; Zhu, F.; Yang, T.; Yang, T.; Luo, J.; Wu, M.; Yu, D. Ultrastable MOF-based foams for versatile applications. *Nano Res.* **2021**, 1–10. [CrossRef]
10. Qian, J.; Gao, X.; Pan, B. Nanoconfinement-Mediated Water Treatment: From Fundamental to Application. *Environ. Sci. Technol.* **2020**, *54*, 8509–8526. [CrossRef]
11. Elabbas, S.; Ouazzani, N.; Mandi, L.; Berrekhis, F.; Perdicakis, M.; Pontvianne, S.; Pons, M.-N.; Lapicque, F.; Leclerc, J.-P. Treatment of highly concentrated tannery wastewater using electrocoagulation: Influence of the quality of aluminium used for the electrode. *J. Hazard. Mater.* **2016**, *319*, 69–77. [CrossRef]
12. Cherdchoo, W.; Nithettham, S.; Charoenpanich, J. Removal of Cr(VI) from synthetic wastewater by adsorption onto coffee ground and mixed waste tea. *Chemosphere* **2019**, *221*, 758–767. [CrossRef]
13. Bharathi, K.S.; Ramesh, S.P.T. Fixed-bed column studies on biosorption of crystal violet from aqueous solution by Citrullus lanatus rind and Cyperus rotundus. *Appl. Water Sci.* **2013**, *3*, 673–687. [CrossRef]
14. Jafari, S.A.; Jamali, A. Continuous cadmium removal from aqueous solutions by seaweed in a packed-bed column under consecutive sorption-desorption cycles. *Korean J. Chem. Eng.* **2016**, *33*, 1296–1304. [CrossRef]
15. Haroon, H.; Ashfaq, T.; Gardazi, S.M.H.; Sherazi, T.A.; Ali, M.; Rashid, N.; Bilal, M. Equilibrium kinetic and thermodynamic studies of Cr(VI) adsorption onto a novel adsorbent of Eucalyptus camaldulensis waste: Batch and column reactors. *Korean J. Chem. Eng.* **2016**, *33*, 2898–2907. [CrossRef]

16. Nordin, N.; Asmadi, N.A.A.; Manikam, M.K.; Halim, A.A.; Hanafiah, M.M.; Hurairah, S.N. Removal of Hexavalent Chromium from Aqueous Solution by Adsorption on Palm Oil Fuel Ash (POFA). *J. Geosci. Environ. Prot.* **2020**, *08*, 112–127. [CrossRef]
17. Villabona-Ortíz, A.; Tejada-Tovar, C.; Ruiz-Paternina, E.; Frías-González, J.D.; Blanco-García, G.D. Optimization of the Effect of Temperature and Bed Height on Cr (VI) Bioadsorption in Continuous System. *Rev. Fac. Ing.* **2020**, *29*, e10477. [CrossRef]
18. Haq, A.U.; Saeed, M.; Anjum, S.; Bokhari, T.H.; Usman, M.; Tubbsum, S. Evaluation of Sorption Mechanism of Pb (II) and Ni (II) onto Pea (Pisum sativum) Peels. *J. Oleo Sci.* **2017**, *66*, 735–743. [CrossRef]
19. Mohammadpour, E.; Yaftian, M.R.; Zamani, A.; Gharbani, P. Optimization of Continuous Flow Adsorption of Heavy Metal Ions on Continuous System Column by Peganum Harmala Seeds. *Arch. Hyg. Sci.* **2017**, *6*, 32–38. [CrossRef]
20. Borna, M.O.; Pirsaheb, M.; Niri, M.V.; Mashizie, R.K.; Kakavandi, B.; Zare, M.R.; Asadi, A. Batch and column studies for the adsorption of chromium(VI) on low-cost Hibiscus Cannabinus kenaf, a green adsorbent. *J. Taiwan Inst. Chem. Eng.* **2016**, *68*, 80–89. [CrossRef]
21. Abdolali, A.; Ngo, H.H.; Guo, W.; Zhou, J.L.; Zhang, J.; Liang, S.; Chang, S.W.; Nguyen, D.D.; Liu, Y. Application of a breakthrough biosorbent for removing heavy metals from synthetic and real wastewaters in a lab-scale continuous fixed-bed column. *Bioresour. Technol.* **2017**, *229*, 78–87. [CrossRef]
22. Ruiz-Paternina, E.B.; Villabona-Ortiz, Á.; Tejada-Tovar, C.; Ortega-Toro, R. Thermodynamic Study of the Removal of Nickel and Chromium in Aqueous Solution using Adsorbents of Agro-Industrial Origin. *Inf. Tecnol.* **2019**, *30*, 3–10. [CrossRef]
23. Naik, R.M.; Ratan, S.; Singh, I. Use of orange peel as an adsorbent for the removal of Cr (VI) from its aqueous solution. *Indian J. Chem. Technol.* **2018**, *25*, 300–305.
24. Abiaziem, C.V.; Williams, A.B.; Inegbenebor, A.I.; Onwordi, C.T.; Ehi-Eromosele, C.O.; Petrik, L.F. Adsorption of lead ion from aqueous solution unto cellulose nanocrystal from cassava peel. *J. Phys. Conf. Ser.* **2019**, *1299*, 012122. [CrossRef]
25. Tejada-Tovar, C.; Ruiz-Paternina, E.; Gallo-Mercado, J.; Moscote-Bohorquez, J. Evaluation of the biosorption with african palm bagasse for the removal of Pb (II) in solution. *Rev. Prospect.* **2015**, *13*, 59–67.
26. Tejada-Tovar, C.; Gonzalez-Delgado, A.D.; Villabona-Ortiz, A. Characterization of Residual Biomasses and Its Application for the Removal of Lead Ions from Aqueous Solution. *Appl. Sci.* **2019**, *9*, 4486. [CrossRef]
27. Asuquo, E.D.; Martin, A.D. Sorption of cadmium (II) ion from aqueous solution onto sweet potato (*Ipomoea batatas* L.) peel adsorbent: Characterisation, kinetic and isotherm studies. *J. Environ. Chem. Eng.* **2016**, *4*, 4207–4228. [CrossRef]
28. Manirethan, V.; Gupta, N.; Balakrishnan, R.M.; Raval, K. Batch and continuous studies on the removal of heavy metals from aqueous solution using biosynthesised melanin-coated PVDF membranes. *Environ. Sci. Pollut. Res.* **2019**, *27*, 24723–24737. [CrossRef]
29. Rinaldi, R.; Yasdi, Y.; Hutagalung, W.L.C. Removal of Ni (II) and Cu (II) ions from aqueous solution using rambutan fruit peels (*Nephelium lappaceum* L.) as adsorbent. *AIP Conf. Proc.* **2018**, *2026*, 020098. [CrossRef]
30. Cai, J.; He, Y.; Yu, X.; Banks, S.; Yang, Y.; Zhang, X.; Yu, Y.; Liu, R.; Bridgwater, T. Review of physicochemical properties and analytical characterization of lignocellulosic biomass. *Renew. Sustain. Energy Rev.* **2017**, *76*, 309–322. [CrossRef]
31. Mahdi, Z.; Yu, Q.J.; El Hanandeh, A. Investigation of the kinetics and mechanisms of nickel and copper ions adsorption from aqueous solutions by date seed derived biochar. *J. Environ. Chem. Eng.* **2018**, *6*, 1171–1181. [CrossRef]
32. Wu, Y.; Fan, Y.; Zhang, M.; Ming, Z.; Yang, S.; Arkin, A.; Fang, P. Functionalized agricultural biomass as a low-cost adsorbent: Utilization of rice straw incorporated with amine groups for the adsorption of Cr(VI) and Ni(II) from single and binary systems. *Biochem. Eng. J.* **2016**, *105*, 27–35. [CrossRef]
33. Dai, Y.; Sun, Q.; Wang, W.; Lu, L.; Liu, M.; Li, J.; Yang, S.; Sun, Y.; Zhang, K.; Xu, J.; et al. Utilizations of agricultural waste as adsorbent for the removal of contaminants: A review. *Chemosphere* **2018**, *211*, 235–253. [CrossRef]
34. Tejada-Tovar, C.; Villabona, A.; Cabarcas, A.; Benitez, C.; Acevedo, D. Optimization of variables in fixed-bed column using the response surface methodology. *Contemp. Eng. Sci.* **2018**, *11*, 1121–1133. [CrossRef]
35. Labied, R.; Benturki, O.; Hamitouche, A.-E.; Donnot, A. Adsorption of hexavalent chromium by activated carbon obtained from a waste lignocellulosic material (*Ziziphus jujuba* cores): Kinetic, equilibrium, and thermodynamic study. *Adsorpt. Sci. Technol.* **2018**, *36*, 1066–1099. [CrossRef]
36. Srivastava, S.; Agrawal, S.B.; Mondal, M.K. Fixed bed column adsorption of Cr(VI) from aqueous solution using nanosorbents derived from magnetite impregnated Phaseolus vulgaris husk. *Environ. Prog. Sustain. Energy* **2019**, *38*, S68–S76. [CrossRef]
37. Deng, Y.; Huang, S.; Laird, D.A.; Wang, X.; Dong, C. Quantitative mechanisms of cadmium adsorption on rice straw- and swine manure-derived biochars. *Environ. Sci. Pollut. Res.* **2018**, *25*, 32418–32432. [CrossRef] [PubMed]
38. Deng, Y.; Huang, S.; Dong, C.; Meng, Z.; Wang, X. Competitive adsorption behaviour and mechanisms of cadmium, nickel and ammonium from aqueous solution by fresh and ageing rice straw biochars. *Bioresour. Technol.* **2020**, *303*, 122853. [CrossRef] [PubMed]
39. Kuppusamy, S.; Thavamani, P.; Megharaj, M.; Venkateswarlu, K.; Lee, Y.B.; Naidu, R. Oak (*Quercus robur*) Acorn Peel as a Low-Cost Adsorbent for Hexavalent Chromium Removal from Aquatic Ecosystems and Industrial Effluents. *Water Air Soil Pollut.* **2016**, *227*, 62. [CrossRef]
40. Medellin-Castillo, N. Biosorption of lead (II) in aqueous solution onto residues of natural fibers from the ixtle industry (*Agave lechuguilla* torr. and *Yucca carnerosana* (trel.) Mckelvey). *Rev. Int. Contam. Ambient.* **2017**, *33*, 269–280. [CrossRef]
41. Rodriguez Narvaez, O.M.; Peralta-Hernandez, J.M.; Goonetilleke, A.; Bandala, E.R. Treatment technologies for emerging contaminants in water: A review. *Chem. Eng. J.* **2017**, *323*, 361–380. [CrossRef]

42. Chen, Y.; An, D.; Sun, S.; Gao, J.; Qian, L. Reduction and Removal of Chromium VI in Water by Powdered Activated Carbon. *Materials* **2018**, *11*, 269. [CrossRef]
43. Basu, H.; Saha, S.; Mahadevan, I.A.; Pimple, M.V.; Singhal, R.K. Humic acid coated cellulose derived from rice husk: A novel biosorbent for the removal of Ni and Cr. *J. Water Process Eng.* **2019**, *32*, 100892. [CrossRef]
44. Ranasinghe, S.; Navaratne, A.; Priyantha, N. Enhancement of adsorption characteristics of Cr(III) and Ni(II) by surface modification of jackfruit peel biosorbent. *J. Environ. Chem. Eng.* **2018**, *6*, 5670–5682. [CrossRef]
45. Akpomie, K.G.; Eluke, L.O.; Ajiwe, V.I.E.; Alisa, C.O. Attenuation kinetics and desorption performance of *artocarpus altilis* seed husk for Co (II), Pb (II) And Zn (II) Ions. *Iran. J. Chem. Chem. Eng.* **2018**, *37*, 171–186.
46. Tejada-Tovar, C.; Herrera-Barros, A.; Ruiz-Paternina, E. Utilization of Biosorbents for Nickel and Lead Removal in Binary Systems. *Cienc. Desarro.* **2016**, *7*, 31–36. [CrossRef]
47. Oguz, E.; Ersoy, M. Biosorption of cobalt(II) with sunflower biomass from aqueous solutions in a fixed bed column and neural networks modelling. *Ecotoxicol. Environ. Saf.* **2014**, *99*, 54–60. [CrossRef]
48. Mishra, A.; Tripathi, B.D.; Rai, A.K. Packed-bed column biosorption of chromium(VI) and nickel(II) onto Fenton modified Hydrilla verticillata dried biomass. *Ecotoxicol. Environ. Saf.* **2016**, *132*, 420–428. [CrossRef]
49. Xavier, A.L.P.; Adarme, O.F.H.; Furtado, L.M.; Ferreira, G.M.D.; da Silva, L.H.M.; Gil, L.F.; Gurgel, L.V.A. Modeling adsorption of copper(II), cobalt(II) and nickel(II) metal ions from aqueous solution onto a new carboxylated sugarcane bagasse. Part II: Optimization of monocomponent fixed-bed column adsorption. *J. Colloid Interface Sci.* **2018**, *516*, 431–445. [CrossRef]
50. Banerjee, M.; Bar, N.; Basu, R.K.; Das, S.K. Removal of Cr(VI) from Its Aqueous Solution Using Green Adsorbent Pistachio Shell: A Fixed Bed Column Study and GA-ANN Modeling. *Water Conserv. Sci. Eng.* **2018**, *3*, 19–31. [CrossRef]
51. Tejada-Tovar, C.; Gallo-Mercado, J.; Moscote, J.; Villabona-Ortíz, A.; Acevedo-Correra, D. Competitive adsorption of lead and nickel ont yam husk and palm bagasse in continuous system. *Rev. Biotecnol. Sect. Agropecu. Agroind.* **2018**, *16*, 52–61. [CrossRef]
52. Pradhan, P.; Arora, A.; Mahajani, S.M. Pilot scale evaluation of fuel pellets production from garden waste biomass. *Energy Sustain. Dev.* **2018**, *43*, 1–14. [CrossRef]
53. Amro, A.N.; Abhary, M.K.; Shaikh, M.M.; Ali, S. Removal of Lead and Cadmium Ions from Aqueous Solution by Adsorption on a Low-Cost Phragmites Biomass. *Processes* **2019**, *7*, 406. [CrossRef]
54. Blanes, P.S.; Bordoni, M.E.; González, J.C.; García, S.I.; Atria, A.M.; Sala, L.F.; Bellú, S.E. Application of soy hull biomass in removal of Cr(VI) from contaminated waters. Kinetic, thermodynamic and continuous sorption studies. *J. Environ. Chem. Eng.* **2016**, *4*, 516–526. [CrossRef]
55. Shahnaz, T.; Padmanaban, V.C..; Narayanasamy, S. Surface modification of nanocellulose using polypyrrole for the adsorptive removal of Congo red dye and chromium in binary mixture. *Int. J. Biol. Macromol.* **2020**, *151*, 322–332. [CrossRef] [PubMed]
56. Muhammad, A.; Shah, A.U.H.A.; Bilal, S. Effective Adsorption of Hexavalent Chromium and Divalent Nickel Ions from Water through Polyaniline, Iron Oxide, and Their Composites. *Appl. Sci.* **2020**, *10*, 2882. [CrossRef]

water

MDPI

Article

Development of Cotton Linter Nanocellulose for Complexation of Ca, Fe, Mg and Mn in Effluent Organic Matter

Vinícius de Jesus Carvalho de Souza [1], José Cláudio Caraschi [1,2], Wander Gustavo Botero [3], Luciana Camargo de Oliveira [4] and Danielle Goveia [1,2,*]

1 Campus of Itapeva, São Paulo State University (UNESP), São Paulo 18409-010, Brazil; viniciusjcsouza@gmail.com (V.d.J.C.d.S.); j.caraschi@unesp.br (J.C.C.)
2 Faculty of Pharmaceutical Sciences, Highway Araraquara-Jaú, São Paulo 14800-903, Brazil
3 Institute of Chemistry and Biotechnology, Federal University of Alagoas (UFAL), Maceió 57072-970, Brazil; wanderbotero@gmail.com
4 Department of Physics, Chemistry and Mathematics, Graduate School of Biotechnology and Environmental Monitoring, Federal University of São Carlos (UFSCar), Rodovia João Leme dos Santos, SP-264, São Paulo 18052-780, Brazil; lcamargo@ufscar.br
* Correspondence: danielle.goveia@unesp.br

Citation: de Jesus Carvalho de Souza, V.; Caraschi, J.C.; Botero, W.G.; de Oliveira, L.C.; Goveia, D. Development of Cotton Linter Nanocellulose for Complexation of Ca, Fe, Mg and Mn in Effluent Organic Matter. *Water* **2021**, *13*, 2765. https://doi.org/10.3390/w13192765

Academic Editors: Cristina Palet and Julio Bastos-Arrieta

Received: 2 September 2021
Accepted: 28 September 2021
Published: 06 October 2021

Publisher's Note: MDPI stays neutral with regard to jurisdictional claims in published maps and institutional affiliations.

Abstract: Effluent organic matter (EfOM) is present in different domestic and industrial effluents, and its capacity to hold metallic ions can interfere in the wastewater treatment process. Due to the low quality of water, new sustainable technologies for this purpose have become extremely important, with the development of renewable-source nanomaterials standing out in the literature. Nanocellulose (NC) deserves to be highlighted in this context due to its physicochemical characteristics and its natural and abundant origin. In this context, the interactions between NC extracted from cotton linter, organic matter fraction (humic substances) and metal ions have been evaluated. Free metal ions (Ca, Fe, Mg and Mn) were separated by ultrafiltration and quantified by atomic absorption spectrometry. The nanomaterial obtained showed potential for the treatment of effluents containing iron even in the presence of organic matter. The probable interaction of organic matter with NC prevents the efficient removal of calcium, magnesium and manganese. For these elements, it is desirable to increase the interaction between metal and NC by modifying the surface of the nanomaterial.

Keywords: aquatic humic substances; nanomaterials; metallic ions; water; bioavailability

1. Introduction

Dissolved organic matter (DOM) is the dominant form of organic matter in aquatic systems. It consists of dissolved molecules as well as aggregates of molecules. These aggregates can be seen as colloids and are very heterogeneous. The main constituents of DOM are the humic substances (HS). As they are complex mixtures of organic compounds with relatively unknown structures, they are operationally defined according to their isolation procedure [1]. The IHSS (International Humic Substances Society) advises the isolation of humic substances from water by adsorption on an XAD-8 column, at pH 2, followed by elution with sodium hydroxide solution. Humic substances are produced from the degradation of biomass and, therefore, are natural compounds present in all environments.

However, the discharge of industrial and municipal effluents may introduce humic materials and humic precursors into surface waters, denominated humic-like substances [2]. Effluent organic matter (EfOM) is characterized by the combination of natural organic matter (NOM), the particulate material in suspension associated with microbial products, and trace harmful chemicals. Most NOM comes from water and is the main component of wastewater. This material, in general, is not considered a toxic pollutant; however, the EfOM can significantly contribute to effluent treatment processes, such as the obstruction of membranes, the limited reuse of wastewater, or the association with other compounds.

Humic substances have abundant functional groups, mainly carboxylic, carbonyl and phenolic, which confer a high capacity for complexation with metallic ions. Because of these interactions, humic substances have an important effect that need to be considered in wastewater treatment processes [3].

Interactions with nutrients and metal ions substantially affect the biogeochemical cycles of these elements, as they are not degraded, remaining in the environment for long periods of time [4]. Even metals considered essential and fundamental for animal and plant metabolisms may pose a risk to the development of the living being in situations of high or prolonged exposure.

Thus, the treatment of wastewater containing metals remains an important issue to be resolved, considering the presence of humic-like substances. There is a growing need to develop methodologies that involve solutions with low environmental impact, where nanotechnology and the use of materials from renewable sources have been highlighted. Tom [5] brought in a detailed review presenting the possible uses of nanotechnology in the field of sustainable water treatment, noting that nanotechnology has great potential in the field of sustainable water treatment and that concerns about the potential toxicity of nanomaterials themselves can be resolved by the use of natural and carbon-based nanomaterials with low toxicity. Natural nanomaterials and biopolymers, with their unique properties, high efficiency, cost-effectiveness and eco-friendly, are considered a good choice for water treatment, especially biobased nanomaterials in cellulose.

Cellulose is a macromolecule responsible for providing some of the important characteristics of plant cells, combining high mechanical strength and low density with good flexibility [6]. All these properties, combined with the fact that cellulose is a non-toxic, renewable and biodegradable material, have caused a great increase in the scientific interest in this material.

When the cellulose macromolecule is broken into nanofibers, its properties are greatly enhanced, in addition to increasing its surface area, making the material attractive for obtaining new composites. Nanocellulose (NC) has high resistance, low specific gravity, a non-abrasive nature, no toxicity, biocompatibility and biodegradability [7]. These characteristics mean NC is a material that is widely used in applications and produced on a large scale.

Thus, for the application of nanocellulose in the removal of metallic ions from wastewater containing organic matter, a very important point to be studied is how this nanomaterial will interact in EfOM and its components, such as metallic ions. It is essential to perform studies on the interactions between metal and organic matter and nanocellulose.

Organic matter plays an important role in the balance of the aquatic system, since its macromolecules, composed of complex aromatic rings and functional groups, allow the absorption of metals and other pollutants, preventing them from being absorbed by living organisms [8]. In this sense, nanomaterials can interact in these systems, causing some type of change in this absorption mechanism due to their small size and differentiated physical-chemical properties. One way to study the interactions between metal ions and natural organic matter is using aquatic humic substances in controlled conditions.

In this context, it is possible to elaborate three hypotheses for the interaction mechanism of NC with the aquatic system: EfOM can complex metals, preventing them from being absorbed by living organisms in a competition process with the nanomaterial. NC can adsorb these metals in a competitive process, increasing their lability or even serving as a transport route for metals into the cells of living organisms. Alternatively, NC can cause changes in the complexing sites of EfOM, decreasing their complexing capacity and releasing the labile metal and bioavailable material into the aquatic system.

To assess the hypotheses presented, it is necessary to study the speciation and distribution of metals in the environment using models that simulate natural conditions. One of the biggest challenges in the speciation analysis of metal ions is in the development of an analytical procedure capable of maintaining the chemical balance of a natural system.

Therefore, it is essential to develop methods to avoid undesirable disturbances in the aquatic system under study [9].

Tangential flow ultrafiltration is a method that has shown good results in differentiating complexes (EfOM + metals) from free metals in aquatic systems. This method consists of using a membrane that allows only the passage of ions, while the larger molecules of organic matter, as well as the EfOM + metal complexes and the NCs and their complexes, are retained in the system. The filtered solution corresponds to the fraction of free metals. Knowing the initial fraction of the solution and the total metals, it is possible to determine the fraction of complexed metals by the difference between total metals and free metals (filtered). The fraction of free metals present in the filtered solution can be determined using spectrometric methods [9,10]. In this sense, the objective of this work is to evaluate the interaction mechanisms between the NC and the metal complexation system (Ca, Fe, Mg and Mn) in the presence or absence of the organic matter, using HS in concentration-controlled experiments.

2. Materials and Methods

2.1. Bleaching Process and Prehydrolysis of the Cotton Linter

The NC used in this study was obtained from cotton linter (Gossypium hirsutum). The linter underwent acid prehydrolysis with hydrochloric acid (HCl) at a concentration of 0.3 mol L^{-1}, a temperature of 123 °C and a pressure of 2.3 kg cm^{-2}. The treatment was performed in a PHOENIX AV-30 model. The ratio of linter cellulose to acid was 1:20 (g linter: mL of acid). The treatment time was 60 min, which was required for the autoclave to reach the specified conditions, followed by another 30 min of cooking at constant temperature and pressure. After the hydrolysis process, the samples were filtered, separating the solid material (prehydrolyzed linter) from the liquor. The filtering process was carried out until the liquor reached a neutral pH.

The prehydrolyzed material underwent a bleaching process based on the holocellulose content determination procedure proposed by Lamaming [11]. For 4.00 g of sample, 140 mL of distilled water, 3 mL of glacial acetic acid, 3.3 g of sodium chlorite and 4.3 g of sodium acetate were used. The reaction medium was placed in a thermostatic bath at 70 °C. After 30 min from the start of treatment, a second addition of reagents (excluding water) was made, maintaining the reaction for another 60 min. At the end of the 90 min process, the samples were filtered, washed with distilled water until reaching a neutral pH and the total removal of chlorine dioxide (yellow in color), and then placed in the oven at 105 ± 3 °C for drying.

2.2. Preparation of Nanomaterial

To obtain the NC, an adaptation of the procedure proposed by Orts et al. [12] with 1:10 sulfuric acid 64% w/w (pulp (g):acid (mL)) was used for the acid hydrolysis of the pulp. The temperature of this process was 60 °C (temperature maintained by a hot plate); the hydrolysis time was 30 min. At the end of the process, 100 g of ice was added and the sample was decanted for 24 h. After decanting, the supernatant was discarded, while the rest went on to the centrifugation process. Centrifugation was performed in 4 cycles of 5 min, and at the end of each cycle the supernatant was discarded for the addition of water.

After centrifugation, the dialysis process was carried out using a 21 mm diameter SERVAPOR membrane. Before use, preparation was necessary, as the membrane contains substances with traces of heavy metals (<50 ppm) that prevent it from drying out. One of the procedures suggested by the manufacturer to remove these substances is to heat the membrane in metal-free water at a temperature between 70 and 80 °C for 2 h, stirring occasionally. The procedure is repeated three times, allowing the water to decant in each repetition.

After removing the metals, the NC solutions obtained were inserted into the membrane, attaching the ends with appropriate clips. The set was inserted into 10 L of deionized water. Changes were made to the water in the set until the water in the container reached a neutral pH.

The characterization of the NC was made by high-resolution electron microscopy (SEM-FEG) using the high-resolution field-emission electron microscope (MEV-FEG), brand JEOL, model JSM-7500 F. The samples were previously dehydrated and mounted directly in stubs, and then metallized in a BAL-TEC® coating system (model SCD-50). The distance between the target and the sample was maintained at 50 mm for all carbon deposits. During deposition, the vacuum was maintained at 0.2 mbar. A conductive carbon layer is generally necessary to reduce the effect of the electric charge on the surface caused by the interaction between the electron beam and the sample, which can result in image distortion [13].

2.3. *Effluent Organic Matter (EfOM) Simulation with Humic Substances (HS)*

Effluent organic matter (EfOM) was simulated using humic substances. The HS was extracted with 100 L of water collected from the Juréia River in the city of São Sebastião, São Paulo, Brazil. The extraction was carried out using the methodology adopted by most researchers associated with the International Humic Substances Society (IHSS), using DAX-8 superlite microporous resin [14]. The EfOM was simulated using deionized water, mass of extracted HS (powder) and pH adjustments to 5.5.

2.4. *Evaluation of the Interaction of NC with Metals*

The verification of the interaction of NC with metallic ion was carried out by preparing the initial solutions of Fe (II)/Mn (II) and Ca (II)/Mg (II) with approximately 2 mg L^{-1}. The pH of the solutions was adjusted to 5.5 in order to approach the pH of natural water, bringing the system closer to real conditions. The NC suspension were added (50–1200 μL) into 250 mL of the multielementar solution. After a period of 24 h (period for the formation of metal-nano complexes) the tangential flow ultrafiltration process was carried out.

The tangential ultrafiltration system used is composed of a peristaltic pump, tygon tubes and a porosity membrane of 1 kDa and 47 mm in diameter. To carry out the filtration, pressure controllers were used, so that when the aliquots were collected, the free metals were separated from the metals complexed to NC. The concentrations of free ions were determined according to item 2.7.

2.5. *Evaluation of the Influence of NC on the Metal-EfOM Complexes*

Into 200.0 mL of EfOM 100 mg L^{-1} solution, 2 mg L^{-1} of the metals Ca, Fe, Mg and Mn were added, and the pH was adjusted to 5.5 to approximate the pH of natural waters. A sample of the solution was collected to determine the exact concentration of the metal in the solution (total metal). After the 24 h period, the tangential ultrafiltration of the solution was performed to determine the free metals not complexed with organic matter, before adding NC (time 0 min).

The NC (300 μL) was then added, and samples of the filtrate were collected from time to time over the 24 h period. The free metals, separated by means of the tangential flow ultrafiltration process, were quantified using the atomic absorption spectrometer (Figure 1). The concentration of complexed metals was evaluated using Equation (1).

$$\text{Complexed metal} = \text{Total Metal} - \text{Free Metal} \quad (1)$$

The Figure 1 represents all the steps described in this section.

2.6. *Evaluation of the Influence of EfOM on Metal—NC Complexes*

Solutions were prepared with the metals at a concentration of 2 mg L^{-1}, and 300 μL of the suspension containing NC was added. This mixture was left to stand for a period of 24 h to form metal—NC bonds. A sample was collected to quantify the total metal present in each solution, the tangential ultrafiltration process was performed, and then EfOM was added at a concentration of 100 mg L^{-1}. Aliquots of the filtrate were collected from time to time during the 24 h period.

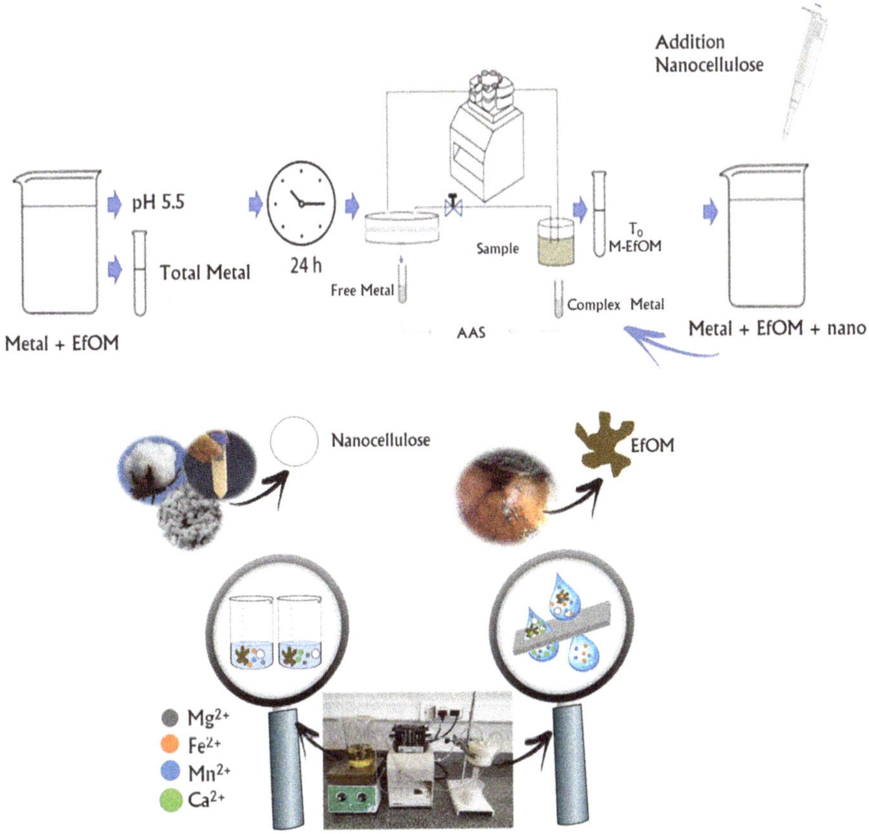

Figure 1. Ultrafiltration system used to evaluate the influence of effluent organic matter in the removal of metals from aquatic systems by nanocellulose. AAS—atomic absorption spectroscopy; EfOM—effluent organic matter; M—metal.

2.7. Determination of Metal Concentration

The analyses for determining the metals present in the aliquot were performed on an atomic absorption spectrometer Analytik Jena flame spray NovAA400. Table 1 represents the parameters used for the determination of metals. For calibrations and metal determinations, the appropriate synthetic standard (atomic absorption spectroscopy (AAS) multielement standard solution, Merck, Darmstadt, Germany) was employed.

Table 1. Parameters used for the determination of ionic metals in the atomic absorption spectrometer.

Elements	Fe	Mn	Ca	Mg
Lamp current	7.5 mA	10 mA	6.5 mA	6.5 mA
Flame	Air/acetileno	Air/acetileno	Air/acetileno	Air/acetileno
Spectral resolution	0.8 nm	1.2 nm	1.2 nm	1.2 nm
Wavelength	248.3 nm	279.8 nm	422.7 nm	202.6 nm
Combined gas flow	55 L h^{-1}	55 L h^{-1}	55 L h^{-1}	55 L h^{-1}
concentration unit	mg L^{-1}	mg L^{-1}	mg L^{-1}	mg L^{-1}

3. Results

3.1. NC Preparation and Interaction with Metals

The NC was obtained by a chemical process referred to as cellulose nanowhiskers. The structure of NC was characterized by scanning electron microscopy (SEM). SEM observations showed that the original cotton linter was successfully acid-hydrolyzed into cellulose nanowhiskers and/or nanocrystalline cellulose with a needle- or short rod-like structure and an average diameter of 15 nm (Figure 2). The fibers have a very wide size distribution, as is often observed for cellulose nanocrystalline structures [15–18]. The NC presented particle agglomeration and consequent stacking, mainly due to its very small size and its large surface area. Furthermore, the hydroxyl group of the surface of the nanowhisker facilitates hydrogen bond formation, which itself facilitates the aggregation.

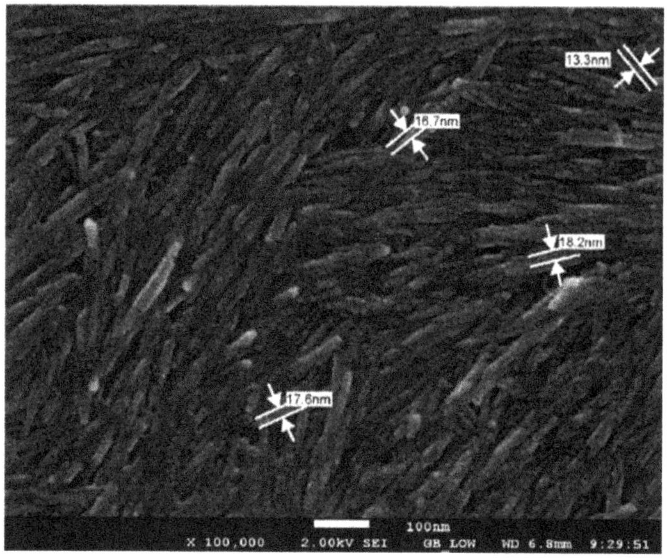

Figure 2. Characterization of cotton linter NC by scanning electron microscopy (SEM).

Figure 3 shows the interaction of nanocellulose with approximately 2 mg L^{-1} of Fe and Mn ions as a function of the volume of NC suspension. Fe has a greater ability to complex with NCs than Mn. Most of the Fe ions were already complexed with the addition of 50 μL of NC suspension, and the increase in concentration did not lead to significant variations. As regards the Mn with the addition of 100 μL of NC suspension, a level of complexed metal was already obtained, which remained constant even with the addition of larger volumes of NC.

This confirms that the interactions between Mn and NC do not depend on the amount of NC present in the system. What probably occurs in this system is the arising of labile manganese, where there is the formation of weak bonds that constantly alternate in association and disassociation with the NC. Considering that even with the addition of nanocellulose, free ions remain, the analysis of all metallic ions proceeded with a fixed volume of nanocellulose suspension and the evaluation of the contact time.

In the evaluation of the interaction of NC with Ca and Mg, it was observed that Ca had a low capacity to complex with NC. The proportion of complexed ions varied between 7% and 12% throughout the analyzed period. Compared to results from the literature, it is observed that Ca generally has a lower affinity for humic material than other metallic species [19]. However, it is possible to observe a small Ca—NC interaction (Figure 4). Mg compared to Ca showed a greater complexing capacity in the first 30 min with 28% of the metal forming complexes with the NC. However, after 24 h, the percentage of complexed

Mg dropped to 4%. This behavior indicates that the bonds between NC and Mg can be weak and unstable.

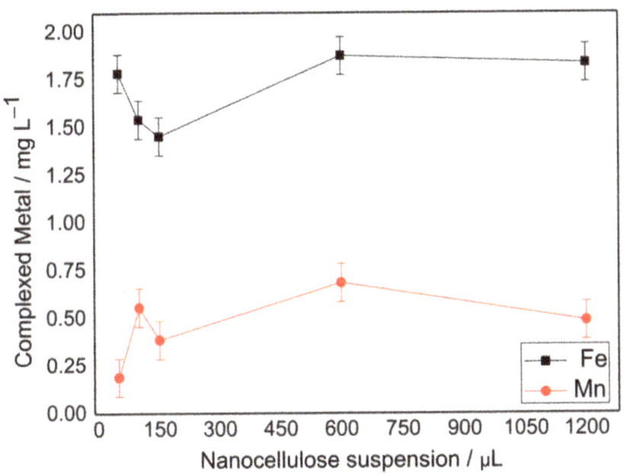

Figure 3. Metals complexed with nanocellulose as a function of the addition of the nanoparticle suspension in the absence of organic matter after 24 h of complexation. pH 5.5. Metal: Fe(II), Mn(II) (2.0 mg L^{-1}). Nanocellulose: 0–1200 µL.

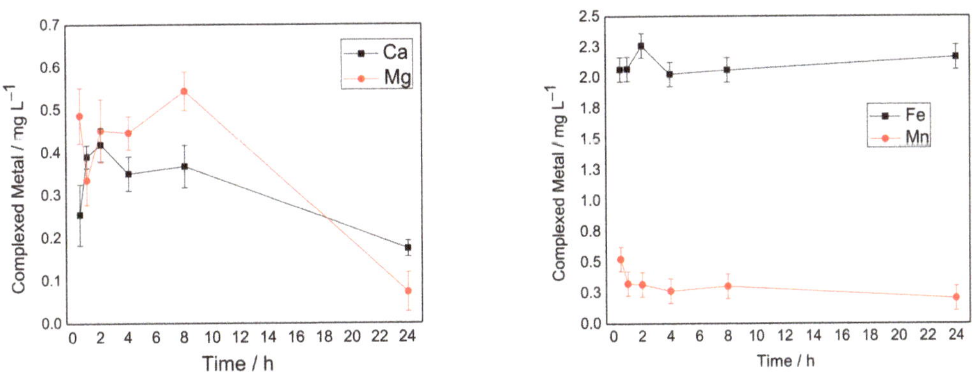

(**a**) (**b**)

Figure 4. Complexed metals to NC in relation to the contact time (**a**) Metal: Ca(II), Mg(II) (2.0 mg L^{-1}). (**b**) Metal: Fe(II), Mn(II) (2.0 mg L^{-1}). pH 5.5. Nanocellulose: 300 µL.

When the interaction of Fe with the nanomaterial was evaluated, it was observed that practically all the Fe present in the solution was complexed with the NC, and there were no changes over the analyzed period, indicating that Fe—NC bonds are strong and stable (Figura 4b). Regarding Mn, the graph suggests a low interaction between the metal and the NC, as observed with Ca and Mg. It is likely that weak and unstable bonds form, so there was less complex Mn during the entire period analyzed.

3.2. Influence of NC on Metals—EfOM Complexes

The values for total metal, free metal (filtrate) and complexed metal (difference between total metal and free metal) with the addition of only EfOM without NC are shown below. The fractionation in tangential flow indicated that 44% of Ca ions, 53% of Mg ions, 75% of Fe ions and 88% of Mn ions were complexed with EfOM (Table 2). These results were obtained before the addition of NC to the system.

Table 2. Fractionation of metal ions complexed to EfOM before adding NC to the system.

	Concentration of Metals (mg L^{-1})		
	Total Metal ± σ (M + M—EfOM)	Free Metal ± σ (M)	Comp. Metal ± σ (M—EfOM)
Ca	2.199 ± 0.074	1.229 ± 0.061	0.971 ± 0.134
Mg	2.284 ± 0.010	0.987 ± 0.036	1.212 ± 0.238
Fe	2.231 ± 0.153	0.569 ± 0.017	1.662 ± 0.170
Mn	2.228 ± 0.044	0.269 ± 0.029	1.959 ± 0.073

With the addition of NCs to the system, competition between ligands occurs. Complexed metal is represented as M—NC and/or M—EfOM. The influence of the addition of NC to the metal—EfOM complex indicated that iron ions are still complexed, probably to EfOM. Depending on the time of concentration of the complexed Ca, the Mg and Mn decreased, indicating that the metal was supplied in a solution (Figure 5a).

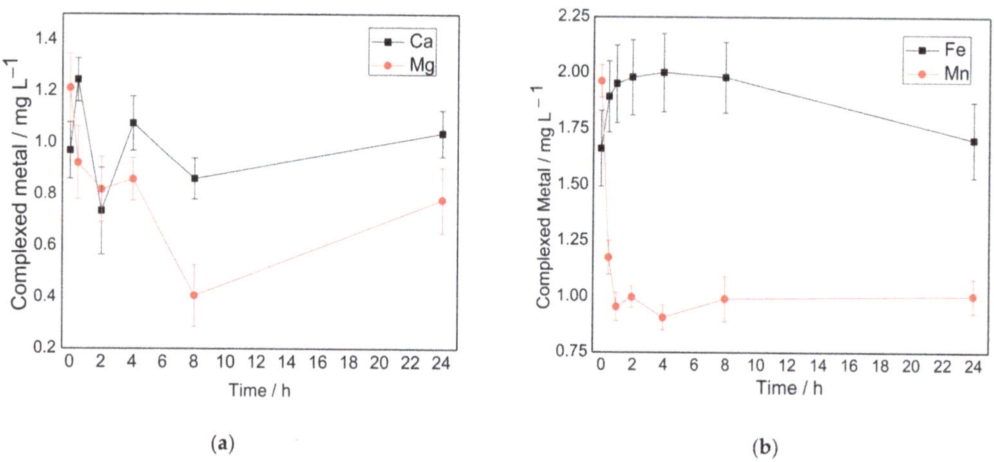

Figure 5. Analysis of the influence of NC on M—EfOM complexes (metals—effluent organic matter). (**a**) Metal: Ca(II), Mg(II) (2.0 mg L^{-1}). (**b**) Metal: Fe(II), Mn(II) (2.0 mgL^{-1}). pH 5.5. EfOM: humic substances (100 mg L^{-1}). Nanocellulose addition: 300 µL.

For Ca, only 44% of the ions complexed with EfOM before the addition of the NC; after 30 min of the addition this value increases to 57%. However, the next points indicate a very large variation in relation to the number of complexed Ca ions. As previously noted, NC forms weak bonds with Ca and tends to release the labile metal into the system, so Ca is sometimes linked with NC, and sometimes it is free, which explains the variation. In addition, the points at which the number of complexed Ca is less than t = 0 may suggest that part of the complexed metal, EfOM, migrates to the NC, which then tends to release the metal due to the low quality of the interaction. Regarding Mg ions, 53% complexed with EfOM, and after the addition, all points indicate a decrease in the number of complexed ions, so the metal may have migrated from EfOM to NC, just as occurred with Ca.

After the addition of NC to the system, there was a small increase in the amount of Fe complexed in the system, which may indicate that part of the metal that was free at time zero was complexed with the NC. However, Fe has the ability to form strong bonds with the two ligands, so it is plausible that there has been an increase in the number of complexed ions. After the addition of NC to the Mn—EfOM complexes, there was a reduction in the complexed Mn ions. This may have happened due to the competition between NC and

EfOM. In this system, the complex of Mn and EfOM migrates to the NC, and ends up forming weaker interactions, releasing the metal more easily (Figure 5b).

3.3. Influence of Organic Matter on M—NC Complexes

The values for total metal, free metal (filtrate) and complexed metal (difference between total metal and free metal) with only NC addition and without organic matter are shown in Table 3. It is observed that, except for Fe, the metals evaluated showed low complexation with NC.

Table 3. Fractionation of metal ions complexed to NC before adding EFOM to the system.

	Concentration of Metals (mg L^{-1})		
	Total Metal ± σ (M + M—NC)	**Free Metal ± σ (M)**	**Comp. Metal ± σ (M—NC)**
Ca	3.400 ± 0.018	3.290 ± 0.017	0.109 ± 0.035
Mg	1.720 ± 0.035	1.647 ± 0.011	0.073 ± 0.046
Fe	2.468 ± 0.013	0.196 ± 0.011	2.272 ± 0.024
Mn	2.236 ± 0.202	2.107 ± 0.036	0.129 ± 0.238

Figure 6 represents the influence of EfOM on the M—NC complexes. In this case, the first point on the graph (0 h) corresponds to the moment before the addition of EfOM to the M—NC complex.

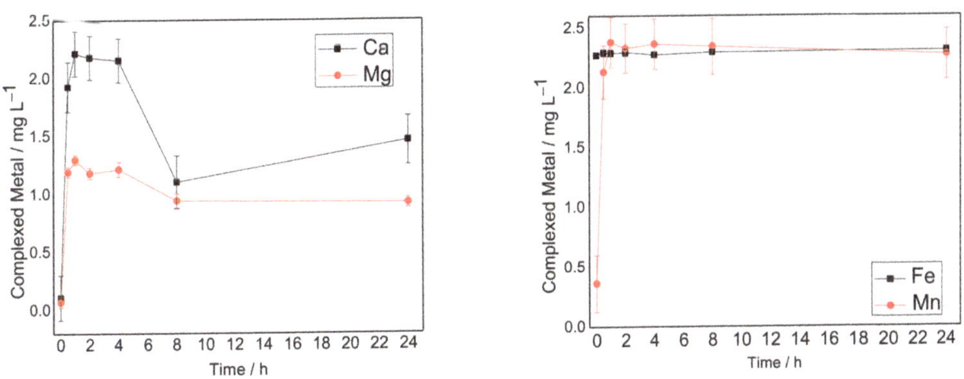

(a) (b)

Figure 6. Analysis of the influence of EfOM on M—NC complexes (metal—nanocellulose). (**a**) Metal: Ca(II), Mg(II) (2.0 mg L^{-1}). (**b**) Metal: Fe(II), Mn(II) (2.0 mgL^{-1}). pH 5.5. Nanocellulose: 300 μL. EfOM addition: humic substances (100 mg L^{-1}).

For Ca, the addition of EfOM into the system caused a large increase in the portion of complexed ions, which went from 3% to 57% 30 min after the addition of EfOM into the system. The next points indicate stability, but 8 h after the addition there was a decrease in the portion of complexed ions. The EfOM was shown to have a greater capacity to complex Ca compared to NC; however, the presence of the two ligands in the system can cause a competition scenario, where Ca passes to the NC forming less stable complexes. The Mg initially presented a behavior very similar to the Ca—before the addition of EfOM into the system, only 4% of the Mg had formed complexes with the NC, and after 30 min from the addition of EfOM into the system, the portion of complexed ions rose to 69%; after 480 min, a slight drop in the number of complexed ions was observed, which rose to 55%, and at the end of 24 h, the portion increased to 54%. The same considerations were made as for Ca, the fall can be justified by a competition between EfOM and NC.

For Fe, the addition of EfOM did not change the proportion of the complexed metal to NC; this is because Fe was shown to be able to form strong bonds with both NC and EfOM (Figure 6b). As for Mn, it can be observed that the proportion of the metal bound to NC was very low before the addition of organic matter (time 0 h); however, after the addition of EfOM, practically all Mn was complexed, and there were practically no changes in this time. This may have happened because, in a competitive scenario among binders, EfOM stands out by forming stronger bonds with Mn.

4. Discussion

The absorbance values found in UV-VIS spectrophotometry at wavelengths of 465 and 665 nm by Vieira et al. [20] indicate that HS does not present such a high degree of structural condensation, which explains the interaction between metal and EfOM.

The competition scenario may cause an increase in the bioavailability of metals due to their weaker bonds with NC. In addition, the competitive elements can also collaborate, so that the metal—NC bonds are absorbed by living beings due to the reduced size of the nanomaterial.

There is probably an interaction between EfOM and nanocellulose. In the first case, where the M—EfOM complexes are in equilibrium, upon coming into contact with the nanomaterial, manganese is released into the environment. In addition, its removal from the effluent is ineffective, increasing its toxicity and impact on the environment, due to its availability in the free metal format.

In the second case, where the M and NC are in equilibrium, when exposed to organic matter, the complexed ions increase. As regards the effluent treatment, the organic matter carrying the ion would probably be incorporated into the NC carrying the metallic ion, overloading the complexation capacity of the NC and blocking the interaction of M with NC.

In the case of Fe, removal is effective even in the presence of organic matter, this being an interesting option for iron removal in effluents containing organic matter, such as wastewater treatment plants. Regarding Ca, Mg and Mn ions, it is possible to verify that there was competition between the complexing groups present in the EfOM and the NC releasing the metal into the solution (free metal). To increase the NC—metal interaction, it could be interesting to develop an organomodification process through the incorporation of selective groups on the surface of the nanomaterial.

Several studies have reported the use of nanocellulose in emerging applications, and focused on the characteristics of this material, such as its high surface area to volume ratio, low environmental impact (mainly for not having associated metals that can be released into the environment), high strength, functionalization and sustainability, and its ability to be applied as a film on support surfaces in the remediation of aquatic environments.

5. Conclusions

This work proposes the valorization of an agroindustrial residue, a fibrous cotton residue, in a nanotechnological input, to be applied in several segments of the technology industry. The results presented in this study show that NC causes changes in the complexation mechanism of metals mixed with organic matter present in water (EfOM). Fe ions showed strong interactions with NC. In relation to Ca, Mg and Mn ions, it is possible to verify that there was competition between EfOM and NC, such that the metal is free in the solution. To increase the NC—metal interaction, it may be interesting to develop an organomodification process, through the incorporation of selective groups onto the surface of the nanomaterial. Nanocellulose has showed high potential for use in the development of new technologies for the treatment of effluents containing metals, mainly iron.

Author Contributions: Conceptualization, J.C.C.; methodology, D.G.; investigation, V.d.J.C.d.S.; data curation, V.d.J.C.d.S.; writing—original draft preparation, V.d.J.C.d.S.; writing—review and editing,

W.G.B. and L.C.d.O.; project administration, D.G.; funding acquisition, D.G. All authors have read and agreed to the published version of the manuscript.

Funding: This research was funded by São Paulo Research Foundation (FAPESP), #2019/24684-2 and #2018/21747-0.

Institutional Review Board Statement: Not applicable.

Informed Consent Statement: Not applicable.

Data Availability Statement: The study did not report any data.

Acknowledgments: Vicunha Rayon Ltd. a for the donation of cellulosic cotton linter fibers.

Conflicts of Interest: The authors declare no conflict of interest. The funders had no role in the design of the study; in the collection, analyses, or interpretation of data; in the writing of the manuscript, or in the decision to publish the results.

References

1. Santos, E.B.H.; Duarte, A.C. The Influence of Pulp and Paper Mill Effluents on the Composition of the Humic Fraction of Aquatic Organic Matter. *Water Res.* **1998**, *32*, 597–608. [CrossRef]
2. Lyu, C.; Liu, R.; Li, X.; Song, Y.; Gao, H. Degradation of Dissolved Organic Matter in Effluent of Municipal Wastewater Plant by a Combined Tidal and Subsurface Flow Constructed Wetland. *J. Environ. Sci.* **2021**, *106*, 171–181. [CrossRef] [PubMed]
3. Tranvik, L.J.; von Wachenfeldt, E. Interactions of Dissolved Organic Matter and Humic Substances in Freshwater Systems. In *Reference Module in Earth Systems and Environmental Sciences*; Elsevier: Amsterdam, The Netherlands, 2014. [CrossRef]
4. Goveia, D. *Metais em Sistemas Aquáticos: Biodisponibilidade e Influência da Matéria Orgânica*, 1st ed.; Editora Appris: Curitiba, Brazil, 2020; p. 136.
5. Tom, A.P. Nanotechnology for Sustainable Water Treatment—A Review. *Mater. Today Proc.* **2021**, *10*, 16–78. [CrossRef]
6. Dufresne, A. Nanocellulose: A New Ageless Bionanomaterial. *Mater. Today* **2013**, *16*, 220–227. [CrossRef]
7. Kim, J.; Shim, B.S.; Kim, H.S.; Lee, Y.; Min, S.; Jang, D.; Abas, Z.; Kim, J. Review of Nanocellulose for Sustainable Future Materials. *Int. J. Prec. Eng. Manuf.-Green Technol.* **2015**, *2*, 197–213. [CrossRef]
8. Tadini, A.M.; Constantino, I.C.; Nuzzo, A.; Spaccini, R.; Piccolo, A.; Moreira, A.B.; Bisinoti, M.C. Characterization of Typical Aquatic Humic Substances in Areas of Sugarcane Cultivation in Brazil Using Tetramethylammonium Hydroxide Thermochemolysis. *Sci. Total Environ.* **2015**, *518*, 201–208. [CrossRef] [PubMed]
9. Goveia, D.; Lobo, F.A.; Burba, P.; Fraceto, L.F.; Dias Filho, N.L.; Rosa, A.H. Approach Combining On-Line Metal Exchange and Tangential-Flow Ultrafiltration for In-Situ Characterization of Metal Species in Humic Hydrocolloids. *Anal. Bioanal. Chem.* **2010**, *397*, 851–860. [CrossRef] [PubMed]
10. Botero, W.G.; Souza, S.O.; Santos, O.S.; Oliveira, L.C.; Amarante, C.B. Influence of sediment humic substances on the bioavailability of metals in the aquatic system. *Quím. Nova* **2014**, *37*, 929–1109. [CrossRef]
11. Lamaming, J.; Rokiah, H.; Cheu, P.L.; Othman, S. Properties of Cellulose Nanocrystals from Oil Palm Trunk Isolated by Total Chlorine Free Method. *Carbohydr. Polym.* **2017**, *156*, 409–416. [CrossRef] [PubMed]
12. Orts, W.J.; Shey, J.; Imam, S.H.; Glenn, G.M.; Guttman, M.E.; Revol, J.F. Application of Cellulose Microfibrils in Polymer Nanocomposites. *J. Polym. Environ.* **2005**, *13*, 301–306. [CrossRef]
13. Barreto, J.; Tita, D.; Orlandi, M. Development of an automated method to perform a quantitative study of particle size distribution and the effect of a conductive layer in scanning electron microscopy. *Quím. Nova* **2019**, *42*, 447–452. [CrossRef]
14. Thurman, E.M.; Malcolm, R.L. Preparative Isolation of Aquatic Humic Substances. *Environ. Sci. Technol.* **2015**, *4*, 463–466. [CrossRef] [PubMed]
15. Satyamurthy, P.; Prateek, J.; Balasubramanya, R.H.; Vigneshwaran, N. Preparation and Characterization of Cellulose Nanowhiskers from Cotton Fibres by Controlled Microbial Hydrolysis. *Carbohydr. Polym.* **2011**, *83*, 122–129. [CrossRef]
16. Nascimento, J.H.O.; Luz, R.F.; Galvão, F.M.G.; Melo, J.D.D.; Oliveira, F.R.; Ladchumananandasivam, R.; Zille, A. Extraction and Characterization of Cellulosic Nanowhisker Obtained from Discarded Cotton Fibers. *Mater. Today Proc.* **2015**, *2*, 1–7. [CrossRef]
17. Sucinda, E.F.; Abdul, M.S.M.; Ridzuan, M.J.M.; Sultan, M.T.H.; Gibson, A.G. Analysis and Physicochemical Properties of Cellulose Nanowhiskers from Pennisetum Purpureum via Different Acid Hydrolysis Reaction Time. *Int. J. Biol. Macromol.* **2020**, *155*, 241–248. [CrossRef] [PubMed]
18. Dashtimoghadam, E.S.; Rasool, N.H.; Knudsen, K.D.; Mirzadeh, H.; Nyström, B. Tunable viscoelastic features of aqueous mixtures of thermosensitive ethyl(hydroxyethyl)cellulose and cellulose nanowhiskers. *Colloids Surf. A Physicochem. Eng. Asp.* **2020**, *590*, 124489. [CrossRef]
19. Botero, W.G.; Oliveira, L.C.; Rocha, J.C.; Rosa, A.H.; Santos, A. Peat humic substances enriched with nutrients for agricultural applications: Competition between nutrients and non-essential metals present in tropical soils. *J. Haz. Mat.* **2010**, *177*, 307–311. [CrossRef] [PubMed]
20. Vieira, D.F.; Nicolas, H.F.; Botero, W.G.; Oliveira, L.C.; Goveia, D. Influence of the diameter of nanoparticles in complexes metal-aquatic humic substances. *Eclética Quím. J.* **2018**, *43*, 44–50. [CrossRef]

water

Article

Coffee Husk and Lignin Revalorization: Modification with Ag Nanoparticles for Heavy Metals Removal and Antifungal Assays

Daniel Fernando Guevara-Bernal [1], Marlon Yesid Cáceres Ortíz [2], Jorge Andrés Gutiérrez Cifuentes [3], Julio Bastos-Arrieta [4], Cristina Palet [5] and Angélica María Candela [1,*]

1 Grupo de Investigación en Nuevos Materiales y Energías Alternativas, Facultad de Química Ambiental, Universidad Santo Tomás, Floridablanca 681003, Colombia; daniel.guevara@ustabuca.edu.co
2 Grupo de Investigación en Bioquímica y Microbiología, Escuela de Química, Facultad de Ciencias, Universidad Industrial de Santander, Bucaramanga 680002, Colombia; mayecaor@correo.uis.edu.co
3 Facultad de Ciencias de la Salud, Universidad del Quindío, Armenia 630004, Colombia; jgutierrez@uniquindio.edu.co
4 Grup d'Electroanàlisi, Department of Chemical Engineering and Analytical Chemistry, Universitat de Barcelona, Carrer Martí i Franqués 1-11, 08028 Barcelona, Spain; julio.bastos@ub.edu
5 Grup de Tècniques de Separació en Química, Unitat Química Analítica, Departament de Química, Facultat de Ciències, Universitat Autònoma de Barcelona, 08193 Bellaterra, Spain; cristina.palet@uab.cat
* Correspondence: angelica.candela01@ustabuca.edu.co

Abstract: This study presents the use of the modified coffee husk and coffee lignin as sorbents in the heavy metal ions sorption of Pb(II), Cd(II), Cr(III), and Cu(II) in an aqueous solution. The modification of sorbents was carried out by the impregnation method, using silver nitrate ($AgNO_3$) and sodium borohydride ($NaBH_4$) as a nanoparticles' (NPs) precursor, and reducing agent, respectively. The obtained nanocomposite material was morphologically characterized by electron microscopy. In addition, an evaluation of metal ions' sorption, pseudo-first-order, and pseudo second-order kinetics modeling was performed. Finally, antifungal activity was evaluated on different *Candida* species. Coffee and lignin modified with AgNPs increased the extraction capacity with the highest sorption for Pb ions with 2.56 mg/g and 1.44 mg/g, respectively.

Keywords: coffee husk; lignin; heavy metals; silver nanoparticles; sorption; antifungal activity

Citation: Guevara-Bernal, D.F.; Cáceres Ortíz, M.Y.; Gutiérrez Cifuentes, J.A.; Bastos-Arrieta, J.; Palet, C.; Candela, A.M. Coffee Husk and Lignin Revalorization: Modification with Ag Nanoparticles for Heavy Metals Removal and Antifungal Assays. *Water* **2022**, *14*, 1796. https://doi.org/10.3390/w14111796

Academic Editor: Mehrab Mehrvar

Received: 5 April 2022
Accepted: 28 May 2022
Published: 2 June 2022

1. Introduction

Globally, freshwater use has increased by a factor of 6 in the last 100 years, and it is estimated that the world will face a 40% water deficit by 2030. Although the actual increase in global water use is not known, scientific articles agree that we will face competition for the growing demand of the industrial and energy sectors based on industrial development and the increased coverage of water and sanitation services [1–6].

In developing countries, part of the population has suffered from health conditions or illnesses caused by the direct consumption of contaminated water. Drinking water is limited, and its quality is under constant pressure due to the presence of infectious agents from microbial agents such as bacteria and fungus, toxic chemicals, and even radiation, which means it is a global challenge that has increased in both developed and developing countries, weakening the economic growth as well as the socio-environmental sustainability and health of billions of people [7,8]. However, the proliferation of heavy metals in water sources is also a growing concern due to limited access to the resources required for implementing effective techniques in pollutant removal, and it is crucial due to their acute toxicity, long-term accumulation, and persistence [5,9,10].

Additionally, the arrangement of large quantities of biomass waste that are continuously produced is one of the primary sources of pollution [11]. Approximately 100 billion metric tons of biomass waste are generated worldwide [12]. These include forest and agricultural waste, fruit, and other food processing waste. On top of this agricultural waste, the coffee industry provides 7.4 million tons of spent coffee beans and an additional amount of pulp, shell, and silver skin (husk) [13,14]. The outer layer of the coffee bean is called silver skin.

Agricultural waste is commonly burned, emitting pollutants into the atmosphere, such as carbon monoxide, nitrous oxide, nitrogen dioxide, and particulate matter. This is accompanied by the formation of ozone and nitric acid, which contribute to acid deposition, bringing risks to human and environmental health, in addition to methane and CO_2, which are greenhouse gases, all of this resulting in global warming [13,15,16].

In this context, biomass revalorization is considered a technological and sustainable solution. Different applications for biomass waste have been proposed, such as the development of new products for energy generation, sorption, and biosorption of pollutants, among others, replacing or reducing the use of hazardous chemicals, as well as lowering costs in industrial processes [12]. Likewise, the valorization of various materials has also involved their modification with nanoparticle systems (NPs), connecting interdisciplinary areas such as physics, biology, engineering, and medicine in corresponding applications [17–19].

In this study, we present the preparation and characterization of nanocomposite based on revalorized materials: coffee husk, coffee lignin, and coffee husk and lignin modified with silver nanoparticles (AgNPs). In addition, the suitability for the sorption capacity towards different metal ions of Pb(II), Cd(II), Cr(III), and Cu(II) and antifungal activity with Candida fungi species were properly evaluated, with modified materials, kinetic modeling and contact time experiments [20–26]. The results obtained contribute with the revalorization of these Colombian agroindustrial wastes due to their modification with silver nanoparticles. The improvement of traditional waste materials in the extraction of heavy metal ions in an aqueous solution with antifungal activity is an alternative to their disposal.

2. Materials and Methods

2.1. Chemical Reagents and Solutions

All the chemicals were of analytic grade. A multiple heavy metals 1000 mg/L stock solution of Pb(II), Cd(II), Cr(III), and Cu(II) is prepared from their nitric salts (all 99% from Panreac, Barcelona, Spain). H_2SO_4 (>99%), $AgNO_3$ salt (>99%), and $NaBH_4$ salt (>98%) are supplied by Sigma Aldrich (Bucaramanga, Colombia). Standards of heavy metal ions were properly diluted in acidified water with 2% of nitric acid to prepare the calibration set for analysis at a concentration range between 0.01 and 1 μg/L.

2.2. Coffee Husk Preparation and Lignin Obtention

The coffee husk has cellulose, hemicellulose, and lignin contents on a dry basis, as is well known. Lignin provides structural rigidity to the cell walls of many plant species [27]. Lignin has been used as a potential adsorbent to remove heavy metals due to its unique polyphenolic structure and physicochemical properties [28].

For that reason, here, lignin is also checked and compared with the original coffee husk biomass.

The coffee husk is obtained from the Café Santander factory in the municipality of Curití, Santander-Colombia. The biomass is obtained in two ways, the first when leaving the coffee roasting production plant, whose shape is an irregular and undefined particle size, and a second that is homogenized in a mill and sieved to a diameter size less than 250 μm. From the sieved coffee husk, acid hydrolysis is performed in H_2SO_4 to 72% (in a ratio of 3 mL of acid per 300 mg of the sample) to extract insoluble lignin. Afterward, it is carried into a thermostatic water bath at 30 °C for one hour, stirring every five minutes. The acidic solution was diluted a concentration of 4%. This solution is autoclaved for two h at 120 °C with a pressure of 103 kPa. Afterward, this mixture is filtered and washed to a neutral pH [29]. Finally, the solid phase lignin material is obtained [30].

2.3. Sorbents Modification with AgNPs

The synthesis of AgNPs is carried out using $AgNO_3$ as a precursor and $NaBH_4$ as a reducing agent [31,32]. For the nano-silver modification of the coffee husk and lignin, this impregnation method is followed in the presence of the biomasses. Thus, the coffee husk and lignin are deposited in falcon tubes, then a 0.1 M silver nitrate solution is added. The biomass and precursor contact time was fixed at 45 min with vigorous agitation. It is then centrifuged at 5000 revolutions per minute (rpm) for 10 min, and then the remaining silver nitrate solution and solid are separated by centrifugation. Once the biomass is impregnated with $AgNO_3$, 0.3 M is added slowly at room temperature. The volume used for the impregnation with the precursor and the reducing agent was always twice the volume of the biomass used, with a contact time of 45 min. Finally, the liquid and solid phases are separated by centrifugation and filtrated, then the solid phase obtained is washed until a neutral pH is obtained. In this way, both coffee husk and lignin were properly modified with silver nanoparticles (AgNPs). The obtained materials' morphology is characterized by using Scanning Electron Microscopy (SEM).

2.4. Sorption Experiments

Sorption experiments were carried out at room temperature with aqueous solutions that are prepared at pH 4.0, containing the metals at a concentration of 0.18 mM each, following conditions of previous studies [25,26,33,34]. The pH is measured with a standardized potentiometer [35].

Batch sorption experiments were performed with the following steps: 25 mg of each sorbent were placed in falcon tubes, and then 2.5 mL of a heavy metal aqueous solution was added into the tube. Later, samples are placed on a rotary mixer (CE 2000 ABT-4, SBS Instruments SA, Barcelona, Spain) and shaken at 25 rpm. The two phases are separated by decantation and later the aqueous supernatant phase is filtered through 0.22 μm Millipore filters (Millex-GS, Millipore, Burlington, MA, USA), diluted, and analyzed, ensuring a concentration range into the calibration range (as mentioned at 0.01–1 μg/L). Finally, the multielement analysis is carried out by ICP-MS (XSERIES 2 ICP-MS, Thermo Scientific, Waltham, MA, USA) [35].

All the results are expressed as the mean value of minimum duplicate experiments, and the standard deviation (SD) is used to analyze data errors.

To optimize sorption conditions, and mainly the contact time, the corresponding experiments were carried out by using all biomass materials (the coffee husk and lignin, and modified ones both with AgNPs) [27–29]. A properly weighed amount of each biomass system (25 mg) with corresponding metal ions solutions (2.5 mL) was shaking both at different contact times (10, 15, 30, 45, 60 min, and 2, 4, 12, and 24 h) [25].

2.5. Kinetic Modeling

The sorption mechanism of the metal ions' sorption process is evaluated using first- and second-order kinetic models with the experimentally obtained data [36]. The Lagergren pseudo-first-order model is based on the range of change in the solute uptake with time, which is generally applicable over the initial stage of an adsorption process (Equation (1)) [36–39].

$$\frac{1}{q_t} = \frac{(k_1/Q_1)}{t} + \frac{1}{Q_1} \tag{1}$$

where Q_1 is the equilibrium adsorption capacity, q_t is the adsorption capacity at the time t, and k_1 (min^{-1}) is the adsorption rate of the pseudo-first-order kinetic model.

The pseudo-second-order kinetic model can be represented in the linear form as (Equation (2)) [40]:

$$\frac{1}{q_t} = \frac{1}{k_2 q_e^2} + \frac{1}{q_e} t \tag{2}$$

where k_2 ($g \cdot mg^{-1} \cdot min^{-1}$) is the rate constant, q_e is the amount of solute adsorbed at equilibrium and q_t is the amount of solute adsorbed ($mg \cdot g^{-1}$) at time t.

2.6. Antifungal Assays

Antifungal assays of each material are performed with *Candida* fungi species such as *Candida albicans*, *Candida parapsilosis*, *Candida glabrata*, *Candida krusei* and *Candida guilliermondii*. These fungus species here selected are well known as opportunistic pathogens, are frequently found in different anatomical sites, and clinical samples could induce systemic, superficial, and nosocomial infections under optimal environmental conditions by their relative resistance to the external environment [30–33].

The antifungal evaluation of the biomass materials is carried out using the Time Kill method. This method is appropriate to evaluate the bactericidal and/or fungicidal activity of certain compounds and allows obtaining information on the dynamic interaction between the antifungal agent and the fungal strain. RPMI 1640 culture medium for different species of *Candida* are inoculated and incubated for 24 h in a shaker at 37 ± 1 °C. Then, 1 mL of this inoculum is deposited in a glass container together with the different biomass materials under the study and at different biomass relative concentrations (0.37, 0.75, 1.5 mg/mL). After 24 h of contact, its viability is evaluated by sowing on the agar surface and evaluating the decrease in the logarithm of CFU (Colony Forming Units), as a measure of the antifungal activity of the biomass materials. The CFU/mL was determined following the procedure described in previous work [41].

2.7. Infrared Spectroscopy

FT-IR analyses were carried out to establish which functional groups are present in the biomass, and which could take part in the modification of coffee and lignin with AgNPs.

2.8. Electron Microscopy Characterization

The morphology of the different nanocomposite materials is analyzed by scanning electron microscopy (SEM) coupled with energy dispersive X-ray (EDX). The samples are coated with gold for the surface study. Images were obtained with the Zeiss EVO® MA10 (Carl Zeiss, Jena, Germany), Zeiss Merlin™ (Carl Zeiss, Germany), and Quanta™ 650 FEG units (FEI Company, Hillsboro, AL, USA).

3. Results and Discussion

3.1. Characterization of Sorbents

Scanning Electron Microscopy (SEM) was performed to analyze the morphological features and evaluate the changes on the material surfaces across the material's modification and sorption experiments (see Figure 1).

It is observed that the coffee material changed its morphology through the different treatments received, such as modification with AgNPs and the possible accumulation of metal ions on the surface of the material after the metal sorption experiments.

The coffee husk sample, as observed in Figure 1a, has a rough and amorphous surface (corrugated). The modification of coffee with AgNPs leads to identifying the presence of scattered particles on the surface, as can be shown in Figure 1b, and they also do not show any morphological change, except in the deposition of AgNPs. Thus, as expected, nanoparticles of Ag were properly immobilized onto the coffee husk.

The black patches on Figure 1a,b are probably pore apertures and cavities, which may influence the increase in sorption kinetics [42–44]. On the other hand, Figure 1c shows some changes in the coffee husk morphology. It looks like the roughness and porosity decreased after the heavy metal ions' adsorption experiment. Notice that there is an evident morphological difference between the surface of the material before (Figure 1a) and after the adsorption of metals (Figure 1c). It is observed that the surface of the coffee husk after the sorption of metal ions became smoother than it was before, which is usually related to being in contact with an acidic aqueous solution, as is the case here (pH 4). These results are consistent with those demonstrated previously in the literature [45], where the same behavior is observed working with copper, zinc, and cadmium metal ion aqueous solutions at pH 4, obtaining different degrees of smoothness of the surface and systems with removal efficacy [26]. The elemental analysis confirmed the presence of Ag on the surface of Figure 1b. The presence of synthesized AgNPs was confirmed by the peak at 3.0 keV in EDS [46–48].

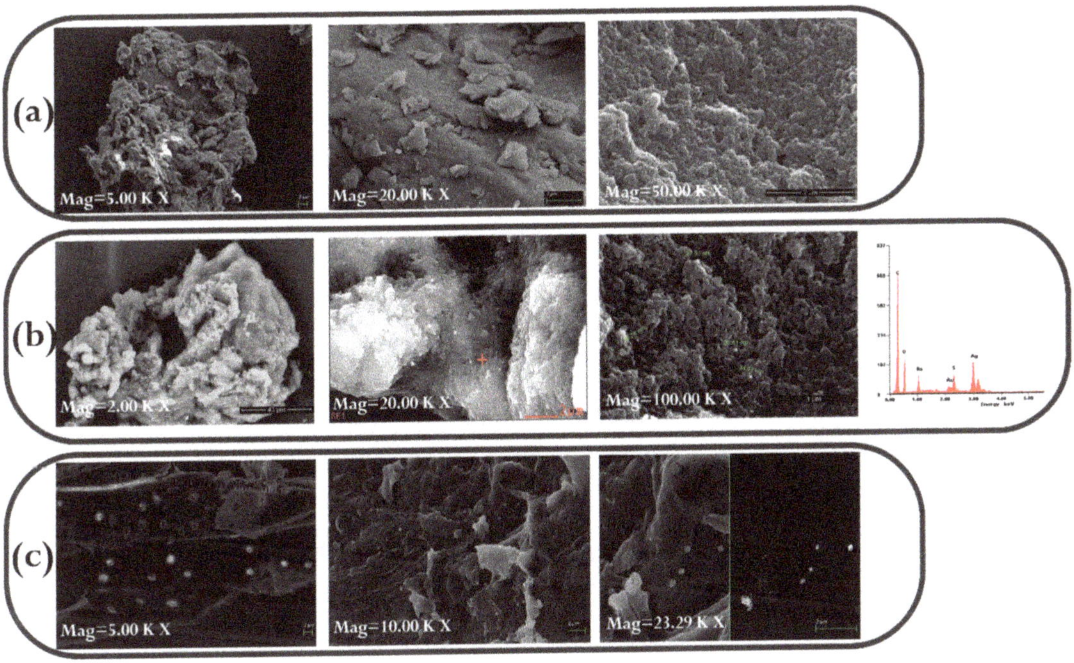

Figure 1. SEM micrographs showing the morphology transformation: (**a**) Coffee husk, (**b**) AgNPs–coffee husk and, (**c**) AgNPs–coffee husk after heavy metal ions' sorption experiment. EDX analysis of AgNPs-coffee husk (at red + sign) is presented in figure (**b**), which includes Ag peak.

FT–IR measurements were used to understand the role of the AgNPs' modification on the biomass materials. The FT–IR spectra is presented in Figure 2. The peak around 3328 cm^{-1} is attributed to the −OH stretching as a result of functional groups of cellulose, hemicellulose, and lignin [44,49]. The peak of 2919 cm^{-1} corresponds to C–H groups, 1731 cm^{-1}, 1639 cm^{-1} C=O of aldehyde and ketone groups, 1458 cm^{-1} from aromatic C=C stretching groups and CH_2 and CH_3 groups, around 1371 cm^{-1} from aromatic C–H stretching and carboxyl−carbonate structures, 1238 cm^{-1} from CHOH groups, and around 1029 cm^{-1} Si–O–Si group [42,44,49,50].

3.2. Sorption Kinetics of Raw Materials and Nanocomposite Materials

The sorption kinetics experimental data for multi-metal aqueous solutions (Pb, Cd, Cr, and Cu) by using the raw and the nanocomposite biomass materials are collected in Figure 3.

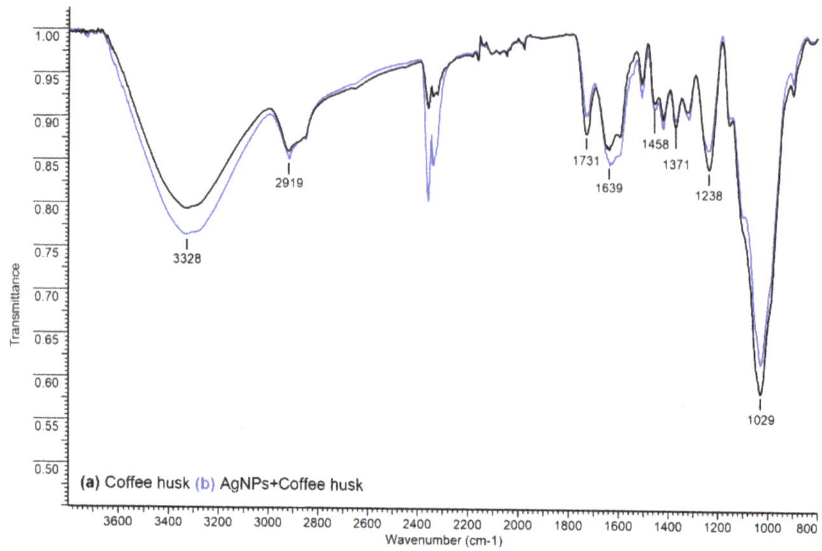

Figure 2. FT–IR spectra of (**a**) coffee husk and (**b**) AgNPs+coffee husk.

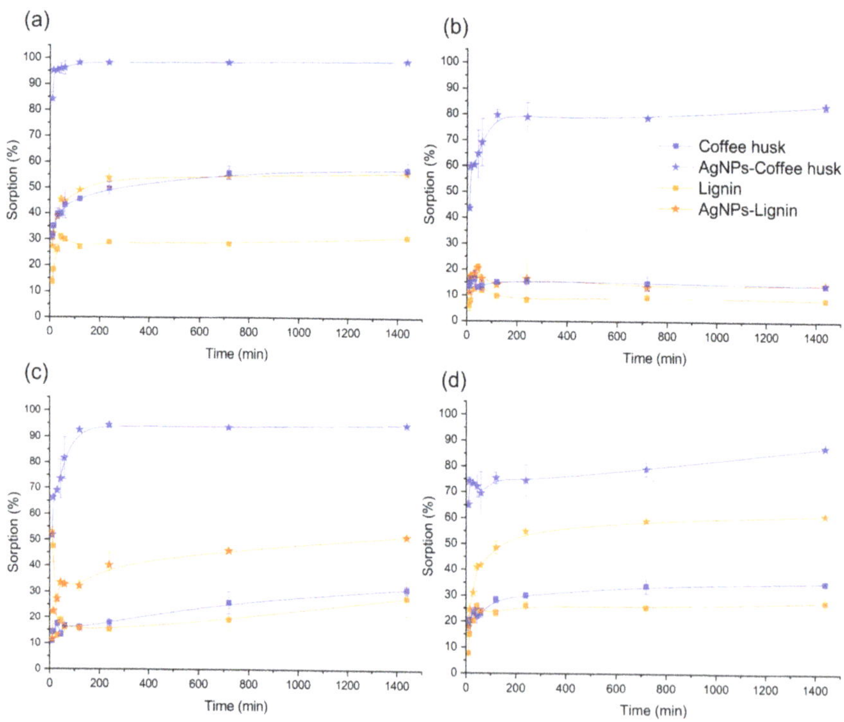

Figure 3. Sorption kinetics for (**a**) lead, (**b**) cadmium, (**c**) chromium, and (**d**) copper.

As can be observed in Figure 3a, in the case of lead comparing lignin and nanocomposite AgNPs' modified lignin materials, a potentiating effect in the metal sorption in the presence of AgNPs is observed, reaching around 46% of Pb sorption at 45 min and with a maximum sorption of 56% after 24 h (with raw lignin of just 31% reached). In the case of coffee husk and AgNPs' modified coffee husk materials, it is observed that the AgNPs' potentiating effect is even higher than with lignin-based materials. In this case, the Pb sorption percentage increases from around 57% with raw coffee husk up to 99% with the nanocomposite AgNPs' modified coffee husk. These results demonstrate that the best material for the sorption of Pb is the coffee husk modified with AgNPs, which also highlights that just 4 h of the experiment are needed under the experimental conditions here studied for all cases (2 h for AgNPs–coffee husk).

In the case of the sorption kinetics experiments for cadmium, the results collected in Figure 3b also show the maximum sorption percentage for the AgNPs' modified coffee husk, reaching 83% after 24 h. The sorption of Cd with the other biomass materials under study does not give significant results. Both the raw lignin and coffee husk, and the AgNPs' modified lignin, have Cd extraction percentages not higher than 20%.

Like Pb, the improved sorption results for chromium are observed with the AgNPs modified coffee husk reaching 95% of chromium sorption after 24 h, highlighting that from around 2 h values above 90% are obtained (as can be observed in Figure 3c). The addition of silver nanoparticles in the coffee husk shows an evident enhancing effect in the extraction of Cr. On the other hand, the modification of the lignin with AgNPs yields a maximum adsorption value of around 50% after 24 h of the sorption experiment, almost double that of raw lignin (around 25%).

Cooper's sorption behavior (see Figure 3d) is similar to those of Pb and Cr, even with less efficient sorption when using coffee husk silver nanocomposite (which is AgNPs–coffee husk), just around 80%. It was also observed that the use of nanocomposite lignin with AgNPs reached an extraction percentage after 24 h of around 60%.

Thus, both coffee husk and lignin modified with AgNPs show an enhancement in the extraction of metal ions compared with corresponding raw materials, probably due to the presence of the silver nanoparticles. This can be deduced since the sorption percentages found for both raw coffee husk and lignin are under 30% of metal sorption, except for Pb than raw coffee husk gives up to 50%. Furthermore, the nanocomposite AgNPs–coffee husk reaches quite a high metal sorption just with 2 h of adsorption experiments for Pb, Cd, and Cr ions. The behavior of the sorption capacity of this AgNPs–coffee husk shows the following decreasing order: Pb(II) $\approx$ Cr(III) > Cu(III) $\approx$ Cd(II). The AgNPs' modification enhances the affinity towards the heavy metal ions due to nanocomposite electrostatic charges.

3.3. Evaluation of Sorption Capacity

The results shown in Table 1 confirm that the modification of the coffee husk and lignin with AgNPs achieves an enhancing effect on the adsorption of metal ions at pH 4 and after 24 h of the sorption experiments. It should be noticed that the coffee husk with AgNPs is the material that achieves the highest values of sorption capacity for each metal ion evaluated, as mentioned previously.

Table 1. Sorption capacity in mg/g.

Sorbent	Heavy Metal Ion Sorption Capacity (mg/g)				Reference
	Pb(II)	Cd(II)	Cr(III)	Cu(II)	
Coffee husk	1.47	0.174	0.188	0.259	This study
Coffee–AgNPs	2.560	1.02	0.575	0.644	
Lignin	0.807	0.103	0.168	0.204	
Lignin–AgNPs	1.440	0.176	0.312	0.451	
Coffee ground	0.628	3.450	-	0.616	[51,52]
Milled olive stones	0.581	0.300	2.340	0.557	[53]
Banana	20.898	3.658	6.855	-	[54]
Corn cob	29.168	13.577	18.782	-	[54]
Sunflower	22.644	11.404	12.206	-	[54]
Limestone	0.0128	0.016	0.016	0.013	[55]
Coconut coir	2.760	-	-	2.25	[56]
Tangerine peel	1.556	0.659	1.480	1.616	[57]

The results evidenced that the modification of the coffee husk with AgNPs has a potentiating effect on the sorption of the studied metal ions. Thus, from the above results, it can be inferred that the increase in the sorption of heavy metals of the AgNPs' modified materials could be due to two reasons. The first can be related to the use of $NaBH_4$ as a reducing agent in the synthesis of AgNPs on biomass, inducing a kind of pretreatment on the surface and increasing the availability of active sites; and the second, the possible effect of AgNPs on the biomass surface to increase the sorption. The AgNPs' modification process could be considered alkaline, modifying the biomasses moieties, and improving the heavy metal ions' sorption system [58]. However, this assumption cannot be verified in the morphological characterization carried out [58,59]. According to [60], the nanocomposites have enhanced active sites that demonstrate stronger sorption on Pb ions than Cr, Cd, and Cu ions.

3.4. Kinectic Modeling

Figure 4 shows the pseudo-first-order kinetics modeling with all the four materials here under study for the metal sorption [28]. It is highlighted that AgNPs–coffee husk achieved better values of $1/q_t$ (see corresponding equation from the Materials and Methods), where the order of best sorption was Pb > Cd > Cu > Cr, which is like those obtained for the other materials (mainly for AgNPs–coffee husk and AgNPs–lignin).

The adjustment of the pseudo-second-order model demonstrates behavior that tends towards linearity, regardless of the material used, except for Cr with both raw biomass systems (as can be seen from Figure 5), where the AgNPs–coffee husk shows the best correlation coefficient (values collected in Table 2. The pseudo-second-order kinetic model assumes that the rate-limiting step is chemical sorption or chemisorption [36].

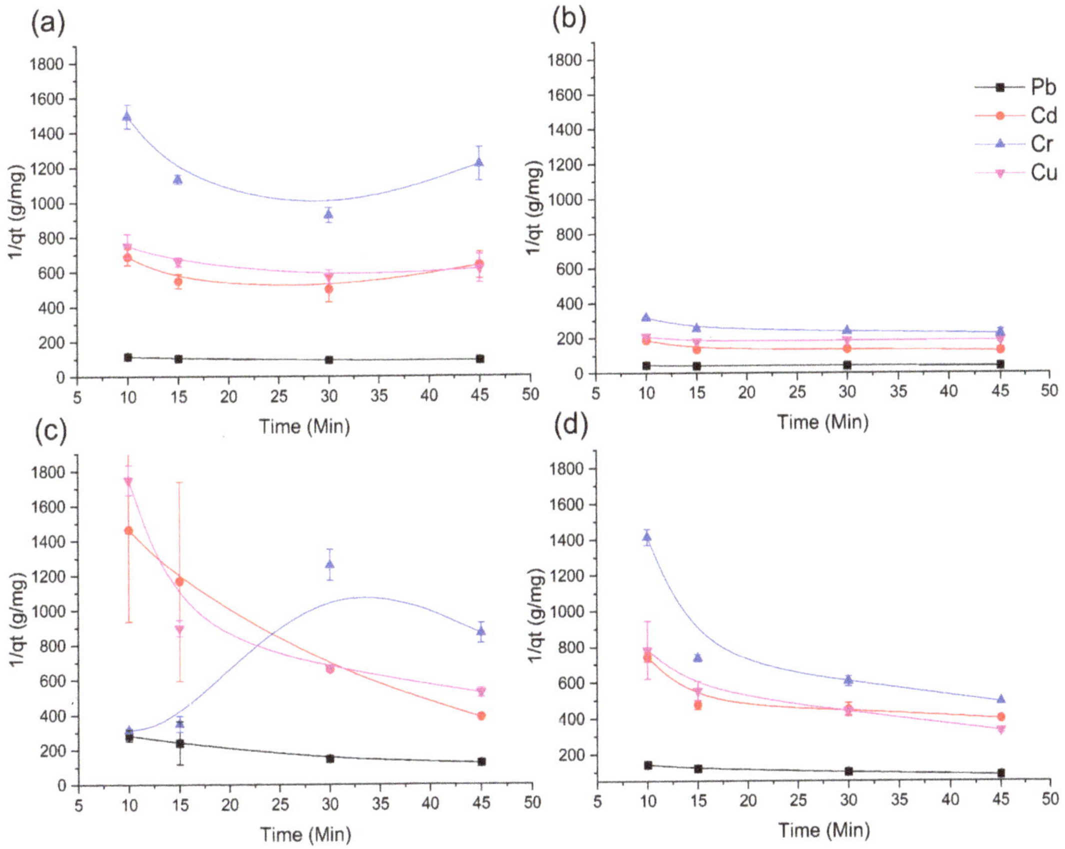

Figure 4. Pseudo-first-order model: (**a**) Coffee husk, (**b**) AgNPs–coffee husk, (**c**) lignin, and (**d**) AgNPs–lignin.

However, the correlation coefficient values show that the sorption mechanisms of the metals on all the sorption materials used do not follow the pseudo-first-order kinetic model. The results of the correlation coefficients of both the pseudo-first-order model and the pseudo-second-order model are presented in Table 2.

The experimental data exhibits good fitting with the pseudo-second-order model with R^2 values ranging from 0.9579 to 1.0000, better than those of the pseudo-first-order model (from 0.0007 to 0.9884) (Table 2). Although such negative values (Table 2) are not usually observed, as in another study reported previously, this may be associated with the electrostatic nature of the adsorption process [61] In the case of the pseudo-first-order model, the correlation coefficients are quite low, which can probably be related to the adsorption of the metals not occurring exclusively by the ion exchange mechanism, which is mostly explained when experimental data can be adjusted by this model. The pseudo-second-order model assumes that the adsorption is controlled by the chemical adsorption or chemisorption process based on valence forces by either sharing or exchanging electrons between the adsorbent and the metal ions of Pb, Cd, Cr, and Cu [62] The pseudo-first-order and the pseudo-second-order kinetic models both assume that the metal ions' adsorption may be [63].

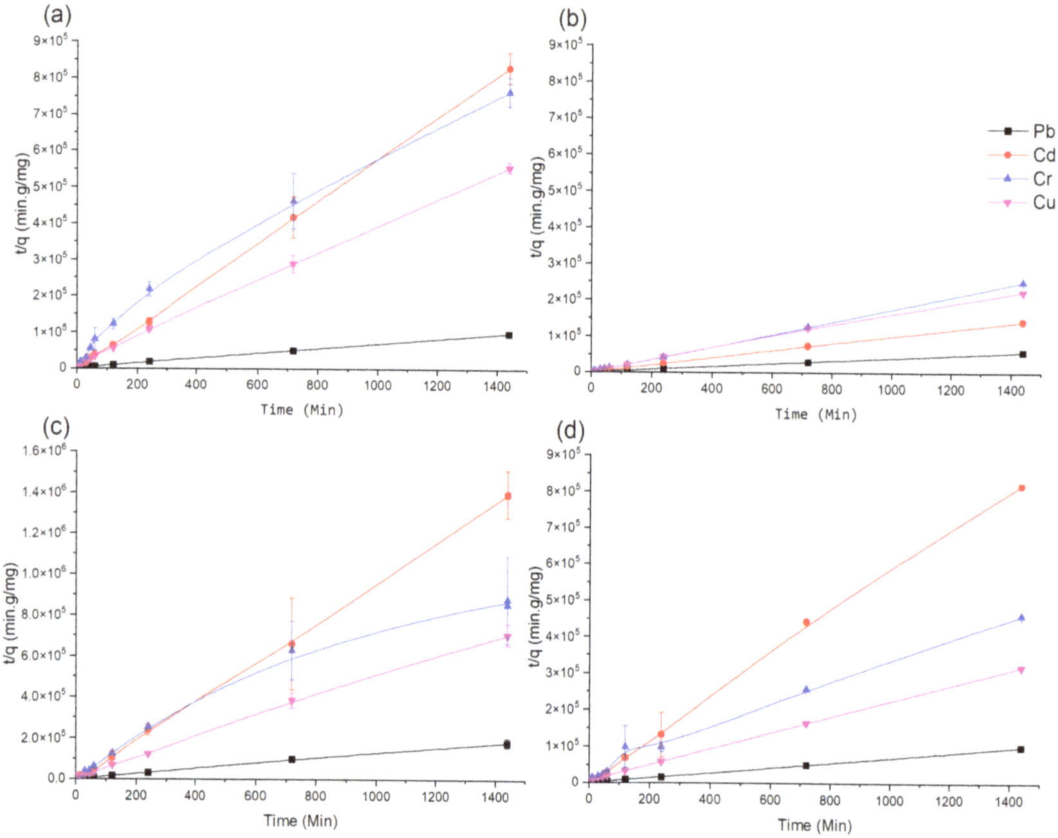

Figure 5. Pseudo-second-order model: (**a**) Coffee husk, (**b**) AgNPs–coffee husk, (**c**) lignin, and (**d**) AgNPs–lignin.

The materials used as adsorbents have a variety of functional groups that allow inducing chemical adsorption together with physical adsorption on the adsorbent surface. Taking that into account, in the case of the use of modified materials, such as AgNPs–coffee husk or AgNPs–lignin, the obtained improvement in the adsorption rate can be related to the modification itself. The modification of the surface can increment the electrostatic interactions between the adsorbent surface and the metal ions [62]. In addition, the kinetics is here checked with a mixed solution of metals that may imply a competition between them. The adsorption of metals could be related to the ionic radius and the electronegativity of each metal ion [34]. Thus, Pb ions with higher ionic radius and electronegativity show the best R^2 and the highest rate constant k_2 of the pseudo-second-order kinetic model with AgNPs–coffee husk as sorbent, and this correlates to being the one with the highest adsorption rate.

Table 2. Pseudo-first- and pseudo-second-order kinetic model values for sorption experiments of Pb(II), Cd(II), Cr(III), and Cu(II), all with an initial concentration of 0.18 mmol/L. The value k_2 is a velocity constant from the pseudo-second-order kinetic model.

Material	Ion	Pseudo-First-Order	Pseudo-Second-Order	
		R^2	k_2 (g/mg·min)	R^2
Coffee husk	Pb(II)	0.8525	3.767	0.9996
	Cd(II)	0.0172	−198.2 *	0.9998
	Cr(III)	0.4268	7.296	0.9841
	Cu(II)	0.8121	15.90	0.9994
AgNPs–coffee husk	Pb(II)	0.8247	22.90	1.000
	Cd(II)	0.9623	7.308	0.9994
	Cr(III)	0.9490	18.79	0.9999
	Cu(II)	0.6747	9.290	0.9976
Lignin	Pb(II)	0.9103	10.05	0.9982
	Cd(II)	0.2986	−77.61 *	0.9973
	Cr(III)	0.4431	10.44	0.9579
	Cu(II)	0.9573	28.06	0.9987
AgNPs–lignin	Pb(II)	0.9884	5.102	0.9999
	Cd(II)	0.0007	−110.4 *	0.9978
	Cr(III)	0.8474	6.089	0.9922
	Cu(II)	0.9657	6.263	0.9988

* The negative values shown in the table are considered non-relevant values. In this sense, the negative coefficients do not explain the sorption kinetics of the system shown [61].

3.5. Antifungal Assays

As in all *Candida* species checked (*C. albicans*, *C. parapsilosis*, *C. glabrata*, *C. krusei* and *C. guillicrmondii*) lignin and its modification give less antifungal activity; thus, just coffee husk and its modifications will be presented here.

Thus, the antifungal behavior of coffee husk and its modification with silver nanoparticles against the species of *Candida albicans* and *Candida parapsilosis* can be observed in Figure 6, respectively. As can be observed, the modified and unmodified coffee husk's best performance is observed at a biomass relative concentration of 1.5 mg/mL, when a fungistatic effect is achieved. In turn, it can be evidenced that the AgNPs' modified coffee husk presents a potentiating effect on the antifungal activity of both *Candida* species, especially when comparing results of experiments with biomass relative concentration of 0.75 mg/mL, in terms of a material to be considered bacteriostatic.

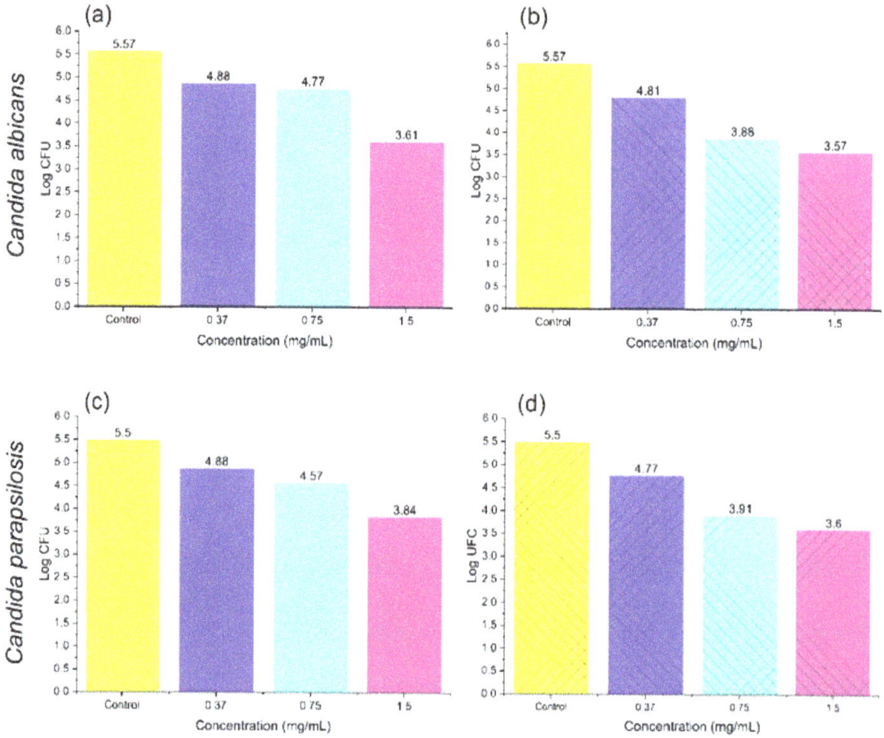

Figure 6. Antifungal activity on *Candida albicans* and *Candida parapsilosis* of (**a**,**c**) coffee husk, (**b**,**d**) AgNPs–coffee husk.

Figure 7 presents the results of *Candida glabrata* that evidence a higher antimicrobial activity in the case of the modified coffee husk AgNPs–coffee husk, in comparison with the previous mentioned species. In this case, the bacteriostatic effect of the modified coffee husk is also present at low relative biomass concentrations.

In Figure 8, greater sensitivity of the *Candida krusei* and *Candida guilliermondii* species to both materials are evidenced, respectively. Especially at a biomass relative concentration of 1.5 mg/mL, where a bacteriostatic effect is obtained. In these cases, as for previous ones, the use of AgNPs–coffee husk shows higher values of logarithmic reduction, reflected in a higher bacteriostatic effect, with higher effectivity at 1.5 mg/mL. *Candida krusei* was more affected than *Candida guilliermondii*, as can be observed.

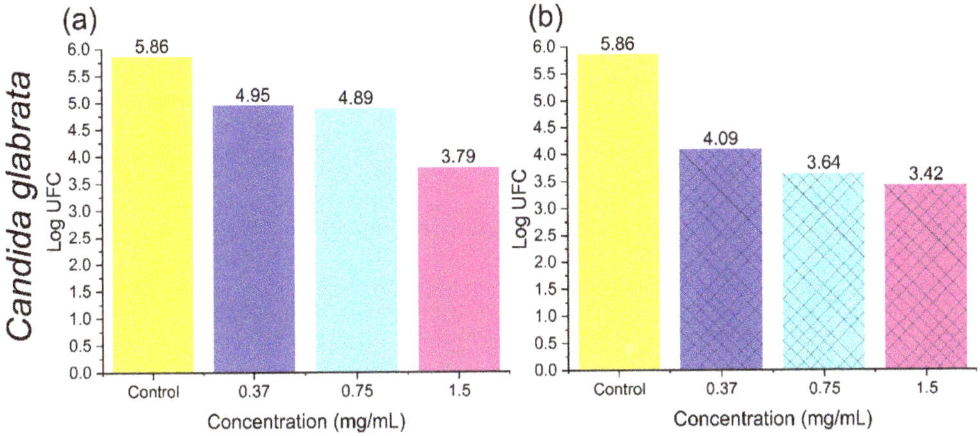

Figure 7. Antifungal activity on *Candida glabrata* of (**a**) coffee husk, and (**b**) AgNPs–coffee husk.

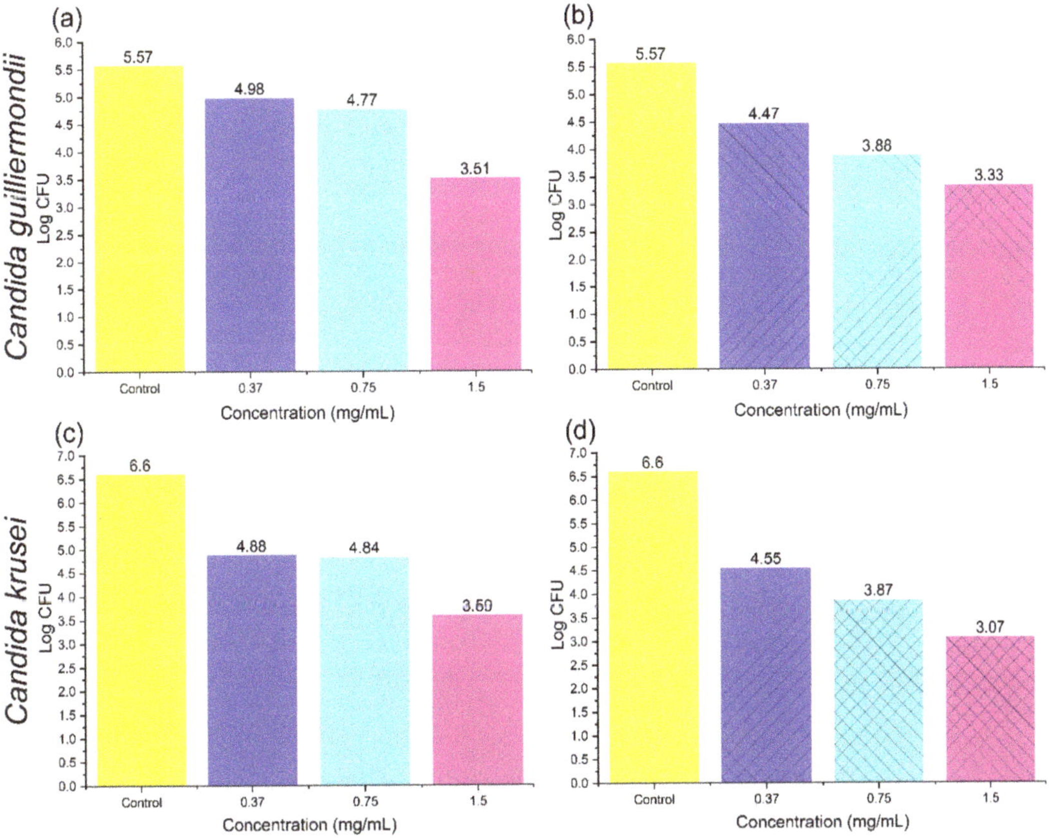

Figure 8. Antifungal activity on *Candida guilliermondii* and *Candida krusei* of (**a**,**c**) coffee husk, and (**b**,**d**) AgNPs–coffee husk.

In summary, the modification of the coffee husk with AgNPs demonstrates an improving effect on the antifungal activity; however, using a 1.5 mg/mL concentration that exceeds two and sometimes reaches three logarithmic reductions could be considered fungistatic and antifungal activity, respectively. It should be noted that the use of AgNPs nowadays is increasing, and should also be applied for the reduction of *Candida* classification species since they have been demonstrated to improve their microbial reduction and achieve biocompatibility in some cases [64].

The materials' concentration at which the highest fungistatic effect is achieved is 1.5 mg/mL. The material with the highest logarithmic reduction values is the modified coffee husk, which demonstrates a higher effect on the reduction of the microbial activity of the evaluated species. *C. krusei* is the species more sensitive to the general presence of both materials under study (coffee husk and its modifications with AgNPs), with the corresponding higher fungicidal effects, followed by *C. glabrata*.

4. Conclusions

The use of four different biomass materials as sorbents in this study demonstrates the viability of coffee husk and lignin to be able to remove heavy metals such as Pb, Cd, Cr, and Cu from aqueous solutions. Their modification with silver nanoparticles is here presented as a way of biomass nanocomposites' preparation and as a comparison of the corresponding raw materials. It is demonstrated that the nanocomposites with silver are an option for the removal of the metals under study, under the conditions here checked (such as pH 4.0 and an initial concentration of 0.18 mmol/L), giving sorption percentages above 90% for Pb and Cr, and around 80% for Cd and Cu. When trying to model the sorption process, the pseudo-second model is the one that best described the heavy metal sorption under the working experimental conditions (24 h of sorption experiments), thus assuming that the adsorption is controlled by the chemisorption process.

When comparing all the sorbents here checked (coffee husk, lignin, and their modifications with AgNPs), the heavy metal sorption efficiency follows the order of AgNPs–coffee husk > coffee husk > AgNPs–lignin > lignin, with sorption capacities of 2.56 mg/g for Pb, 1.02 mg/g for Cd, 0.644 mg/g for Cu, and 0.575 mg/g for Cr with AgNPs–coffee husk.

Finally, the antifungal evaluation demonstrates that the modified and unmodified coffee husk have fungistatic activity on *C. albicans*, *C. glabrata*, *C. krusei*, and *C. guilliermondii* mainly with a concentration of 1.5 mg/mL, as they do not exceed the two logarithms of reduction, therefore demonstrating some fungicidal effect.

In this research, the implementation of the coffee husk modified with AgNPs was developed, and constitutes a promising sorbent in the sorption of heavy metal ions such as lead, cadmium, copper, and chromium from aqueous solutions, and can also inhibit the growth of fungus pathogens at the same time. To achieve a realistic application with wastewater using these materials, it is necessary to further study the reuse of the material and the cost of the modified materials to produce nanoparticle-modified biomass systems.

Author Contributions: Conceptualization, A.M.C., J.A.G.C., J.B.-A. and C.P.; data curation: M.Y.C.O. and D.F.G.-B.; methodology, D.F.G.-B., A.M.C., J.A.G.C. and M.Y.C.O.; formal analysis, D.F.G.-B. and M.Y.C.O.; investigation, D.F.G.-B., A.M.C., J.A.G.C., J.B.-A. and C.P.; writing—original draft preparation, D.F.G.-B.; writing—review and editing, A.M.C., J.B.-A. and C.P.; visualization, J.B.-A. and C.P.; supervision, A.M.C., J.B.-A. and C.P.; project administration, A.M.C.; Project administration: A.M.C.; funding acquisition: A.M.C., J.A.G.C., C.P. and M.Y.C.O. All authors have read and agreed to the published version of the manuscript.

Funding: This study was financially supported by institutional funds for the development of the research, within the framework of project No. 01946233 Act No. 16-VII-2019 of Universidad Santo Tomás–Bucaramanga, and by one Spanish research project No. CTM2015-65414-C2-1-R.

Institutional Review Board Statement: Not applicable.

Informed Consent Statement: Not applicable.

Data Availability Statement: Not applicable.

Acknowledgments: Acknowledgment to the *Grup de Tècniques de Separació en Química* (GTS) for allowing and financing the sample analysis, as employees in the study of kinetics modeling. Thanks to the *Servei de Microscòpia* from *Universitat Autónoma de Barcelona* for allowing morphological studies of the samples. Moreover, thanks to the *Grupo de Investigación en Bioquímica y Microbiología* for allowing the microbiology analysis of samples; and thanks to the *Grupo en Nuevos Materiales y Energías Alternativas* (GINMEA) of the *Universidad Santo Tomás* for directing the project (GIFQCAAMBUISUPCP62017).

Conflicts of Interest: The authors declare no conflict of interest.

References

1. WHO Drinking-Water. Available online: https://www.who.int/news-room/fact-sheets/detail/drinking-water (accessed on 27 January 2021).
2. OECD. *OECD Environmental Outlook to 2050: The Consequences of Inaction*; OECD Publishing: Paris, France, 2012; ISBN 978-92-64-122161.
3. Burek, P.; Satoh, Y.; Fischer, G.; Kahil, M.T.; Scherzer, A.; Tramberend, S.; Nava, L.F.; Wada, Y.; Eisner, S.; Hanasaki, N.; et al. *Water Futures and Solution—Fast Track Initiative*; IIASA: Laxenburg, Austria, 2016.
4. *IEA Water Energy Nexus, Excerpt from the World Energy Outlook 2016*; IEA: Paris, France, 2016.
5. Genç-Fuhrman, H.; Mikkelsen, P.S.; Ledin, A. Simultaneous Removal of As, Cd, Cr, Cu, Ni and Zn from Stormwater Using High-Efficiency Industrial Sorbents: Effect of PH, Contact Time and Humic Acid. *Sci. Total Environ.* **2016**, *566–567*, 76–85. [CrossRef] [PubMed]
6. United Nations. *The United Nations World Water Development Report 2021: Valuing Water*; UNESCO: Paris, France, 2021.
7. Picetti, R.; Deeney, M.; Pastorino, S.; Miller, M.R.; Shah, A.; Leon, D.A.; Dangour, A.D.; Green, R. Nitrate and Nitrite Contamination in Drinking Water and Cancer Risk: A Systematic Review with Meta-Analysis. *Environ. Res.* **2022**, *210*, 112988. [CrossRef]
8. Sagasta, M.; Zadeh, S.M.; Turral, H. *More People, More Food, Worse Water?: A Global Review of Water Pollution from Agriculture*; FAO: Rome, Italy, 2018.
9. Joseph, L.; Jun, B.-M.; Flora, J.R.V.; Park, C.M.; Yoon, Y. Removal of Heavy Metals from Water Sources in the Developing World Using Low-Cost Materials: A Review. *Chemosphere* **2019**, *229*, 142–159. [CrossRef] [PubMed]
10. Ma, Y.; Egodawatta, P.; McGree, J.; Liu, A.; Goonetilleke, A. Human Health Risk Assessment of Heavy Metals in Urban Stormwater. *Sci. Total Environ.* **2016**, *557–558*, 764–772. [CrossRef] [PubMed]
11. Muralikrishna, I.v.; Manickam, V. Introduction. In *Environmental Management*; Elsevier, 2017; pp. 1–4.
12. TerraGreen Global Waste—Solvable Problem as a Renewable Energy Resource. Available online: https://medium.com/@support_61820/global-waste-solvable-problem-as-a-renewable-energy-resource-5d8f05cc1a7d (accessed on 27 January 2021).
13. Alatzas, S.; Moustakas, K.; Malamis, D.; Vakalis, S. Biomass Potential from Agricultural Waste for Energetic Utilization in Greece. *Energies* **2019**, *12*, 1095. [CrossRef]
14. Kondamudi, N.; Mohapatra, S.K.; Misra, M. Spent Coffee Grounds as a Versatile Source of Green Energy. *J. Agric. Food Chem.* **2008**, *56*, 11757–11760. [CrossRef] [PubMed]
15. Singh, C.K.; Kumar, A.; Roy, S.S. Quantitative Analysis of the Methane Gas Emissions from Municipal Solid Waste in India. *Sci. Rep.* **2018**, *8*, 2913. [CrossRef]
16. Sabiiti, E. Utilising Agricultural Waste to Enhance Food Security and Conserve the Environment. *Afr. J. Food Agric. Nutr. Dev.* **2011**, *11*, 3–5.
17. Radhakrishnan, R.; Patra, P.; Das, M.; Ghosh, A. Recent Advancements in the Ionic Liquid Mediated Lignin Valorization for the Production of Renewable Materials and Value-Added Chemicals. *Renew. Sustain. Energy Rev.* **2021**, *149*, 111368. [CrossRef]
18. Becker, J.; Wittmann, C. A Field of Dreams: Lignin Valorization into Chemicals, Materials, Fuels, and Health-Care Products. *Biotechnol. Adv.* **2019**, *37*, 107360. [CrossRef]
19. Cao, L.; Yu, I.K.M.; Liu, Y.; Ruan, X.; Tsang, D.C.W.; Hunt, A.J.; Ok, Y.S.; Song, H.; Zhang, S. Lignin Valorization for the Production of Renewable Chemicals: State-of-the-Art Review and Future Prospects. *Bioresour. Technol.* **2018**, *269*, 465–475. [CrossRef]
20. Mohanpuria, P.; Rana, N.K.; Yadav, S.K. Biosynthesis of Nanoparticles: Technological Concepts and Future Applications. *J. Nanoparticle Res.* **2008**, *10*, 507–517. [CrossRef]
21. Królikowska, A.; Kudelski, A.; Michota, A.; Bukowska, J. SERS Studies on the Structure of Thioglycolic Acid Monolayers on Silver and Gold. *Surf. Sci.* **2003**, *532–535*, 227–232. [CrossRef]
22. Kumar, A.; Mandal, S.; Selvakannan, P.R.; Pasricha, R.; Mandale, A.B.; Sastry, M. Investigation into the Interaction between Surface-Bound Alkylamines and Gold Nanoparticles. *Langmuir* **2003**, *19*, 6277–6282. [CrossRef]
23. Chandrasekharan, N.; Kamat, P.v. Improving the Photoelectrochemical Performance of Nanostructured TiO 2 Films by Adsorption of Gold Nanoparticles. *J. Phys. Chem. B* **2000**, *104*, 10851–10857. [CrossRef]
24. Pető, G.; Molnár, G.L.; Pászti, Z.; Geszti, O.; Beck, A.; Guczi, L. Electronic Structure of Gold Nanoparticles Deposited on SiOx/Si(100). *Mater. Sci. Eng. C* **2002**, *19*, 95–99. [CrossRef]
25. Zhang, L.; Webster, T.J. Nanotechnology and Nanomaterials: Promises for Improved Tissue Regeneration. *Nano Today* **2009**, *4*, 66–80. [CrossRef]

26. Zhang, H.; Carrillo-Navarrete, F.; Palet-Ballús, C. Human Hair Biogenic Fiber as a Biosorbent of Multiple Heavy Metals from Aqueous Solutions. *J. Nat. Fibers* **2022**, *19*, 2018–2033. [CrossRef]
27. Suárez-Forero, S.J.; Candela-Soto, A.M.; Henao-Martínez, J.A.; Bayona-Ayala, O.L. Evaluation of the Performance of the Preteretment with the Hydrogen Peroxide on Sugar Cane Bagasse for Removing Lignin. *ITECKNE* **2019**, *16*, 21–28. [CrossRef]
28. Tokay, B.; Akpınar, I. A Comparative Study of Heavy Metals Removal Using Agricultural Waste Biosorbents. *Bioresour. Technol. Rep.* **2021**, *15*, 100719. [CrossRef]
29. Sluiter, A.; Hames, B.; Ruiz, R.; Scarlata, C.; Sluiter, J.; Templeton, D.; Crocker, D. *Determination of Structural Carbohydrates and Lignin in Biomass*; NREL: Golden, CO, USA, 2008.
30. Burbano Patiño, A.A. *Síntesis y Caracterización de Nanopartículas Magnéticas Del Tipo Core-Shell Fe_3O_4@Ag Soportadas Sobre Lignina y Cascarilla de Café*; Universidad Santo Tomás: Bucaramanga, Colombia, 2018.
31. Zhang, X.-F.; Liu, Z.-G.; Shen, W.; Gurunathan, S. Silver Nanoparticles: Synthesis, Characterization, Properties, Applications, and Therapeutic Approaches. *Int. J. Mol. Sci.* **2016**, *17*, 1534. [CrossRef] [PubMed]
32. Wiley, B.; Sun, Y.; Mayers, B.; Xia, Y. Shape-Controlled Synthesis of Metal Nanostructures: The Case of Silver. *Chem.—A Eur. J.* **2005**, *11*, 454–463. [CrossRef] [PubMed]
33. Zhao, J.; Shen, X.-J.; Domene, X.; Alcañiz, J.-M.; Liao, X.; Palet, C. Comparison of Biochars Derived from Different Types of Feedstock and Their Potential for Heavy Metal Removal in Multiple-Metal Solutions. *Sci. Rep.* **2019**, *9*, 9869. [CrossRef] [PubMed]
34. Zhao, J.; Boada, R.; Cibin, G.; Palet, C. Enhancement of Selective Adsorption of Cr Species via Modification of Pine Biomass. *Sci. Total Environ.* **2021**, *756*, 143816. [CrossRef]
35. Zhang, H. Biosorption of Heavy Metals from Aqueous Solutions Using Keratin Biomaterials. Ph.D. Thesis, Universitat Autònoma de Barcelona, Barcelona, Spain, 2014.
36. Sahoo, T.R.; Prelot, B. Adsorption Processes for the Removal of Contaminants from Wastewater. In *Nanomaterials for the Detection and Removal of Wastewater Pollutants*; Elsevier: Amsterdam, The Netherlands, 2020; pp. 161–222.
37. Ho, Y.S.; McKay, G. Pseudo-Second Order Model for Sorption Processes. *Process Biochem.* **1999**, *34*, 451–465. [CrossRef]
38. Zur Theorie Der Sogenannten Adsorption Gelöster Stoffe. *Z. Für Chem. Und Ind. Der Kolloide* **1907**, *2*, 15. [CrossRef]
39. Zheng, Y.-M.; Li, N.; Zhang, W.-D. Preparation of Nanostructured Microspheres of Zn–Mg–Al Layered Double Hydroxides with High Adsorption Property. *Colloids Surf. A Physicochem. Eng. Asp.* **2012**, *415*, 195–201. [CrossRef]
40. Rout, S.; Kumar, A.; Mana Ravi, P.; Mangal Tripathi, R. Pseudo Second Order Kinetic Model for the Sorption of U (VI) onto Soil: A Comparison of Linear and Non-Linear Methods. *Int. J. Environ. Sci.* **2015**, *6*, 145–154. [CrossRef]
41. Loo, Y.Y.; Rukayadi, Y.; Nor-Khaizura, M.-A.-R.; Kuan, C.H.; Chieng, B.W.; Nishibuchi, M.; Radu, S. In Vitro Antimicrobial Activity of Green Synthesized Silver Nanoparticles Against Selected Gram-Negative Foodborne Pathogens. *Front. Microbiol.* **2018**, *9*, 1555. [CrossRef]
42. Daffalla, S.B.; Mukhtar, H.; Shaharun, M.S. Preparation and Characterization of Rice Husk Adsorbents for Phenol Removal from Aqueous Systems. *PLoS ONE* **2020**, *15*, e0243540. [CrossRef]
43. Jahirul, M.; Rasul, M.; Chowdhury, A.; Ashwath, N. Biofuels Production through Biomass Pyrolysis—A Technological Review. *Energy* **2012**, *5*, 4952–5001. [CrossRef]
44. Konneh, M.; Wandera, S.M.; Murunga, S.I.; Raude, J.M. Adsorption and Desorption of Nutrients from Abattoir Wastewater: Modelling and Comparison of Rice, Coconut and Coffee Husk Biochar. *Heliyon* **2021**, *7*, e08458. [CrossRef]
45. Oliveira, W.E.; Franca, A.S.; Oliveira, L.S.; Rocha, S.D. Untreated Coffee Husks as Biosorbents for the Removal of Heavy Metals from Aqueous Solutions. *J. Hazard. Mater.* **2008**, *152*, 1073–1081. [CrossRef]
46. Sharma, A.; Sagar, A.; Rana, J.; Rani, R. Green Synthesis of Silver Nanoparticles and Its Antibacterial Activity Using Fungus Talaromyces Purpureogenus Isolated from Taxus Baccata Linn. *Micro Nano Syst. Lett.* **2022**, *10*, 2. [CrossRef]
47. Chandrasekharan, S.; Chinnasamy, G.; Bhatnagar, S. Sustainable Phyto-Fabrication of Silver Nanoparticles Using Gmelina Arborea Exhibit Antimicrobial and Biofilm Inhibition Activity. *Sci. Rep.* **2022**, *12*, 156. [CrossRef]
48. Kaidi, S.; Belattmania, Z.; Bentiss, F.; Jama, C.; Reani, A.; Sabour, B. Synthesis and Characterization of Silver Nanoparticles Using Alginate from the Brown Seaweed Laminaria Ochroleuca: Structural Features and Antibacterial Activity. *Biointerface Res. Appl. Chem.* **2021**, *12*, 6046–6057. [CrossRef]
49. Fu, B.; Ge, C.; Yue, L.; Luo, J.; Feng, D.; Deng, H.; Yu, H. Characterization of Biochar Derived from Pineapple Peel Waste and Its Application for Sorption of Oxytetracycline from Aqueous Solution. *BioResources* **2016**, *11*, 9017–9035. [CrossRef]
50. Kaur, P.; Kaur, P.; Kaur, K. Adsorptive Removal of Imazethapyr and Imazamox from Aqueous Solution Using Modified Rice Husk. *J. Clean. Prod.* **2020**, *244*, 118699. [CrossRef]
51. Huang, M.-L.; Yen, P.-L.; Liao, V.H.-C. A Combined Approach to Remediate Cadmium Contaminated Sediment Using the Acidophilic Sulfur-Oxidizing Bacterial SV5 and Untreated Coffee Ground. *Chemosphere* **2021**, *273*, 129662. [CrossRef]
52. Seniūnaitė, J.; Vaiškūnaitė, R.; Bazienė, K. Mathematical Modelling for Copper and Lead Adsorption on Coffee Grounds. In Proceedings of the Proccedings of 10th International Conference "Environmental Engineering", VGTU Technika, Vilnius, Lithuania, 27 April 2017.
53. Ben Amar, M.; Walha, K.; Salvadó, V. Evaluation of Olive Stones for Cd(II), Cu(II), Pb(II) and Cr(VI) Biosorption from Aqueous Solution: Equilibrium and Kinetics. *Int. J. Environ. Res.* **2020**, *14*, 193–204. [CrossRef]
54. Mahmood-ul-Hassan, M.; Suthor, V.; Rafique, E.; Yasin, M. Removal of Cd, Cr, and Pb from Aqueous Solution by Unmodified and Modified Agricultural Wastes. *Environ. Monit. Assess.* **2015**, *187*, 19. [CrossRef] [PubMed]

55. Aziz, H.A.; Adlan, M.N.; Ariffin, K.S. Heavy Metals (Cd, Pb, Zn, Ni, Cu and Cr(III)) Removal from Water in Malaysia: Post Treatment by High Quality Limestone. *Bioresour. Technol.* **2008**, *99*, 1578–1583. [CrossRef] [PubMed]
56. Yusoff, S.N.M.; Kamari, A.; Putra, W.P.; Ishak, C.F.; Mohamed, A.; Hashim, N.; Isa, I.M. Removal of Cu(II), Pb(II) and Zn(II) Ions from Aqueous Solutions Using Selected Agricultural Wastes: Adsorption and Characterisation Studies. *J. Environ. Prot.* **2014**, *05*, 289–300. [CrossRef]
57. Abdić, Š.; Memić, M.; Šabanović, E.; Sulejmanović, J.; Begić, S. Adsorptive Removal of Eight Heavy Metals from Aqueous Solution by Unmodified and Modified Agricultural Waste: Tangerine Peel. *Int. J. Environ. Sci. Technol.* **2018**, *15*, 2511–2518. [CrossRef]
58. Kumar, P.; Barrett, D.M.; Delwiche, M.J.; Stroeve, P. Methods for Pretreatment of Lignocellulosic Biomass for Efficient Hydrolysis and Biofuel Production. *Ind. Eng. Chem. Res.* **2009**, *48*, 3713–3729. [CrossRef]
59. Morales De La Rosa, S. Hidrólisis Ácida de Celulosa y Biomasa Lignocelulósica Asistida Con Líquidos Iónicos. Ph.D. Thesis, Universidad Autónoma de Madrid, Madrid, Spain, 2015.
60. Ituen, E.; Yuanhua, L.; Verma, C.; Alfantazi, A.; Akaranta, O.; Ebenso, E.E. Synthesis and Characterization of Walnut Husk Extract-Silver Nanocomposites for Removal of Heavy Metals from Petroleum Wastewater and Its Consequences on Pipework Steel Corrosion. *J. Mol. Liq.* **2021**, *335*, 116132. [CrossRef]
61. Kul, A.R.; Caliskan, N. Equilibrium and Kinetic Studies of the Adsorption of Zn(II) Ions onto Natural and Activated Kaolinites. *Adsorpt. Sci. Technol.* **2009**, *27*, 85–105. [CrossRef]
62. Xiao, G.; Wang, Y.; Xu, S.; Li, P.; Yang, C.; Jin, Y.; Sun, Q.; Su, H. Superior Adsorption Performance of Graphitic Carbon Nitride Nanosheets for Both Cationic and Anionic Heavy Metals from Wastewater. *Chin. J. Chem. Eng.* **2019**, *27*, 305–313. [CrossRef]
63. Cuong Nguyen, X.; Thanh Huyen Nguyen, T.; Hong Chuong Nguyen, T.; Van Le, Q.; Yen Binh Vo, T.; Cuc Phuong Tran, T.; Duong La, D.; Kumar, G.; Khanh Nguyen, V.; Chang, S.W.; et al. Sustainable Carbonaceous Biochar Adsorbents Derived from Agro-Wastes and Invasive Plants for Cation Dye Adsorption from Water. *Chemosphere* **2021**, *282*, 131009. [CrossRef]
64. Torres Acosta, L.; Mendieta, I.; Hernández, G.; Núñez, R.; Castaño, V. Citotoxicidad y Genotoxicidad de AgNPs Para Disminuir La Adherencia de Candida Albicans En Prótesis Dentales. *Mundo Nano. Rev. Interdiscip. Nanocienc. Nanotecnol.* **2011**, *5*.

water

MDPI

Article

Removal of Various Hazardous Materials Using a Multifunctional Biomass-Derived Hydroxyapatite (HAP) Catalyst and Its Antibacterial Effects

Sanha Jang [1], Kyeongmun Park [1], Sehwan Song [2], Haksoo Lee [3], Sungkyun Park [2], Buhyun Youn [3,*] and Kanghyun Park [1,4,*]

1 Department of Chemistry, Pusan National University, Busandaehak-ro 63beon-gil, Geumjeong-gu, Busan 46241, Korea; jangs0522@naver.com (S.J.); pkmkm951@naver.com (K.P.)
2 Department of Physics, Pusan National University, Busandaehak-ro 63beon-gil, Geumjeong-gu, Busan 46241, Korea; sehwansong465@gmail.com (S.S.); pak@pusan.ac.kr (S.P.)
3 Department of Biological Sciences, Pusan National University, Busandaehak-ro 63beon-gil, Geumjeong-gu, Busan 46241, Korea; hs950303@gmail.com
4 SoulDot Co., Ltd., Pusan National University, Busandaehak-ro 63beon-gil, Geumjeong-gu, Busan 46241, Korea
* Correspondence: bhyoun72@pusan.ac.kr (B.Y.); chemistry@pusan.ac.kr (K.P.); Tel.: +82-51-510-2264 (B.Y.); +82-51-510-2238 (K.P.)

Citation: Jang, S.; Park, K.; Song, S.; Lee, H.; Park, S.; Youn, B.; Park, K. Removal of Various Hazardous Materials Using a Multifunctional Biomass-Derived Hydroxyapatite (HAP) Catalyst and Its Antibacterial Effects. *Water* **2021**, *13*, 3302. https://doi.org/10.3390/w13223302

Academic Editors: Cristina Palet and Julio Bastos-Arrieta

Received: 28 October 2021
Accepted: 19 November 2021
Published: 22 November 2021

Abstract: In the present study, oyster shells, a cause of environmental pollution, were employed effectively to synthesize hydroxyapatite (HAP) by facile oxidation and phosphorylation. The ability of HAP to adsorb various metal cations and inhibit bacterial growth was validated. The biomass-derived HAP catalyst exhibited high metal cation adsorption in water at room temperature and under various acidic conditions (M = Cr, Mn, Ni, Cu, Cd, Ba, and Pb). HAP was demonstrated to have a maximum removal efficiency of 92.8% for the heavy metal Pb. Even under different pH conditions, HAP was demonstrated to be effective for the removal of three harmful heavy metals, Cr, Cd, and Pb, with a particularly high removal efficiency demonstrated for Pb under all conditions (average removal efficiency of Cr: 63.0%, Cd: 59.9%, and Pb: 91.6%). In addition, HAP had a significant influence on phosphate ion adsorption in aqueous solution, eliminating 98.1% after 3 min. Furthermore, biomass-derived HAP was demonstrated to have significant antibacterial activity against *E. coli* and *S. aureus* (5 mM: 74% and 78.1%, 10 mM: 89.6% and 96.0%, respectively).

Keywords: biomass-derived material; nanomaterial; environmental chemistry; multifunctional catalyst; antibacterial; metal ion adsorption; phosphate ion adsorption

1. Introduction

Environmental pollution, especially in the ocean and groundwater, is increasing rapidly because of the organic and inorganic pollutants in hazardous wastewater [1–10]. These pollutants come from industrial wastewater and electrical devices containing heavy metals, such as Pb, Cd, and Cr. Pb, in particular, is a hazardous metal that causes osteoporosis, and Cd is a hazardous metal that causes itai-itai disease [2,11–13]. The elimination of heavy metals is of particular interest to scientists and engineers because of the negative impact of these pollutants on a wide range of living organisms [14,15]. As a result, various techniques for heavy metal removal from wastewater, such as ion exchange, chemical precipitation, membrane separation, and adsorption, have been developed [16,17]. Precipitation accompanied by a chemically or externally mediated transition to non-labile forms is a popular method for removing metals from water and wastewater. The long-lasting matrix to which the metals are transferred must be stable under various conditions, especially pH and temperature [11]. Auaok et al. reported the effectiveness of Pb (II) adsorption using polypyrrole-based activated carbon [18]. Li et al. was successful in adsorbing heavy metals through zeolite, a material with a large surface area and pores [17].

In an earlier study, we successfully used Sb_2S_5 nanofibers to selectively adsorb Pb in water at low temperatures [19]. According to the literature, heavy metals such as Pb, Ni, Cu, and Cd, can be quickly eliminated from wastewater using hydroxyapatite (HAP, $Ca_{10}(PO_4)_6(OH)_2$) [20–22].

Phosphorus (P) is an important nutrient for all species and is required for the healthy operation of ecosystems [23]. Phosphoric acid (H_3PO_4) is decomposed in the aqueous phase at various pK_a values to yield phosphate species ($H_2PO_4^-$ at pK_a = 2.15, HPO_4^{2-} at pK_a = 7.20, and PO_4^{3-} at pK_a = 12.35). Thus, the total phosphate is the sum of all anionic phosphates [24]. However, excess P that is not digested in the biogeochemical P cycle is released into the waterways, causing eutrophication [25]. In aquatic environments, eutrophication causes toxic algal blooms, biodiversity loss, and dead zones [26]. As a result, effective and practical techniques to remove phosphate from water bodies are critical [27]. HAP has been used to confirm the replacement of phosphate anions in previous studies [21,28,29].

Microbial toxins such as total coliform, Escherichia coli (*E. coli*), fecal coliform, and Salmonella are often found in groundwater where treatment facilities are lacking. This microbial contamination results in acute and chronic diseases, such as diarrhea and high fever [30–32]. The high volume of sewage dumped into water bodies is a major issue in the industrial era [1]. Significant research efforts have been aimed at developing a catalyst that utilizes chemical and physical adsorption to solve this problem. In addition, many engineers and scientists have worked tirelessly to eliminate bacteria. Some materials, both organic and inorganic, have antibacterial properties [33–40]. Lamkhao et al. recently reported that a microwave-assisted combustion process can be used to produce HAP with antibacterial properties [38].

Consumption of seafood produced by the aquaculture sector has increased in recent years as the world's population has grown, and interest in the recycling of industrial waste from seashells has also grown [41–43]. According to data from the National Assembly Research Service in South Korea [44], approximately 40% of fish and shellfish output was discarded each year between 2010 and October 2019 (Figure S1). Seashells have been employed as raw materials for a variety of applications. Nakanishi [45], Shon [46], and Koopan [47] et al. reported on the recycling of oyster shells for various uses. Furthermore, Yeom et al. reported successful phosphate ion removal using a crab shell [48]. Oysters are a popular seafood worldwide. In particular, oyster shells made of calcium carbonate are aquaculture waste that causes serious disposal problems worldwide, as well as on the southeastern coast of South Korea (Figure S2). The oyster shells are easy to collect for recycling purposes and are therefore can be utilized as a calcium supply.

Herein, we report the preparation of HAP catalysts from oyster shells using facile methods such as heat treatment and phosphorylation, and characterize their metal cation adsorption efficiency and antibacterial performance. In this investigation, Pb, an extremely toxic heavy metal, was adsorbed at a rate of 92.8% by the HAP catalyst. Furthermore, as a result of experiments with various pH conditions that are representative of actual wastewater, a high removal efficiency was demonstrated for Pb, while Cr and Cd demonstrated maximum effectiveness at pH 7. When HAP was used to remove phosphate ions over a period of 9 min, the removal efficiency was found to be 98.1% at 3 min, and remained constant thereafter. Bacteria are one of the most dangerous organisms for humans; thus, we evaluated the antibacterial properties of *E. coli* and *S. aureus*. We validated the antibacterial capabilities of HAP against *E. coli* and *S. aureus* (In the case of 5 mM of HAP, *E. coli*: 74.4% and *S. aureus*: 74.9%. in the case of 10 mM of HAP, *E. coli*: 90.9% and *S. aureus*: 96%, respectively).

2. Results and Discussion

2.1. Preparation and Characterization of HAP Catalyst

CaO was produced from oyster shells according to Equation (1) using a simple heat treatment at 900 °C. We used existing research as a guide and synthesized the HAP catalyst according to Equation (2) [45].

$$CaCO_3 \rightarrow CaO + CO_2 \quad (1)$$

$$10CaO + 6H_3PO_4 \rightarrow Ca_{10}(PO_4)_6(OH)_2 + 8H_2O \quad (2)$$

Transmission electron microscopy (TEM) images revealed the calcium oxide formed by oxidation, and the HAP formed by phosphorylation, in the initial calcium carbonate oyster shells (Figure 1a,c). There was no discernible form to the CaO, but there was an indication of a definite rod shape morphology for the HAP. This was caused by the phosphorylation of CaO. The high-resolution TEM (HRTEM) images and Fourier-transform patterns revealed the formation of CaO crystals with a lattice distance of 0.242 nm and HAP crystals with a lattice distance of 0.280 nm, which correspond to the (200) and (211) planes, respectively (Figure 1b,d).

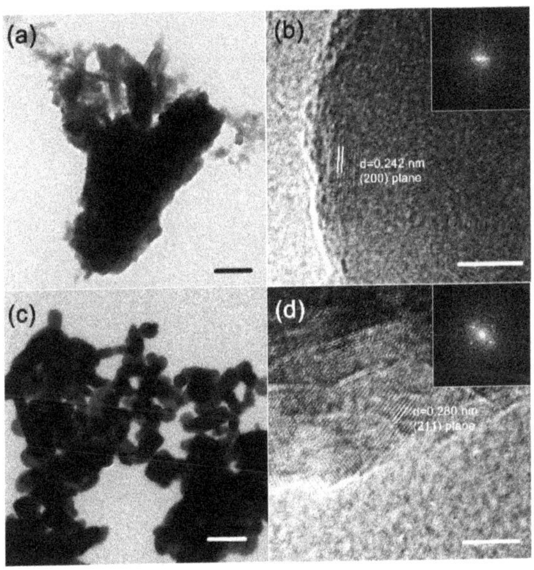

Figure 1. (**a**) TEM and (**b**) HRTEM (inset: SAED pattern) images of CaO. (**c**) TEM and (**d**) HRTEM (inset: SAED pattern) images of hydroxyapatite (HAP). The bars indicate 500 nm (**a**,**c**) and 5 nm (**b**,**d**), respectively.

Figure 2 illustrates the element mapping analysis obtained from high-angle annular dark-field scanning transmission electron microscopy (HAADF-STEM) images. The HAADF image exhibited bright particles (Figure 2a). The bright particles of HAP in the HAADF image are more angular than the brilliant particles of CaO in the CaO HAADF image (Figure S3a). Ca (green), P (cyan), and O (red) elemental maps depict the distribution of Ca, P, and O, which are components of HAP (Figure 2b–d). The existence of P was not verified in the elemental analysis of CaO, while Ca (green) and O (red) were observed, with the strongest signal corresponding to Ca (Figure S3b–d).

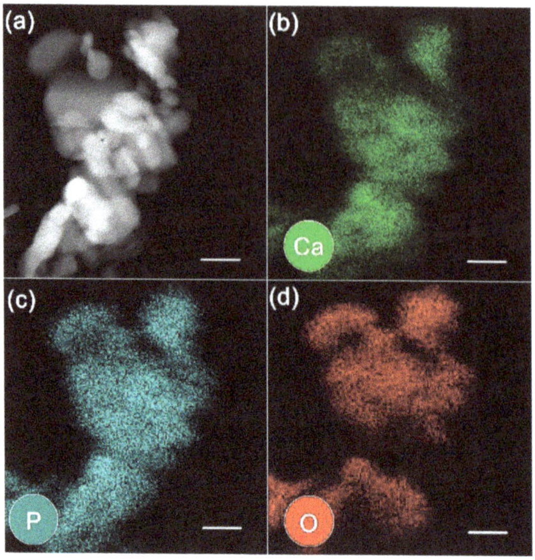

Figure 2. (**a**) HAADF-STEM image and (**b**–**d**) elemental mapping images of HAP. The bars indicate 100 nm.

Figure 3a shows the Fourier-transform infrared (FT-IR) spectra of CaO and HAP. The broad and sharp peaks at 1383–1546 cm^{-1} and 875 cm^{-1} are characteristic CaO peaks. The chemical groups PO_4^{3-}, OH^-, and HPO_4^{2-} that are typical of HAP were observed in the FT-IR spectrum. The PO_4^{3-} group provides a significant IR transmission band at 568 and 602 cm^{-1} and 1000–1100 cm^{-1}. The transmission wavelength of water causes distinct peaks at 3572 cm^{-1} and 633 cm^{-1}. In the XRD pattern, the CaO peak was observed at 2θ = 32.3°, 37.5°, 54.0°, 64.4°, and 67.6°, corresponding to the reflections of the (111), (200), (220), (311), and (222) planes of CaO, respectively (JCPDS No. 82-1691) (Figure 3b). However, when the oyster shell was oxidized, it was not entirely transformed into calcium oxide, and some calcium carbonate remained. After the phosphorylation of HAP, there were numerous peaks other than that of CaO, with particularly prominent peaks appearing at 31.7°, 32.2°, and 32.9°, corresponding to the reflections of the (211), (112), and (300) planes of HAP, respectively (JCPDS No. 09-0432). According to the literature [20], CaO is not entirely phosphorylated, and therefore the results represent a variety of different crystal structures for CaO. As a result, among the HAP XRD peaks, there is a peak at 37.5°. The lattice distance determined by HRTEM was quite consistent with the information interpreted from the XRD peak.

The chemical states of the samples were examined using XPS. The survey scan of CaO and HAP, shown in Figure S4a,b, demonstrates that all samples consisted of N, O, Ca, and C. In addition, P was found in the HAP powder. Figure 3c,d show the C 1*s* and P 2*p* core-level X-ray photoelectron spectra and deconvoluted results of CaO and HAP powders, respectively. All spectra were calibrated by adjusting the C 1*s* peak to 284.5 eV [49]. Furthermore, all the peaks were deconvoluted using a Gaussian-Lorentzian fitting using CASA. The C 1*s* spectra show surface carbon contamination (C-C, C-O-C, and O-C=O) and -CO_3 (carbonate) (Figure 3c). The relative -CO_3 peak, compared to C-C, is less dominant in the HAP powder due to the phosphorylation. Unfortunately, as there is only a slight difference in the binding energy between the source materials, $CaCO_3$ and HAP, it is difficult to separate the two peaks [50]. However, the lack of satellite loss feature, shown in Figure S4c in the Ca 2*p* spectra, provides an indication as to how HAP forms after the reaction [https://xpssimplified.com/elements/calcium.php accessed on 13 May 2021]. Therefore, the CaO powder is composed of $CaCO_3$, while HAP is

$Ca_{10}(PO_4)_6(OH)_2$. Furthermore, the P 2*p* spectrum is only represented in the HAP powder, as shown in Figure 3d. The O 1*s* spectra, shown in Figure S4d, also show additional HAP bonds and O-C=O, C-O-C, and $CaCO_3$ for the HAP powder. However, O-C=O, C-O-C, and $CaCO_3$ bonds were present in the CaO powder spectra.

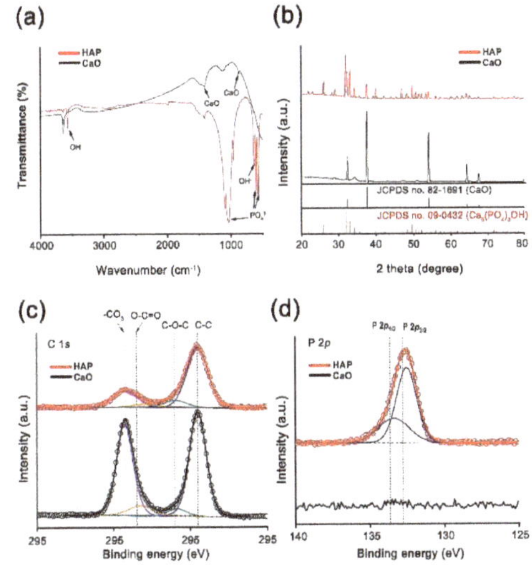

Figure 3. (**a**) FT-IR spectrum of CaO and HAP powder. (**b**) XRD data for CaO and HAP powder. Deconvoluted (**c**) C 1s and (**d**) P 2*p* X-ray photoelectron spectrum of CaO and HAP powder. The vertical dotted lines indicate the reference binding energy [50,51]

2.2. Metal Cation Removal Efficiecny of HAP Catalyst

To test the catalytic performance of HAP as an adsorbent for metal cations in water, we prepared a model solution containing seven metal ions, Cr^{3+}, Mn^{2+}, Ni^{2+}, Cu^{2+}, Cd^{2+}, Ba^{2+}, and Pb^{2+}. To prepare 1 L of 100 ppm metal cation solution, known quantities of metal nitrates were mixed with distilled water. In a typical removal test, biomass-derived HAP catalyst (0.01 mmol) was inserted into the model solution (50 mL), and the HAP catalyst was collected by centrifugation after 24 h of stirring at RT. In addition, we created a pH 6 solution to represent wastewater using HNO_3 (0.1 M) and NaOH solution (0.1 M). Inductively coupled plasma-optical emission spectroscopy (ICP-OES) was used to evaluate the residual aqueous solutions. Through ICP-OES analysis, we discovered that Cr^{3+}, Mn^{2+}, Ni^{2+}, Cu^{2+}, Cd^{2+}, Ba^{2+}, and Pb^{2+} were present at concentrations of 115.39, 85.44, 98.47, 112.60, 99.63, 114.79, and 103.52 ppm, respectively, in the model solution. Furthermore, we demonstrated the removal efficiency of the metal cations (Table S1).

The removal efficiency and the adsorption capacity (q_e) were calculated using Equations (3) and (4):

$$Re\% = \left(\frac{C_0 - C_e}{C_0}\right) \times 100 \tag{3}$$

$$q_e = \frac{C_0 - C_e}{W} \times V \tag{4}$$

where C_0 and C_e ($mg \cdot L^{-1}$) are the liquid-phase concentrations of the metal cations before and after adsorption, respectively. V (L) and W (g) are the volume and mass of each metal ion solution, respectively. Figure 4 illustrates the removal efficiencies of each metal cation. HAP was shown to be capable of removing each cation and a removal efficiency of more than 55% was demonstrated for the extremely harmful Cr, Cd, and Pb (Figure 4a). The elimination efficiency of Pb (92.8%), which causes osteoporosis, was especially high (Cr: 57.9% and Cd: 73.9%, respectively). Furthermore, HAP also exhibits Mn, Ni, Cu, and Ba adsorption characteristics, and the removal rates were 66.3%, 37.1%, 29.1%, and 44.0%, respectively. Equation (3) was used to obtain the adsorption capacities. The HAP adsorption capacities were 66.8, 56.6, 36.6, 32.8, 73.6, 50.5, and 96.0 $L \cdot g^{-1}$ of Cr, Mn, Ni, Cu, Cd, Ba, and Pb, respectively (Figure 4b). A high capacity for the adsorption of Pb, Cu, and Ni and several other metal cations was demonstrated for HAP. Numerous studies have established the potential of HAP to absorb metal ions via two major processes (Equations (5)–(8)). The first is ion exchange, which occurs when calcium ions in the HAP lattice are replaced by adsorbed metal ions via diffusion [52–57].

$$Ca_{10}(PO_4)_6(OH)_2 + xM^{2+} \rightarrow Ca_{10-x}M_x(PO_4)_6(OH)_2 + xCa^{2+} \quad (5)$$

$$Ca_{30}((PO_4)_6(OH)_2)_3 + 20Cr^{3+} \rightarrow Cr_{20}((PO_4)_6(OH)_2)_3 + 30Ca^{2+} \quad (6)$$

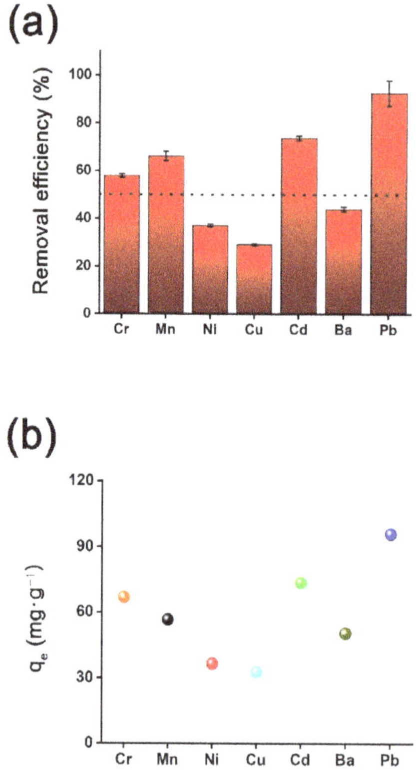

Figure 4. HAP catalytic performance at pH: (**a**) removal efficiency and (**b**) adsorption capacity. Reaction conditions: hydroxyapatite (0.05 g), pH 6, time 24 h, RT, respectively. The concentration of M ion aqueous solution exhibits in Table S1.

The other process involves the precipitation of metal phosphate complex materials in acidic environments. In this state, phosphate interacts with metallic ions according to the equations below, forming a new crystalline structure with limited solubility [52,58–65].

$$Ca_{10}(PO_4)_6(OH)_2 + 14H^+ \rightarrow 10Ca^{2+} + 6H_2PO_4^- + 2H_2O \quad (7)$$

$$10M^{2+} + 6H_2PO_4^- + 2H_2O \rightarrow M_{10}(PO_4)_6(OH)_2 + 14H^+ \quad (8)$$

According to Mobasherpour et al., adsorption is accomplished using a two-step adsorption technique [66]. Initially, Ni ions are rapidly adsorbed through surface interactions with = POH sites, leading to the formation of $Ca_{10-x}M_x(PO_4)_6(OH)_2$. Thus, the initial process involves Ni adsorption by HAP, followed by Ca^{2+} replacement by Ni^{2+}.

The pH of actual wastewater varies based on the surroundings. Thus, we selected, Cr, Cd, and Pb, which are known to be hazardous, to test under various pH conditions (Table S2). The pH of the solution has a considerable impact on heavy metal adsorption because it affects adsorbent characteristics such as surface charge, adsorbate speciation, and degree of ionization in aqueous solutions. Because each pH condition was simulated with solutions derived from the stock solution prepared in Figure 4, no further ICP-OES analysis of the stock solution was carried out. The same quantity of HAP was added to the stock solution, and a 24 h adsorption experiment was performed. Although the data from the experiments performed at pH 6 are already shown in Figure 4, they are also included in Figure 5 for comparison. Figure 5a–c depict the removal efficiency of heavy metal ions by the HAP catalyst at various pH levels. In general, the pH of wastewater is known to be 6 to 8; however, to account for the acidification produced by acid rain, the experiment was performed at pH 4, 5, 6, and 7. At pH 7, a good removal efficiency was demonstrated for the three heavy metal cations. Cr^{3+} ions were removed more efficiently as the pH increased, whereas Pb^{2+} ions were removed efficiently in a wide range of pH conditions. The removal efficiency for the Cd ion (15.6%) was lowest at pH 5, whereas the removal efficiency Cr^{3+} was 44% at this pH level. According to the literature, because the surface of HAP is negatively charged in this range, the most appropriate pH range for Cd^{2+} adsorption is above pH_{pzc} (surface charge of HAP is zero). The pH can impact efficient ion exchange adsorption between cationic species and the Ca^{2+} in HAP [67,68]. In contrast, HAP has a high adsorption capacity at pH 4, 5, 6, and 7 for Pb, which suggests that Pb ion adsorption in wastewater would be efficient (Figure 5d). Although the heavy metal removal efficiency varied with pH, the average removal efficiencies of Cr, Cd, and Pb were 63.0%, 59.9%, and 91.6%, respectively (Table S2).

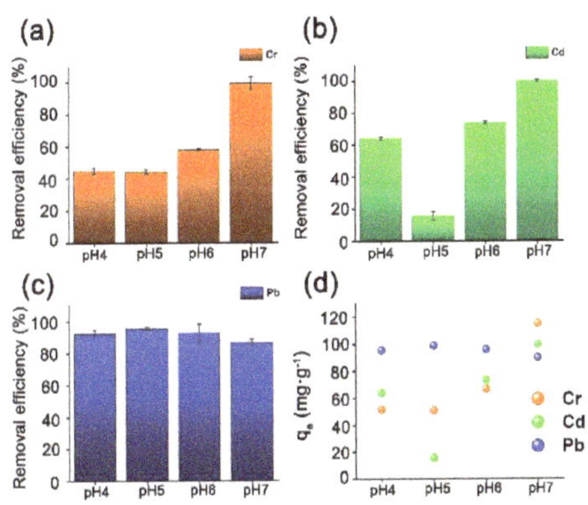

Figure 5. Removal efficiency using HAP under various pH conditions for three major hazardous metal cations: (**a**) Cr, (**b**) Cd, and (**c**) Pb. (**d**) Adsorption capacities. Reaction conditions: hydroxyapatite (0.05 g), pH 4, 5, 6, and 7. Time 24 h and RT, respectively. The concentration of M ion aqueous solution exhibits in Table S2.

2.3. *Phosphate Ion Removal Efficiecny*

According to the literature [69], phosphate ion removal in aqueous solution is caused by the flexibility of the structure surface activity and the non-stoichiometry of HAP. As a result, the phosphate ion adsorbs to the location next to the Ca^{2+} ion on the HAP surface or to the phosphate ion defect. To explore the phosphate ion removal efficiency we applied the ascorbic acid solution method outlined in the experimental procedure with monitoring every 3 min (Figure 6). When performing UV-VIS analyses, we used a wavelength of 625 nm to account for the color reagent. We compared the phosphate ion removal efficiency of HAP with that of CaO. The phosphate ion removal efficiency of HAP was demonstrated to be 98.1% after 3 min, and subsequent testing at 3 min intervals showed that the efficiency remained consistent thereafter. In contrast, CaO was demonstrated to have an ion removal efficiency of 71.1% after 3 min, at which point the solution was saturated. As a consequence, the phosphate ion removal efficiency of HAP was 27% greater than that of CaO, and saturation was achieved very rapidly.

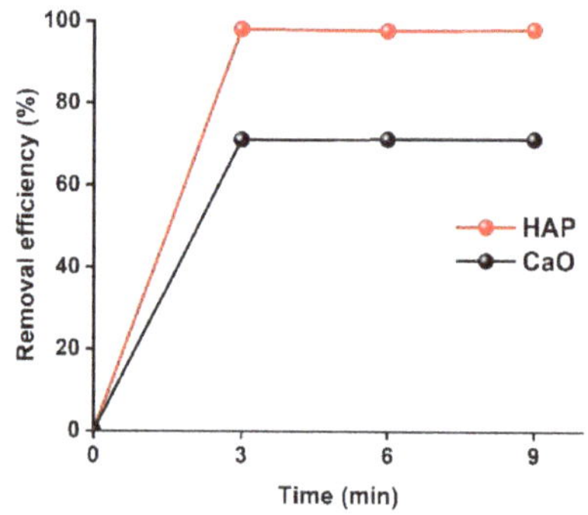

Figure 6. Comparison of phosphate ion removal efficiency using HAP and CAO. Reaction conditions: Hydroxyapatite (2 g) and CaO (2 g). The reaction time 0, 3, 6, and 9 and RT.

2.4. *Antibacteiral Performance of HAP*

The antibacterial performance of HAP (5 mM and 10 mM) was evaluated against Gram-negative and Gram-positive bacteria, *E. coli* and *S. aureus*. Further, 600 nm was chosen as an appropriate wavelength to measure the concentration of bacteria. Equation (9) was used to calculate the antibacterial efficiency.

$$\frac{I_0 - I}{I_0} \times 100 \tag{9}$$

I_0 and I indicate the initial absorbance of the bacteria and the final absorbance, respectively. Before cultivating the bacteria, 100 mL of the HAP solution was added to the culture solution, and the bacteria were cultured for 24 h. Figure 7a shows the estimated antibacterial activity of HAP based on Equation (9). Furthermore, we demonstrated the antimicrobial efficiency of *E. coli* and *S. aureus* (Table S3). The antibacterial activity was 74.4% against *E. coli* and 74.9% against *S. aureus* when 5 mM HAP was used, indicating similar antibacterial effectiveness. Furthermore, when 10 mM HAP was applied, *E. coli* demonstrated 90.9% antibacterial effectiveness, while *S. aureus* demonstrated 96.0% antibacterial effectiveness. We used FE-SEM to examine the damaged bacterial cells (Figure 7b–e). In the case of *E. coli*, the cells appeared to be damaged or melted at 5 mM and 10 mM HAP, respectively (Figure 7b,c). Aggregation occurred with 5 mM HAP in *S. aureus*, and cell surface modification and aggregation occurred with 10 mM (Figure 7d,e). These results indicate that HAP demonstrates successful antibacterial performance against *E. coli* and *S. aureus*.

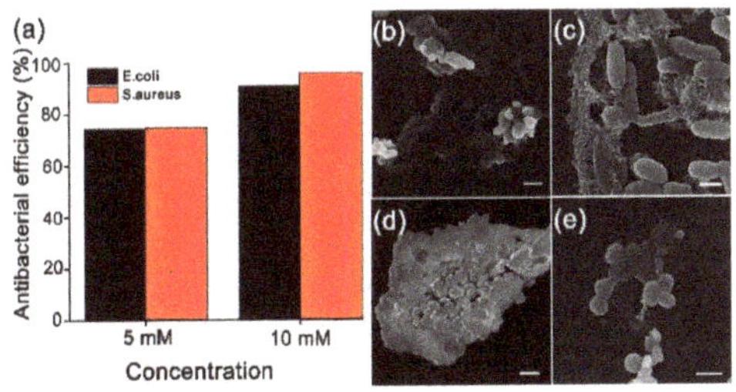

Figure 7. (**a**) Antibacterial efficiency using HAP. SEM images of antibacterial performance against *E. coli* (**b**) 5 mM and (**c**) 10 mM HAP. SEM images showing the antibacterial performance against *S. aureus* with (**d**) 5 mM and (**e**) 10 mM HAP. The bars indicate (**b**–**e**) 1 μm.

3. Materials and Methods

3.1. Chemicals and Characterization

Oyster shells were obtained from Tongyeong, South Korea. Solvents and all other chemicals were obtained from commercial sources and used as received, unless otherwise noted. Biomass-derived CaO and HAP catalysts were characterized using field-emission transmission electron microscopy (TALOS F200X), field emission scanning electron microscopy (FE-SEM), and elemental mapping (ZEISS SUPRA 25). X-ray diffraction (XRD) was performed on a Rigaku RINT 2200 HK diffractometer, and the chemical states were characterized using high-performance X-ray photoelectron spectroscopy (XPS). Fourier-transform infrared spectra (FT-IR) were recorded on a Spectrum Two spectrometer (PerkinElmer). Bacteria were characterized using UV-VIS (UV-1800, Shimadzu UV spectrophotometer).

3.2. Synthesis of Biomass-Derived HAP Catalyst from Oyster Shell

The discarded oyster shells were cleaned with clean water and dried in an oven at 140 °C for 24 h. The oyster shells (10 g) were mechanically ground in a mortar for 10 min under atmospheric conditions. To synthesize CaO powder, the powdered oyster shell was placed in an alumina bowl and heated at a ramping rate of 7.5 °C·min^{-1} up to 900 °C under an air atmosphere, and then held at 900 °C for 4 h. After heating, the resulting yellowish-white powder was cooled to room temperature (RT). HAP powder was synthesized using the method reported by Nakanishi group [45]. To synthesize HAP powder, distilled water (30 mL), H_3PO_4 (30 mmol), and ammonia solution (10 mmol) were added to a 100 mL round-bottom flask and stirred for 10 min. Powdered CaO (2.804 g), H_3PO_4 (30 mmol), and NH_4OH (10 mmol) were added to the round-bottom flask and stirred at 40 °C for 24 h. The subsequent colloidal dispersion was cooled to RT, washed three times with distilled water, and the sample was separated by centrifugation at 3500 rpm for 10 min. The precipitate was dried in an oven at approximately 100 °C. Finally, the powder was placed in an alumina bowl and heat-treated at a ramping rate of 6.7 °C·min^{-1} up to 800 °C under an air atmosphere, and then held at 800 °C for 4 h.

3.3. Removal Efficiency of Various Metal Cation Species

First, we prepared various M nitrate salt solutions containing 100 ppm of M cations at RT (M = Cr^{3+}, Mn^{2+}, Ni^{2+}, Cu^{2+}, Cd^{2+}, Ba^{2+}, and Pb^{2+}). Experiments were carried out under two sets of conditions to determine the metal ion removal efficiency of the biomass-derived HAP catalyst.

1. Standard conditions: The biomass-derived HAP powder (50 mg) was dispersed in a metal ion solution (50 mL) and stirred at RT for 24 h. The pH of the solution was adjusted to 6 using nitric acid (0.1 M) and sodium hydroxide solution (0.1 M), to create a stock solution with conditions comparable to wastewater.

2. pH controlled conditions: pH 4, pH 5, and pH 7 solutions were prepared to recreate severe and realistic wastewater conditions. Each solution was generated in the same manner as in Experiment 1.

3.4. Removal Efficiency for Phosphate Ion

We prepared a phosphate solution containing 100 ppm of KH_2PO_4 at RT. A volumetric flask (1 L) was filled with powdered KH_2PO_4 (4.393 g) and distilled water to make a phosphate solution. To test the phosphate ion adsorption capabilities of HAP, a pre-prepared phosphate solution (20 mL) was placed in a round-bottom flask, and after adding HAP (2 g), solution was collected every 3, 6, and 9 min, respectively. CaO was also tested under the same conditions as those used to test HAP. The phosphate was quantified using the ascorbic acid method, a color development reagent technique, and analyzed using UV-VIS [27]. Before creating a chromogenic reagent for phosphate analysis, three reagents were synthesized.

1. Potassium antimonyl tartrate solution was mixed with potassium antimonyl(III) tartrate hydrate (6.8 mg) in distilled water (50 mL).
2. An ammonium molybdate solution was prepared by combining ammonium molybdate tetrahydrate (1.5 g) with distilled water (50 mL).
3. Ascorbic acid solution was prepared by combining ascorbic acid (0.88 g) with distilled water (50 mL).

To prepare the color reagent, sulfuric acid (14% v/v, 25 mL), potassium antimonyl tartrate solution (2.5 mL), ammonium molybdate solution (7.5 mL), and ascorbic acid solution (15 mL) were added to a 100 mL flask and stirred for 10 min. A phosphate solution (100 ppm of PO^{3-}) was prepared by diluting 1 mL of phosphate solution in a volumetric flask (10 mL) with water. Then, 5 mL of the diluted H_3PO_4 solution and 0.8 mL of pre-prepared color development reagent were combined, stirred, and measured after 5 to 10 min.

All of the above experiments were performed under standard conditions (RT and 1 bar).

3.5. Antimicrobial Test Using Gram-Netaive and Gram-Positive Bacteria

We evaluated the antibacterial effect by growing bacteria in different concentrations of HAP (Scheme S1). To determine the optimal concentration, we prepared 5 and 10 mM HAP in lysogeny broth (LB). Bacteria were incubated in an Erlenmeyer flask with LB (100 mL) and HAP at 37 °C for 24 h. We measured the bacterial concentration using UV-VIS at 600 nm to assess the efficacy of HAP for antibacterial activity. We used FE-SEM to detect the damaged bacterial cells.

4. Conclusions

In this study, we successfully synthesized biomass-derived HAP using low-value oyster shells, which are otherwise considered industrial waste. Due to the top-down approach of oxidation and phosphorylation, the purity of the HAP was low. However, it demonstrated high removal efficiency by metal cation adsorption, particularly for the Pb ion (92.8%). Generally, the pH of wastewater varies from environment to environment, and average removal efficiencies of 63.0% (Cr), 59.9% (Cd), and 91.6% (Pb) were demonstrated at pH 7, 5, and 4. Furthermore, because the phosphate ion removal efficiency (98.1%) was also high, the use of HAP could prevent freshwater and saltwater eutrophication. Moreover, as bacterial contamination is a component of water pollution, we confirmed the antibacterial properties of HAP. When we tested the antimicrobial activity against *E. coli* and *S. aureus* using 5 mM HAP, we achieved antibacterial efficacies of 74.4% and 78.1%, respectively. The antibacterial efficacies of 10 mM HAP were 89.6% and 96.0%, respectively. We expect that application of these biomass-derived products would reduce pollution and contribute to a cleaner environment for future generations.

Supplementary Materials: The Supplementary Materials are available online at https://www.mdpi.com/article/10.3390/w13223302/s1. Figure S1: Histogram of South Korean fishery production and byproducts [44]; Figure S2: Oyster shells that have not been treated in South Korea. The photo was taken by Sanha Jang; Figure S3: (a) HAADF-STEM image of CaO and (b-d) elemental mapping images of CaO. The bars indicate 100 nm, respectively; Figure S4: Survey spectrum of (a) CaO and (b) HAP powders and deconvoluted (c) Ca 2p and (d) O 1s spectrum. The vertical dotted lines indicate the reference binding energy; Table S1: ICP data table for RT condition of various metal cation adsorption; Table S2: ICP data table for various acidic condition of various metal cation adsorption; Table S3: Calculation of antibacterial efficiency of E. coli and S. aureus for 24 h. Scheme S1. Schematic image of antibacterial tests using Gram negative and Gram positive, respectively.

Author Contributions: K.P. (Kanghyun Park), S.P. and B.Y. provided academic direction, conceptualization, validation, funding acquisition, writing—reviewing and editing. S.J. contributed especially in the synthesis and analysis of materials, wrote the introduction, results and discussion, experimental and conclusion parts. K.P. (Kyeongmun Park) contributed regarding adsorption of heavy metals and phosphate of HAP. S.S. and H.L. took part formal analysis such as XPS and antibacterial tests. All authors have read and agreed to the published version of the manuscript.

Funding: This research was supported by the Basic Science Research Program through a National Research Foundation of Korea (NRF) grant funded by the Korean Government (MSIP) (NRF-2020R1I1A3067208) and Korea Basic Science Institute (National research Facilities and Equipment Center) grant funded by the Ministry of Education (2021R1A6C101A429).

Institutional Review Board Statement: Not applicable.

Informed Consent Statement: Not applicable.

Data Availability Statement: Data are contained within the article.

Conflicts of Interest: The authors declare no conflict of interest.

References

1. Sandoval, A.; Hernández-Ventura, C.; Klimova, T.E. Titanate nanotubes for removal of methylene blue dye by combined adsorption and photocatalysis. *Fuel* **2017**, *198*, 22–30. [CrossRef]
2. Briffa, J.; Sinagra, E.; Blundell, R. Heavy metal pollution in the environment and their toxicological effects on humans. *Heliyon* **2020**, *6*, e04691. [CrossRef]
3. Zhou, Q.; Yang, N.; Li, Y.; Ren, B.; Ding, X. Total Concentrations and Sources of Heavy Metal Pollution in Global River and Lake Water Bodies from 1972 to 2017. *Glob. Ecol. Conserv.* **2020**, *22*, e00925. [CrossRef]
4. Aithani, D.; Saurav, D.; Siddiqui, Z.; Kumar, A.; Khillare, P.S. Source Apportionment, Pollution Assessment, and Ecological and Human Health Risk Assessment Due to Trace Metals Contaminated Groundwater along Urban River Floodplain. *Groundw. Sustain. Dev.* **2020**, *11*, 100445. [CrossRef]
5. Rajkumar, H.; Naik, P.K.; Rishi, M.S. A New Indexing Approach for Evaluating Heavy Metal Contamination in Groundwater. *Chemosphere* **2020**, *245*, 125598. [CrossRef] [PubMed]
6. Sun, Z.; Yue, Y.; He, W.; Jiang, F.; Lin, C.; David, Y.; Pui, H.; Liang, Y.; Wang, J. The Antibacterial Performance of Positively Charged and Chitosan Dipped Air Filter Media. *Build. Environ.* **2020**, *180*, 107020. [CrossRef]
7. Liu, Z.; Zhu, Z.; Zhu, Y.; Xu, W.; Li, H. Investigation of Dust Loading and Culturable Microorganisms of HVAC Systems in 24 Office Buildings in Beijing. *Energy Build.* **2015**, *103*, 166–174. [CrossRef]
8. Dechow, M.; Shon, H.; Steinhanses, J. Concentrations of selected contaminants in cabin air of airbus aricrafts. *Chemosphere* **1997**, *35*, 21–31. [CrossRef]
9. Jang, S.; Song, S.; Lim, J.H.; Kim, H.S.; Phan, B.T.; Ha, K. Application of Various Metal-Organic Frameworks (MOFs) as Catalysts for Air and Water Pollution Environmental Remediation. *Catalysts* **2020**, *10*, 195. [CrossRef]
10. Lindgre, T.; Norb ck, D.; Wieslander, G. Perception of cabin air quiality in airline crew related to air humidification on intercontinental flights. *Indoor Air* **2007**, *17*, 204–210. [CrossRef]
11. Srininasan, M.; Ferraris, C.; White, T. Cadmium and Lead Ion Capture with Three Dimensionally Ordered Macroporous Hydroxyapatite. *Environ. Sci. Technol.* **2006**, *40*, 7054–7059. [CrossRef] [PubMed]
12. Amarasinghe, B.M.W.P.K.; Williams, R.A. Tea Waste as a Low Cost Adsorbent for the Removal of Cu and Pb from Wastewater. *Chem. Eng. J.* **2007**, *132*, 299–309. [CrossRef]
13. Yarkandi, N.H. Removal of Lead (II) from Waste Water by Adsorption. *Int. J. Curr. Microbiol. Appl. Sci.* **2014**, *3*, 207–228.
14. Mahmud, H.N.M.E.; Hosseini, S.; Yahya, R.B. Polymer adsorbent for the removal of lead ions from aqueous solution. *Int. J. Technol. Res.* **2014**, *11*, 4–8.

15. Mohammadi, S.Z.; Karimi, M.A.; Afzali, D.; Mansouri, F. Removal of Pb (II) from Aqueous Solutions Using Activated Carbon from Sea-Buckthorn Stones by Chemical Activation. *Deslination* **2010**, *262*, 86–93. [CrossRef]
16. Liu, L.; Zhang, K. Nanopore-Based Strategy for Sequential Separation of Heavy-Metal Ions in Water. *Environ. Sci. Technol.* **2018**, *52*, 5884–5891. [CrossRef]
17. Hong, M.; Yu, L.; Wang, Y.; Zhang, J.; Chen, Z.; Dong, L.; Zan, Q. Heavy Metal Adsorption with Zeolites: The Role of Hierarchical Pore Architecture. *Chem. Eng. J.* **2019**, *359*, 363–372. [CrossRef]
18. Alghamdi, A.A.; Al-Odayni, A.; Saeed, W.S.; Al-kahtani, A.; Alharthi, F.A.; Aouak, T. Efficient adsorption of lead (II) from aqueous phase solutions using polypyrrole-based activated carbon. *Materials* **2019**, *12*, 2020. [CrossRef]
19. Park, K.H.; Choi, J.; Chun, J.; Kim, H.J.; Son, S.U. Low-Temperature Synthesis of Ultrathin Sb_2S_5 Nanofibers and Their Application as Highly Selective Pb-Adsorbents in Water. *Chem. Commun.* **2008**, *14*, 1659–1661. [CrossRef]
20. Dabiri, S.M.H.; Rezaie, A.A.; Moghimi, M.; Rezaie, H. Extraction of Hydroxyapatite from Fish Bones and Its Application in Nickel Adsorption. *BioNanoScience* **2018**, *8*, 823–834. [CrossRef]
21. Ibrahim, M.; Labaki, M.; Giraudon, J.; Lamonier, J. Hydroxyapatite, a Multifunctional Material for Air, Water and Soil Pollution Control: A Review. *J. Hazard. Mater.* **2020**, *383*, 121139. [CrossRef] [PubMed]
22. Gupta, N.; Kushwaha, A.K.; Chattopadhyaya, M.C. Adsorptive removal of Pb^{2+}, Co^{2+} and Ni^{2+} by hydroxyapatite/chitosan composite from aqueous solution. *J. Taiwan Inst. Chem. Eng.* **2012**, *43*, 125–131. [CrossRef]
23. Huang, Y.; Wu, D.; Wang, C.; Du, P.; Cheng, C.; Cheng, C. Survey of Rice Production Practices and Perception of Weedy Red Rice (Oryza Sativa f. Spontanea) in Taiwan. *Weed Sci.* **2020**, *69*, 526–535. [CrossRef]
24. Awual, R.M. Efficient phosphate removal from water for controlling eutrophication using novel composite adsorbent. *J. Clean. Prod.* **2019**, *228*, 1311–1319. [CrossRef]
25. Hautier, Y. General destabilizing effects of eutrophication on grassland productivity at multiple saptial scales. *Nat. Commun.* **2020**, *11*, 5375. [CrossRef]
26. Yang, F.; Zhang, S.; Sun, Y.; Tsang, D.C.W.; Cheng, K.; Sik, Y. Assembling Biochar with Various Layered Double Hydroxides for Enhancement of Phosphorus Recovery. *J. Hazard. Mater.* **2019**, *365*, 665–673. [CrossRef] [PubMed]
27. Palansooriya, K.N.; Kim, S.; Igalavithana, A.D.; Hashimoto, Y.; Choi, Y.; Mukhopadhyay, R.; Sarkar, B.; Ok, Y.S. Fe (III) Loaded Chitosan-Biochar Composite Fibers for the Removal of Phosphate from Water. *J. Hazard. Mater.* **2021**, *415*, 125464. [CrossRef]
28. Ishikawa, K.; Ducheyne, P.; Radin, S. Determination of the Ca/P Ratio in Calcium- Deficient Hydroxyapatite Using X-ray Diffraction Analysis. *J. Mater. Sci. Mater. Med.* **1993**, *4*, 165–168. [CrossRef]
29. Shimabayashi, S.; Fukuda, H.; Aoyama, T.; Nakagaki, M. Adsorption of phosphate ion by hydroxyapatite in water. *Chem. Pharm. Bull.* **1982**, *30*, 3074–3081. [CrossRef]
30. Andrade, L.; Kelly, M.; Hynds, P.; Weatherill, J.; Majury, A.; Dwyer, O. Groundwater Resources as a Global Reservoir for Antimicrobial-Resistant Bacteria. *Water Res.* **2020**, *170*, 115360. [CrossRef]
31. Motlagh, A.M.; Yang, Z.; Saba, H. Groundwater Quality. *Water. Environ. Res.* **2020**, *92*, 1649–1658. [CrossRef] [PubMed]
32. Ighalo, J.O.; Adeniyi, A.G.; Otoikhian, K.S. Recent advances in environmental protection of oil polluted surface and groundwater in the nigerian context. *J. Eng. Exact Sci.* **2020**, *6*, 416–420. [CrossRef]
33. Shi, H.; Liu, F.; Xue, L. Fabrication and characterization of antibacterial PVDF hollow fibre membrane by doping Ag-loaded zeolites. *J. Membr. Sci.* **2013**, *437*, 205–215. [CrossRef]
34. Shen, L.; Zhang, Y.; Yu, W.; Li, R.; Wang, M.; Gao, Q.; Li, J.; Lin, H. Fabrication of hydrophilic and antibacterial poly(vinylidene fluoride) based separation membranes by a novel strategy combining radiation grafting of poly(acryl acid) (PAA) and electroless nickel plating. *J. Colloid Interface Sci.* **2019**, *543*, 64–75. [CrossRef] [PubMed]
35. Li, R.; Wu, Y.; Shen, L.; Chen, J.; Lin, H. A novel strategy to develop antifouling and antibacterial conductive Cu/polydopamine/polyvinylidene fluoride membranes for water treatment. *J. Colloid Interface Sci.* **2018**, *531*, 493–501. [CrossRef]
36. Wu, W.; Zhao, W.; Wu, Y.; Zhou, C.; Li, L.; Liu, Z.; Dong, J.; Zhou, K. Antibacterial Behaviors of Cu_2O Particles with Controllable Morphologies in Acrylic Coatings. *Appl. Surf. Sci.* **2019**, *465*, 279–287. [CrossRef]
37. Zhang, M.; Yu, J.; Zhang, J.; Lan, Q.; Dai, J.; Huang, Y.; Li, G.; Fan, Q.; Fan, X.; Zhou, Z. Shape-Controlled Synthesis of Cu_2O Nanocrystals by One Pot Solution-Phase Reduction Process. *Chem. Phys. Lett.* **2017**, *671*, 154–160. [CrossRef]
38. Lamkhao, S.; Phaya, M.; Jansakun, C.; Chandet, N.; Thongkorn, K.; Rujijanagul, G.; Bangrak, P.; Randorn, C. Synthesis of Hydroxyapatite with Antibacterial Properties Using a Microwave-Assisted Combustion Method. *Sci. Rep.* **2019**, *9*, 4015. [CrossRef]
39. Zhao, X. Antibacterial Bioactive Materials. *Bioact. Mater. Med. Des. Appl.* **2011**, 97–123.
40. Jang, S.; Jung, S.; Song, S.; Lee, S.; Lee, H.; Cho, E.; Lee, H.J.; Park, S.; Youn, B.; Park., K.H. Preparation and characterization of multifunctional nanofibers containing metal-organic frameworks and Cu_2O nanoparticles: Particualte matter capture and antibacterial activity. *Environ. Sci. Nano* **2021**, *8*, 1226–1235. [CrossRef]
41. Silva, T.H.; Mesquita-Guimarães, J.; Henriques, B.; Silva, F.S.; Fredel, M.C. The Potential Use of Oyster Shell Waste in New Value-Added by-Product. *Resources* **2019**, *8*, 13. [CrossRef]
42. Barros, M.C.; Bello, P.M.; Bao, M.; Torrado, J.J. From Waste to Commodity: Transforming Shells into High Purity Calcium Carbonate. *J. Clean. Prod.* **2009**, *17*, 400–407. [CrossRef]
43. Ruviaro, C.F.; Gianezini, M.; Brandão, F.S.; Winck, C.A.; Dewesl, H. Life Cycle Assessment in Brazilian Agriculture Facing Worldwide Trends. *J. Clean. Prod.* **2012**, *28*, 9–24. [CrossRef]

44. Yu, J.; Kim, G. Reform measures for generation and treatment problems of fischery wastes and byproducts. *Natl. Assem. Res. Serv.* **2020**, *118*, 1–17.
45. Onoda, H.; Nakanishi, H. Preparation of Calcium Phosphate with Oyster Shells. *Nat. Resour.* **2012**, *3*, 71–74. [CrossRef]
46. Jung, J.H.; Yoo, K.S.; Kim, H.G.; Lee, H.K.; Shon, B.H. Reuse of Waste Oyster Shells as a SO_2/NO_x Removal Absorbent. *J. Ind. Eng. Chem.* **2007**, *13*, 512–517.
47. Kwon, H.B.; Lee, C.W.; Jun, B.S.; Yun, J.D.; Weon, S.Y.; Koopman, B. Recycling Waste Oyster Shells for Eutrophication Control. *Resour. Conserv. Recycl.* **2004**, *41*, 75–82. [CrossRef]
48. Jeon, D.J.; Yeom, S.H. Recycling Wasted Biomaterial, Crab Shells, as an Adsorbent for the Removal of High Concentration of Phosphate. *Bioresour. Technol.* **2009**, *100*, 2646–2649. [CrossRef]
49. Chen, X.; Wang, X.; Fang, D. A Review on C1s XPS-Spectra for Some Kinds of Carbon Materials. *Fuller. Nanotub. Carbon Nanostruct.* **2020**, *28*, 1048–1058. [CrossRef]
50. Gomes, G.C.; Borghi, F.F.; Ospina, R.O.; López, E.O.; Borges, F.O.; Mello, A. Nd:YAG (532 nm) Pulsed Laser Deposition Produces Crystalline Hydroxyapatite Thin Coatings at Room Temperature. *Surf. Coat. Technol.* **2017**, *329*, 174–183. [CrossRef]
51. Roychowdhury, T.; Bahr, S.; Dietrich, P.; Meyer, M.; Thißen, A.; Linford, M.R. Calcite ($CaCO_3$), by near-Ambient Pressure XPS. *Surf. Sci. Spectra* **2019**, *26*, 014025. [CrossRef]
52. Stötzel, C.; Müller, F.A.; Reinert, F.; Niederdraenk, F.; Barralet, J.E.; Gbureck, U. Ion Adsorption Behaviour of Hydroxyapatite with Different Crystallinities. *Colloids Surf. B Biointerfaces* **2009**, *74*, 91–95. [CrossRef] [PubMed]
53. Suzuki, T.; Ishigaki, K.; Miyake, M. Synthetic hydroxyapatites ans inroganic cation exchanges. *J. Chem. Soc. Faraday Trans.* **1984**, *80*, 3157–3165. [CrossRef]
54. Miyake, M.; Ishigaki, K.; Suzuki, T. Structure Refinements of Pb^{2+} Ion-Exchanged Apatites by X-ray Powder Pattern-Fitting. *J. Solid State Chem.* **1986**, *61*, 230–235. [CrossRef]
55. Sugiyama, S.; Fukuda, N.; Matsumoto, H.; Hayashi, H.; Shigemoto, N.; Hiraga, Y.; Moffat, J.B. Interdependence of Anion and Cation Exchanges in Calcium Hydroxyapatite: Pb^{2+} and Cl^-. *J. Colloid Interface Sci.* **1999**, *220*, 324–328. [CrossRef] [PubMed]
56. Shashkova, I.L.; Rat'ko, A.I.; Kitikova, N.V. Removal of Heavy Metal Ions from Aqueous Solutions by Alkaline-Earth Metal Phosphates. *Colloids Surf. A Physicochem. Eng. Asp.* **1999**, *160*, 207–215. [CrossRef]
57. Laperche, V.; Logan, T.J.; Gaddam, P.; Traina, S.J. Effect of apatite amedments on plant uptake of lead from contaminated soil. *Environ. Sci. Technol.* **1997**, *31*, 2745–2753. [CrossRef]
58. Laperche, V.; Traina, S.J.; Gaddam, P.; Logan, T.J. Chemical and mineralogical characterizations of Pb in a containted soil: Reactions with synthetic apatite. *Environ. Sci. Technol.* **1996**, *30*, 3321–3326. [CrossRef]
59. Chojnacka, K. Equilibrium and Kinetic Modelling of Chromium(III) Sorption by Animal Bones. *Chemosphere* **2005**, *59*, 315–320. [CrossRef]
60. Ma, Q.Y. Effects of aqueous Al, Cd, Cu, Fe(II), Ni, and Zn on Pb immobilization by hydroxyapatite. *Environ. Sci. Technol.* **1994**, *28*, 1219–1228. [CrossRef]
61. Ma, Q.Y.; Logan, T.J.; Traina, S.J. Lead Immobilization from Aqueous Solutions and Contaminated Soils Using Phosphate Rocks. *Environ. Sci. Technol.* **1995**, *29*, 1118–1126. [CrossRef]
62. Ruby, M.V.; Davis, A.; Nicholson, A. In Situ Formation of Lead Phosphates in Soils as a Method to Immobilize Lead. *Environ. Sci. Technol.* **1994**, *28*, 646–654. [CrossRef] [PubMed]
63. Chen, X.; Wright, J.V.; Conca, J.L.; Peurrung, L.M. Effects of PH on Heavy Metal Sorption on Mineral Apatite. *Environ. Sci. Technol.* **1997**, *31*, 624–631. [CrossRef]
64. Zhang, P.; Ryan, J.A. Formation of Pyromorphite in Anglesite-Hydroxyapatite Suspensions under Varying pH Conditions. *Environ. Sci. Technol.* **1998**, *32*, 3318–3324. [CrossRef]
65. Zhang, P.; Ryan, J.A.; Yang, J. In Vitro Soil Ph Solubility in the Presence of Hydroxyapatite. *Environ. Sci. Technol.* **1998**, *32*, 2763–2768. [CrossRef]
66. Mobasherpour, I.; Salahi, E.; Pazouki, M. Removal of Nickel (II) from Aqueous Solutions by Using Nano-Crystalline Calcium Hydroxyapatite. *J. Saudi Chem. Soc.* **2011**, *15*, 105–112. [CrossRef]
67. Thom, N.T.; Thanh, D.T.M.; Nam, P.T.; Phuong, N.T.; Buess-Herman, C. Adsorption behavior of Cd^{2+} ions using hydroxyapatite (HAP) powder. *Green Process. Synth.* **2018**, *7*, 409–416. [CrossRef]
68. Kadouche, S.; Zemmouri, H.; Benaoumeur, K.; Drouiche, N. Procedia Engineering Metal Ion Binding on Hydroxyapatite (HAP) and Study of the Velocity of Sedimentation. *Procedia Eng.* **2012**, *33*, 377–384. [CrossRef]
69. Kang, J.T.; Jeong, G.H.; Shin, H.G. The Adsorption of Phosphate Son Using Hydroxyapatite Synthesized by Wastewater Sludge of Semiconductor Fabrication Process. *J. Environ. Sci.* **2002**, *11*, 257–262.

MDPI
St. Alban-Anlage 66
4052 Basel
Switzerland
Tel. +41 61 683 77 34
Fax +41 61 302 89 18
www.mdpi.com

Water Editorial Office
E-mail: water@mdpi.com
www.mdpi.com/journal/water

www.ingramcontent.com/pod-product-compliance
Lightning Source LLC
Chambersburg PA
CBHW040827250926
51850CB00007BA/1279
9783036569758